Conservation
in
Perspective

Conservation in Perspective

Edited by
A. WARREN
Department of Geography
University College London, UK
and
F.B. GOLDSMITH
Department of Botany and Microbiology
University College London, UK

A Wiley-Interscience Publication

JOHN WILEY & SONS
Chichester · New York · Brisbane · Toronto · Singapore

Library of Congress Cataloging in Publication Data:
Main entry under title:

Conservation in perspective.

  Includes index.
  1. Nature conservation. 2. Ecology. I. Warren,
Andrew. II. Goldsmith, F.B. (Frank Barrie)
QH75.C666 1983   333.7'2   82-23848
ISBN 0 471 10321 7
ISBN 0 471 10381 0 (pbk.)

British Library Cataloguing in Publication Data:

Conservation in perspective.
  1. Environmental pollution
  I. Warren, A.   II. Goldsmith, F.B.
  333.7'2   TD170

  ISBN 0 471 10321 7
  ISBN 0 471 10381 0 Pbk

Typeset by MHL Typesetting Limited, Coventry
Printed in Great Britain by The Pitman Press, Bath, Avon

Contents

PART II   ECOLOGY AND CONSERVATION IN PRACTICE

PART III   ORGANIZING CONSERVATION

Contents   vii

Foreword

During the 1950s, as the national programme for nature conservation took shape, such far-seeing leaders as Sir Arthur Tansley, Professor W.H. Pearsall, Sir Julian Huxley and Charles Elton were concerned to ensure through concerted research and training the build-up of a first-rate group of pure and applied ecologists as a mainstay for future advances in conservation. The fulfilment of this aim reached a decisive stage in July 1960 with the start at University College London of a pioneer postgraduate Diploma Course in Conservation—it had to begin simply with a diploma in order to by-pass difficult university degree requirements, which became much easier to satisfy once a course was in being. It followed lines which Professor Pearsall and I had worked out 'to provide trained men and women capable of dealing with problems of animal populations, vegetation and land use, both in Britain and overseas', as we described it in the Nature Conservancy's annual report to Parliament.

Experience and knowledge soon accumulated, and by 1974 it was possible to produce an impressive review under the title *Conservation in Practice*, to which I wrote an extended Foreword recounting preceding developments, which need not be repeated here. It is to cover the further notable advances in knowledge and the accompanying fresh perspectives that the present even more impressive and comprehensive work has been produced. It forms at once a benchmark for the latest progress of scientific conservation, an invaluable manual for practitioners and users of such expertise, and an accessible channel to help the many non-ecologists who now speak and write glibly on the subject to understand what they are talking about.

Its publication is particularly opportune as it coincides with the formulation of the United Kingdom contribution to the World Conservation Strategy, contained in a series of Review Group Reports specially prepared by leading specialist rapporteurs on rural, coastal, industrial, urban, international, educational, and ethical elements. This national programme, focused and co-ordinated in an Overview, has the ambitious aim of integrating ecological and conservation principles with other components of public policy on a long-term basis. It rests on the kind of modern knowledge which is set out more comprehensively and in fuller detail within *Conservation in Perspective*.

This new important revised edition of *Conservation in Practice* provides for the 1980s a firm foundation and clear guidelines for one of the most important tasks on this decade's agenda. Although written primarily in British terms it has global relevance and interest, especially since, let it be modestly stated, no other nation is currently in a position to produce anything comparable in comprehensiveness or of greater scientific substance. In these challenging times ecologists and conservationists stand out as knowing the realities from the myths, as the following pages well demonstrate. In congratulating the team responsible for them we may hope that this work will receive the wide recognition and assiduous attention which it merits.

Max Nicholson
October 1982

Preface

The publication of our first collection of essays on nature conservation (*Conservation in Practice*) coincided with a growing groundswell of environmental concern. In some senses that groundswell broke in the late 1970s, in a welter of polemic in support of a manifestly dwindling natural heritage. The movement for nature conservation is now in a much stronger state than it was in 1974: it is based on sounder scientific understanding of ecosystems, on a much improved data base, on much better practical techniques, on far more widespread public involvement, on stronger institutions and on a wider political foundation. No conservationist, however, would claim that any of these elements was anywhere near perfect.

Immodestly we believe that the members (past and present) of the Ecology and Conservation Unit at University College London, who were the first to hear much of the material that was included in this and our earlier volume, have had no small part in this improvement. The movement now has a professional cadre as well as an enthusiastic voluntary wing. It has its own literature, popular and scientific, its own methodology and, inadequate as it is, its own legislative foundation. The current attack on its financial position, though serious, is unlikely to slow its momentum more than marginally.

We believe that the essays in this book reflect the change in the movement over the last decade. The collection shows a greater focusing of interest onto critical issues than the first volume: less on broad environmental—or resource—issues, and more onto nature conservation itself. Inevitably we have had to restrict ourselves to a few habitats and elements of the ecosystem, to a limited number of practical issues and to a small part of the huge range of political and organizational activity that is going on. Nonetheless we believe that we have captured the essence of an increasingly self-confident discipline and movement. Most chapters in this book were submitted by April 1982. Inevitably some material they include has become dated in the interval before publication and for this we must apologise.

We would like to acknowledge the following for their help in preparing this book: Mrs Elsie Banham for typing; Alick Newman, Richard Davidson, Anne Mason, and Sarah Skinner for preparing the artwork; Marian Whittaker and Jacki Chandler for preparing the index; Carolyn Harrison, Richard Munton and Brian Wood for their support in preparing the book as a whole and finally to Mr E.M. Nicholson for his enthusiasm and support.

A. Warren and F.B. Goldsmith
1983

Contributors

R.J. Berry — *Department of Zoology and Comparative Anatomy, University College London, Gower Street, London WC1E 6BT, UK*

J. Bowers — *School of Economic Studies, University of Leeds, Leeds LS 2 9JT, UK*

A.D. Bradshaw — *Department of Botany, University of Liverpool, PO Box 147, Liverpool L69 3BX, UK*

L. Cole — *Land Use Consultants Ltd., 731 Fulham Road, London SW6 6UL, UK*

E. Duffey — *Cergne House, Church Lane, Wadenhoe, Peterborough PE8 5ST, UK*

F.B. Goldsmith — *Ecology and Conservation Unit, University College London, Gower Street, London WC1E 6BT, UK*

J. Gordon — *48 Blenheim Avenue, Stony Stratford, Milton Keynes, Buckinghamshire, UK*

C.M. Harrison — *Department of Geography, University College London, Gower Street, London WC1E 6BT, UK*

P.D. Lowe — *Bartlett School of Planning and Architecture, University College London, Gower Street, London WC1E 6BT, UK*

A. MacEwen — *Manor House, Wootton Courteney, Minehead, Somerset TA24 8RD, UK*

M. MacEwen — *Manor House, Wootton Courteney, Minehead, Somerset, TA24 8RD, UK*

G.R. Miller — *Institute of Terrestrial Ecology, Banchory Research Station, Hill of Brathens, Glassel, Banchory, Kincardineshire AB3 4BY, UK*

N.W. Moore — *The Farm House, Swavesey, Cambridgeshire, UK*

R. Munton — *Department of Geography, University College London, Gower Street, London WC1E 6BT, UK*

F. PERRING   *Royal Society for Nature Conservation, The Green,*
                    Nettleham, Lincoln LN2 2NR, UK

G.F. PETERKEN   *Nature Conservancy Council, P.O. Box 6, Godwin*
                    House, George Street, Huntingdon PE18 6BU, UK

T. PRITCHARD   *Nature Conservancy Council, Plas Penrhos, Ffordd*
                    Penrhos, Bangor, Gwynedd LL57 2LQ, UK

J. SHEAIL   *Institute of Terrestrial Ecology, Monks Wood Experi-*
                    mental Station, Abbots Ripton, Huntingdon PE 17 2LS,
                    UK

C.R. TUBBS   *Nature Conservancy Council, Shrubbs Hill Road, Lynd-*
                    hurst, Hampshire SO4 7DJ, UK

A. WARREN   *Ecology and Conservation Unit, University College*
                    London, Gower Street, London WC1E 6BT, UK

A. WATSON   *Institute of Terrestrial Ecology, Banchory Research Sta-*
                    tion, Hill of Brathens, Glassel, Banchory, Kincardine-
                    shire AB3 4BY, UK

T.C.E. WELLS   *Institute of Terrestrial Ecology, Monks Wood Experi-*
                    mental Station, Abbots Ripton, Huntingdon PE17 2LS,
                    UK

J.B. WOOD   *Ecology and Conservation Unit, University College*
                    London, Gower Street, London WC1E 6BT, UK

CHAPTER 1

An Introduction to Nature Conservation

A. WARREN and F.B. GOLDSMITH

Any book about the conservation of nature must draw on the experience of a broad range of academic and professional disciplines. This is first because nature is valued by many different groups of people: by countrymen as part of their livelihood; by scientists for research; by resource conservationists for the future; by teachers for education; by naturalists to satisfy their curiosity; and by most of us for the opportunities it offers for recreation, for its beauty and for its very naturalness. And second, it is because nature can only be conserved by combining the skills of foresters and agriculturalists, natural and social scientists, economists, journalists, planners, critics, administrators and politicians.

In this introduction we present an account of nature conservation which draws on most of these disciplines. First, we relate nature conservation to the other objectives of conservationists. Then we outline some of its own objectives. We follow this with a review of the 'conservation imperative'—the perception of the need to conserve nature, and with an explanation of some of the meanings that are used in this book for the word 'conservation' itself. Finally we focus onto the practice of nature conservation, particularly in Britain, for the chapters in this book are primarily about these islands.

THE SCOPE OF CONSERVATION

Four facets of the environment concern people today: resources, pollution, aesthetic heritage and nature. Of these the one that receives the most attention in the press and in the academic world is the supply of natural resources—of food, fibre, energy and minerals. Because resources are valued in economic ways, resource conservation is, very properly, the preserve of social scientists and particularly of economists (Cole, 1978; O'Riordan, 1976). Although the value-systems and methods of analysis among resource conservationists have different objectives from those of nature conservationists, as we will explain, they have produced many models that are useful for thinking about the future partly because they share a similar time-scale for their projections. On the other hand, their interests threaten nature because the two objectives are often in conflict over the same things. A wood, for example, has value both as a resource for timber and in other ways that we shall discuss shortly. Some of the authors in this book believe the conflict to be inevitable whereas others, perhaps the majority, believe there to be potential long-term common interests (Peterken, Chapter 6; Miller and Watson, Chapter 7; and MacEwen and MacEwen, Chapter 22).

A second environmental concern that has also been prominent in the last two decades has been about pollution. The main targets here have been unclean air, water and soil, and in this respect the campaign against pollution has shared many goals with nature conservation (as shown by Bradshaw, Chapter 11, in the case of derelict land). In Britain the gross eutrophication of the Norfolk Broads' which has caused the loss of valuable communities of aquatic macrophytes (Moss, 1977) and the problem of acid rain in the uplands (see Goldsmith and Wood, Chapter 17 for references) are further examples of shared interest. Urban nature conservation, a development of the last decade, is in part, a response to growing distaste of visual pollution (see Cole, Chapter 16). But, cleanliness is easier to quantify and the time needed to achieve it is shorter than the time needed to manage whole ecosystems.

Historical buildings and landscapes are the third objective for conservation (Lowenthal and Binney, 1981). Nature conservationists share long time perspectives and difficulties of evaluation with those who wish to conserve our historical and aesthetic heritage and have a very strong common interest with them when they seek to conserve landscapes, for example in ancient deer parks (Rose, 1974); chalk grasslands or coppice woodlands (Rackham, 1980; Tittensor, 1981; see also Sheail, Chapter 18 of this volume). But the two aims are subtly different: a coppice, to a lover of the past, is a record of a different age, perhaps a laboratory in which to discover how people once lived. In a sense, it is the very artificiality of coppices that is valued in historical landscapes as Newby (1979) among others has pointed out. Nature conservationists place different values on the same wood, as we must now explain.

THE OBJECTIVES OF NATURE CONSERVATION

In all environmental arguments there is a large element that is unquantifiable (see, for example, Ashby, 1978); in nature conservation this element is overwhelming. It is true that there are some useful, even commercial, byproducts of nature conservation and these have often been rehearsed (Myers, 1979) and in this sense nature is a resource. Some species of plant and animal may become useful in producing food, fibre, fuel or medicine (Frankel and Soulé, 1981), although this will never involve more than a very small proportion of the global complement of species. The once-popular argument that a diversity of species promoted global or local stability is now discredited. Connell and Orias pointed out as long ago as 1964 that is was more probable that stability promoted diversity, although the problem is now seen to be much more complex (May, 1974). Soil stability and carbon dioxide levels are clearly linked to a plant cover of the earth, but both could be maintained with a relatively small number of common species. A more convincing argument, though hard to quantify, is that in conserving species we are keeping our options open for the future (International Union for Conservation of Nature and Natural Resources, 1980; Goldsmith, 1980). Last among the payoffs in nature conservation is the commercial return from the enjoyment of nature when, for example, it helps to support the economy of Kenya or of the remote parts of Britain, but this is the outcome of individual evaluations that are themselves not strictly economic.

The great majority of those who believe in nature conservation, and to almost all of our authors (and certainly to ourselves) the importance of nature far transcends a narrowly economic definition of utility, and among many the utilitarian arguments are used as a mere disguise for deeper, less well-articulated feelings (Grove, 1981; Sax 1980). But if our goals are not as immediate, transparent or concrete as the supply of coal, wheat, timber or the maintenance of a healthy environment, it is worth recalling that these apparently concrete goals themselves depend ultimately on ephemeral individual evaluations. Nature conservation is therefore not alone in associating scientific analysis and technical management with a poor understanding of people's real values, as some economists acknowledge (Robinson, 1964).

There can be no doubt that nature is very important to most people in ways quite apart from its economic utility. Pre-literate folk, who may be thought to be a better guide to 'natural' enthusiasms than modern men, have been discovered again and again to have had an enormous repertoire of natural knowledge well beyond their material needs, as Lévi-Strauss's review showed (1966, pp. 1–9). In a European context the historian Le Roy Ladurie (1980, pp. 290–294) recounted a detailed knowledge of nature among illiterate fourteenth-century Pyrenean peasants that evidently had no connection with their material subsistence. The well-known record of nature that has been left by literate men, from Egypt in the third millenium BC to the present, and the popularity of these accounts of nature in both aural tradition and in literature are further evidence of the importance of nature (see, for example, Diana Spearman's *Animal Anthology*, 1966). The writers and media-men who purvey nature to us today have a vast market. For example, the BBC programme *Wildlife on One* on 29 January 1981 commanded an audience of over 15 million. Further evidence of the widespread modern interest in nature comes from the opinion poll conducted for the Countryside Commission of England and Wales which revealed that 54 per cent of the sample had made a countryside visit in the summer of the survey and that the average number of visits among this group had been 4.3. On an average summer Sunday in that year, no less than 14 million people were finding recreation in the countryside (Fitton, 1978). Nature was undeniably a very prominent component of their recreation, together with clean air, peace and an interest in history.

An enquiry into the ways that people value nature would involve the whole of our culture. Here, we do no more than list the more prominent values that are applied to nature in contemporary Britain. 'Wild' nature is seen by many people as an economic resource. In Britain most plants and animals are owned by individuals or groups who see them as a source of 'return' on capital investment in terms of grazing, timber or sport. Some economists, following the lead of Clawson (1969) have attempted to extend economic analysis to this 'resource' by placing monetary values on informal recreation, and Helliwell (1973) even tried to value particular species in such a way (see also Goldsmith, Chapter 14).

Scientists value nature as a subject for observation and experiment. They value the 'interest' of an ecosystem, by which they mean that it provides stimulation in the form of surprises, soluble puzzles, or tests for hypotheses. Some believe that a scientific approach, like that of the economists, is necessarily reductionist (Popper, 1974) and as such, others would argue, antipathetic to the more common 'holistic' view of nature (Habermas, 1971). But the real drive for enlightenment in activities such as science probably comes from deep psychological needs and is very widely shared (Thomas, 1974).

Most people value nature in many ways which, in spite of (or perhaps even because of) their imprecision, are touched on in an enormous literature. Simply put, they see nature holistically as 'life-enhancing', something that does them good. Those who wish to conserve nature for this reason are really moralists who believe that most of the population share their opinions 'deep down' if not on the surface. The power that conservationists sometimes acquire—as in institutions or the National Parks in the United States and Great Britain —seems to confirm this belief (Sax, 1980). The moral view of nature is undeniably woven deeply into the fabric of British culture (Williams, 1973), ideology and politics (Sandbach, 1980), and it has many modern exponents (Dillard, 1976, Mabey, 1980, Shoard, 1980). It has only occasionally been studied by social survey methods, but those who have used them have found deep and widespread values in nature (Mostyn, no date).

The reason that nature is so universally valued is another fascinating field that we cannot adequately explore here although we have been debating it elsewhere (Page and Warren, 1982). We noted there that Lévi-Strauss (1966), Douglas (1973) and Leach (1970) are anthropologists who believe that nature is valued as a language of symbols. Thomas (1974) is probably the most accessible of many authorities who have suggested that nature was valued because of our evolutionary connections with it. Lewis (1943) pointed to the widespread religious valuation in nature, and Ashby (1978) and Passmore (1974), among others, have discovered a kind of secular veneration of nature. Philip Lowe (Chapter 19) puts these values in historical perspective.

This great variety in the ways in which nature is valued is both a strength and a weakness in the nature conservation movement. It is a strength because almost everyone has some conception of the value of nature. It is a weakness in that nature conservationists are constantly accused of not being clear about their priorities. The arguments for the conservation of one wood, for example Grass Wood in Wharfedale, flush out at least the following views: a resource of timber, a good shoot, a reservoir of pests, a useful field experiment, a good open-air teaching laboratory, a retreat for renewal from the urban parts of South Yorkshire, a place to overcome alienation from nature, an important influence on local property prices, a place to employ youth, an historic landscape, an example of ancient land use, an essential part of the beauty of the Dales, a place to paint pictures or to photograph, part of a romanticized past, part of the Yorkshire heritage, the habitat of rare species, a species-rich ecosystem, a reservoir of species being made extinct elsewhere and so on. The authors in this book will, at one time or other, adopt all of these views and examples of almost all of these value-systems will be found in the chapters that follow.

THE NEED TO CONSERVE

Conservation has usually been advocated in response to the perception that without long-term and sensitive management of natural systems, resources are likely to become less available in future. Until recently such perceptions usually led to calls for resource conservation, and only rarely for nature conservation. The idea that we owe it to our descendants to care for resources, is said to have been first articulated by Jeremy Bentham (1748–1832) who inspired the foundation of University College London. Specifically in respect of food, the idea was more starkly expressed by Bentham's contemporary Thomas Malthus (1766–1835) and it is his views that are better remembered today: population, he believed, expanded geometrically, while food production could only do so arithmetically. Some nineteenth-century economists such as Ricardo (1772–1823) and Jevons (1835–82, Professor of Economics at University College) had even gloomier views, the first about food, the second about energy (Barnett and Morse, 1963; Simon, 1981), but most of their colleagues in the latter part of the century, including Marx, took a much rosier view of the future of resources. John Stuart Mill (1806–73), although among these optimists, seems to have been one of the first to advocate the conservation of pleasant environments and to note how this might constrain the unfettered exploitation of resources for mere material needs (Barnett and Morse, 1963).

The impetus behind the first conservation movement in the early twentieth century in the United States was the belief that the forest resource was being depleted too quickly, and that behind the Soil Conservation Service in the 1930s was concern that soil was being eroded faster than its renewal (Held and Clawson, 1968). There was a euphoric feeling about resources in the decade after the Second World War but this gave way again to some extreme forms of pessimism in the late 1960s and early 1970s, most memorably in the *Ecologist's Blueprint for Survival* (Goldsmith *et al.*, 1972) and in *The Limits to Growth*, sponsored by the Club of Rome (Meadows *et al.*, 1972).

In the field of resource conservation the most recent models (see the review by Cole, 1978; and the recent 'Cheermonger' polemic by Simon, 1981) eschew extreme pessimism, especially about minerals, but even, in many cases, about energy and food as well (although not about their distribution). But in nature conservation alarm has actually increased to critical proportions in many areas. Indeed, success in other areas of environmental exploitation may be the very reason for the crisis in nature conservation.

In Britain, the headlong, some would say blind, drive towards greater productivity of food and fibre seems to many to have led to the wholesale destruction of nature (see the discussion by Munton, Chapter 20 and Bowers, Chapter 21). There have been many accounts of these changes (Mabey, 1980; Shoard, 1980). Since those publications the Nature Conservancy Council has revealed that 4 per cent of their Sites of Special Scientific Interest were being damaged each year. Quite startling losses of heathland were in progress (see Harrison, Chapter 5). A survey of raised peat bogs in three areas of northern Britain showed that by 1978, 87 per cent of the original area had been used in some way. In Lancashire 99.5 per cent of these mosses had been reclaimed. Of 3188 Scottish deciduous woods in the 1950s only 2436 remained in the late 1970s. Most had been 'coniferized' (Goode, 1981; see also the discussion by Munton, Chapter 20). Parry, Bruce and Harkness (1982) have shown that in the North Yorkshire Moors the rate of primary moorland reclamation (from previously unreclaimed land) rose from about 8 hectares per annum in the 1904–50 period to about 180 ha year^{-1} in the 1963–74 period. The rate of primary loss to afforestation reached over 400 ha year^{-1} in the latter period. Parry's team have shown similar rates of loss in other national parks (Parry, Bruce and Harkness, 1981).

Further afield, change seems to be even more rapid. The tropical rain forest, where 40–50 per cent of the total world complement of species are to be found, is under very severe threat. Myers (1979) believed that as much as a half of the Amazon forest had disappeared, much of it in the last 25 years. He quoted Food and Agriculture Organization (FAO) statistics to the effect that the rate of clearance was 125 000 km^2 year^{-1} (almost the size of Great Britain). In Indonesia resettlement programmes were planned over 180 000 km^2 in the 1976–85 period, and reports suggest that the project is on schedule. Myers believed that over 1000 species of animals and plants were becoming extinct (world wide) every year.

The feeling that nature is seriously threatened in Britain is reflected in the results of the Countryside Commission's questionnaire survey (Fitton, 1978). About a half of the sample thought there had been a change for the worse in the countryside. It is also reflected in the growth in the membership of bodies such as the Royal Society for the Protection of Birds, the County Naturalists Trusts, the National Trust and many other concerned organizations (Perring, Chapter 25 of this volume). Other manifestations of the concern are the prominence given to nature conservation in the press, the large rallies that the Friends of the Earth can summon to causes such as 'Save the Whale', the guerilla tactics of organizations such as Greenpeace, and the length and vehemence of the debate in Parliament over the recent Wildlife and Countryside Act.

CONSERVATION AS A PROCEDURE

'Conservation' came to mean the planned use of functioning natural systems in nineteenth-century England. Calver's *Conservation and Improvement of Tidal Rivers* (1853) used the word to mean 'the scientific adjustment of means to an end'. In forestry the word was introduced into Britain from the Indian Forest Service, where 'the conservator' was the guardian of a resource that needed a long view and a knowledge of the ecology of the system. In North America, this use of the term is said to derive from Gifford Pinchot, one of the founders of the 'First Conservation Movement'. Later, in the 1940s the same meaning attached to the Soil Conservation Service of the US Department of Agriculture.

The ideas of responsibility towards nature are much older than the use of the label 'conservation'. In tracing these ideas back to antiquity, Passmore (1974) distinguished the persistent Judaeo-Christian belief that man had dominion over nature, from an ancient, at base animist, respect and sympathy with nature. He agreed with Lynn White (1967) in assigning modern science and technology to the first tradition, but his book was a search for a more sympathetic western tradition on which to base a more 'conservationist' policy.

Debates about ancient texts are by no means irrelevant today. The present US Secretary of State for the Interior, James Watt, is a born-again Christian, who bases (or at least based) his policies of opening up federal land to exploitation by timber, mining and energy companies on his belief that 'we must occupy [use] the land until Jesus returns'. He has recently retracted to some extent, but he has set the tone of the debate for his opponents in the Sierra Club who, unable to combat such statements with logic, themselves resort to Biblical quotation: 'Beware of false prophets. . . in sheep's clothing', etc. (Reed, 1981).

There were no alternatives to very vague notions of responsibility until the early nineteenth century. Then, Malthus advocated 'moral constraint' or specifically control of human reproduction. The British 'Conservation Society' still adopts this as one of their main policies. Jevons was in favour of some control on the use of coal, Pinchot in the twentieth century advocated careful use of forests, and Bennett, the first director of the US Soil Conservation Service, wanted the federal government to promote certain land-use practices and a land-use policy to control soil erosion.

The first real definitions of conservation as a procedure seem to have come in the writings of American economists in the middle decades of this century (Ciriacy-Wantrup, 1952; Clawson see Held and Clawson, 1968) (Zimmermann, 1951). Most of them were concerned with resources rather than nature conservation but their models are a useful treatment of the conservation process as a whole. Held and Clawson (1968), basing their formulation on the earlier work of the economist Raleigh Barlowe (1958), expressed this in a now well-known diagram (Figure 1.1).

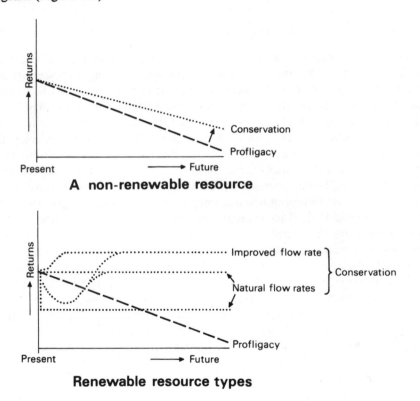

FIGURE 1.1   The conservation of non-renewable and renewable resources. In the lower diagram the uppermost curve represents conservation with no investment (a rare and lucky case); the middle curve represents an investment that only pays off after a few years—hence the initial drop in returns; the lowest curve represents a drop from the present level of use, because the resource cannot sustain such a level.

The conservation of two kinds of resource can be defined using this diagram. First are non-renewable, or stock resources, in other words those of which there is a defined amount, and which are not being recreated. For these conservation can only be what Ciriacy-Wantrup (1952, p. 51) called 'a redistribution of use in the direction of the future'. (He gave a mathematical formulation in his Appendix). In nature conservation unique sites such as Dungeness and Cow Green in Upper Teesdale (Gregory, 1971) or rare species such as an endemic *Sorbus* are really stock resources.

Second are renewable or flow resources, in other words those which are continuously being recreated such as timber, grain, etc. For these, conservationists aim at a steady yield. Most of nature conservation is concerned with flow resources. In the long term the usual target is sustained yield—an idea that has been explored very thoroughly by foresters and range-managers and has also been developed for wild populations of animals by ecologists such as Odum (1971). The idea is further discussed by Wood in Chapter 8 of this volume.

We can use Held and Clawson's model as the basis for explaining three more basic conservationist ideas: carrying capacity, time-horizons and social responsibility.

Some renewable resources are probably being exploited at rates far above their long-term capacity. If production is not drastically reduced soon, then a crash or an even stronger decline can be expected in the medium-term future. The suggested remedy is shown in the lowest curve on the lower diagram on Figure 1.1, that is a reduction in the rate of use. This is needed because of the belief that each primary resource, be it land, grazing or whatever, cannot sustain use above a certain threshold (it has a finite 'carrying capacity'). Production at higher levels may be possible temporarily but such exploitation would exhaust the system. 'Carrying capacity' is a very complex notion (Warren and Maizels, 1977; see also Wood, Chapter 8) and it is often very difficult to decide just what is the limiting primary resource, and, indeed, the limits vary temporally and spatially and with subtly different products. As Holling (1973) pointed out, environmental fluctuations as well as the dynamics of the species concerned make for enormous temporal variability in production. This complicates the notion of 'carrying capacity' and it also means that the 'target' levels of production must be set to allow for natural variation: other resources must not be threatened when the level is too high; and the genetic viability of the population must not be endangered by allowing it to become too low (see Wood, Chapter 15).

Another important part of conservation as a concept is that of a time horizon. If there are cyclical fluctuations in an animal population or in primary production as a result, for example, of climatic cycles, then the manager needs to plan for a period that involves at least one and preferably several complete cycles. If one were conserving a population of eland in the Sahel, for example, one would choose to maintain it in numbers that were adequate to survive a rare, severe drought and not too large to endanger other grazers such as wild gazelle or domestic cattle. Climatic cycles, of course, are rather difficult to detect, but some meteorologists claim that there have been 30-year cycles of drought in the Sahel, and if this is the case, a 30-year management period would be the very least that was adequate for conservation. Many 'natural' cycles are as long or longer than this.

Conservation is necessarily a long-term view of resource management, for another reason as well. Most of the systems that conservationists are concerned with are very slow to reach stability, (they have long 'relaxation times'). Examples are soils (see Warren Chapter 2) tree crops (see Goldsmith and Wood Chapter 17), and animal populations (Wood, Chapter 8). Even whole historic landscapes mature only after many decades (MacEwen and MacEwen, Chapter 22).

Long time-periods of concern necessarily entail wide spatial horizons as well (Harrison and Warren, 1970). For example, concern with the lifetime of a wood inevitably involves concern with the whole ecosystem of which it is a part. One needs to think, first, of a large enough unit to allow for rare tree-throw in some parts of the wood but continuity of cover elsewhere; second, of the external inputs of pollutants from a river catchment or an upwind factory; and, third, of the wood as part of a series of other woodlands, from and to which plants and animals might migrate.

Held and Clawson's (1968) model implies finally, and perhaps most fundamentally, that conservation is a concern more of society than of individuals. This follows from what has been said about temporal and spatial scales, for although each human individual can contribute his small share to everyday conservation of individual species, only a large group of people can be expected to concern itself with periods greater than an individual lifetime and areas greater than an individual estate. The historical growth of complexity in national organizations and the increased democratic diffusion of responsibility in societies over the last century can be seen in this light to have been a necessary precursor to the growth of conservationist ideas. In the United States some define conservation primarily as the long-term management of resources for the public good (Dana, 1958). In Britain a conservation policy could hardly be conceived of for an organization smaller than a Naturalists Trust (Perring, Chapter 24), or perhaps a local authority (Gordon, Chapter 23), or for a period of less than a decade, and more usually several. All of the following chapters envisage that conservation will be achieved only in spatial and temporal units larger than these minimum sizes.

CONSERVATION IN PRACTICE

The discussion above has now descended from Olympian, even transcendental, themes to the everyday practicality of local government. It was intended to place a conceptual framework around the more practical chapters that follow. But they need to be placed in another framework: the historical context of nature conservation in Britain, of its outcome in the organization of conservation today, and of the current debate about where it should go in the next few decades.

The first major revolution in British attitudes and approaches to nature was clinched by the National Parks and Access to the Countryside Act of 1949. With what are really only minor modifications, that Act and the thinking behind it (reported in various seminal background papers), still governs the way that nature is conserved in Britain today. Philip Lowe (Chapter 19) explains some of the detail of the evolution of those Acts and the parallel growth of voluntary groups concerned with nature.

Put very briefly, the Act and some of its precursors introduced Britain to a system that, as Duffey (Chapter 25) explains, is unique in Western Europe and perhaps in the world. It divided the so-called 'scientific' aspects of nature from a 'countryside' or recreational aspect, assigning each to a different arm of central government. It allowed the designation of different types of protection from Areas of Outstanding Natural Beauty (AONB), through national parks and Sites of Special Scientific Interest (SSSI) to national nature reserves. The division of natural science from social policy was further exaggerated by the later division of the Nature Conservancy in 1973 into the executive and managerial Nature Conservancy Council and the research-orientated Institute of Terrestrial Ecology.

Many nature conservationists still believe in the necessity of this 'two-culture' division for the analysis of their field, although it is undeniable that unity is needed for practical projects. It is hard to merge two very different paradigms of study, and our book rather clearly divides itself into three sections: in the first two natural science is the theme (as a background to management or in the actual technology of management), while in the third practical, social and administrative issues are the prime focus. We believe that there are advantages in this division: we see the natural science as an exploration of the options and constraints that nature offers and as a background to evolving a technology of management. It is for others using quite different approaches to interpret these possibilities in their social context and to apply them in practice.

A salient feature of the nature conservation field in Britain today, reflected in almost all the chapters in this book, is that the movement appears to be limbering up to a second revolution moving from a consensus mode, in which it is thought to have lost out, into more of a conflict mode. There have been several pointers to this. First was the recent passage through Parliament of the Wildlife and Countryside Act of 1981. Most conservationists regard the final form of the Act with deep suspicion (Rose and Secrett, 1982) and certainly do not regard it as in any way a closure of debate on some vital issues such as the protection of Sites of Special Scientific Interest, the agricultural grant-support system, the question of planning control in the countryside or even the whole organization of conservation that was inherited from the 1949 Act (see MacEwen and MacEwen, Chapter 22). Indeed, the paltry sums of money that have now been provided for the compensation of the landowners of SSSIs for *not* desecrating them seems to be the most obvious of the ticking timebombs (Rose and Secrett,

1982). It is clear that the landowning and farming lobbies believe this too, for they have cautioned their members to cooperate under the real threat that, if they do not, any future government would almost certainly introduce real controls.

Second, there has been a phenomenal growth in the membership of the relevant voluntary bodies as described by Perring (Chapter 24). Such numbers as a proportion of our total population are unique in a world context, although the other side of the coin may be the surprisingly small amount of interest in political action in Britain as compared to some European countries.

Third, there has been the series of important international studies that have appeared in recent years. These began with the Stockholm Environment Conference of 1972 which lead to the institution of the UN Environment Programme and the series of World Conferences on Population, Desertification (Warren and Maizels, 1977), Water, the Mediterranean and so on. More recently there has been the so-called Brandt Report on 'North–South' problems (Brandt *et al.*, 1980), and most recently the World Conservation Strategy (WCS) published by the International Union for the Conservation of Nature and Natural Resources (IUCN, 1980).

The World Conservation Strategy is of necessity very generalized, aimed as it is at a global audience, but it encourages individual governments to produce more specific national strategies that conform to its own recommendations, and these are now appearing. In Britain a study is proceeding to produce a strategy by the middle of 1983 and this is generating some informed and useful debate about the future of conservation in the United Kingdom.

Several older, more diffuse debates are now being focused onto the idea of this conservation strategy. One is the debate about a national land-use policy, an approach that has long been advocated by the Nature Conservancy Council (Nature Conservancy Council, 1977). This body has even encouraged localized attempts at such a policy, notably in the Berwyn Mountains in north central Wales (Williams and Harding, 1982) but with mixed success. A land-use policy, though, has powerful opponents not least among farmers and foresters and even from the ranks of conservationists themselves. Hall (1981), for example, believed it to be not only quite impractical but also highly undesirable, since it would pass all decisions to a centralized, poorly accountable and probably secretive bureaucracy. Very much the same debate with similar conclusions had taken place in the United States in the early 1970s (see Criley, 1972, and others in the same journal). In Britain some observers fear that nature conservation would come off second best in any secret struggles, although it has not done so in other countries such as the Netherlands.

Another related issue that has simmered ever since the 1959 Act but is now boiling up has been the question of the division of roles between the Nature Conservancy Council and the two Countryside Commissions. There are many who now advocate the amalgamation of the three bodies (Shoard, 1980; see also MacEwen and MacEwen, Chapter 22) for the reasons that they both cater for the management and appreciation of nature, and that they should be more effective as one body in maintaining our natural heritage. There are, however, many others who believe that nature conservation might suffer, and that one body would be an easier target than several.

And finally the issue of Planning Controls in the Countryside that was dominant for a long time after the 1949 Act is surfacing with renewed vigour. Shoard's (1980) book is the strongest advocate of this and others will be found within this book.

The chapters in this book will be approaching these and other questions each from its own particular viewpoint: that of a habitat, a process, an agency, or a type of interest. We have introduced each of the three sections with another, much shorter overview within which to locate their individual approaches.

ACKNOWLEDGEMENTS

We would like to thank the following for their comments and suggestions on earlier drafts of this chapter: David Lowenthal, Richard Munton, Philip Lowe, Sam Berry, and Brian Wood.

REFERENCES

Ashby, E. (Lord) (1978). *Reconciling Man with Environment*, Oxford University Press London, 104 pp.

Barlowe, R. (1958). *Land Resource Economics. The Political Economy of Rural and Urban Land Resource use*, Prentice-Hall, Englewood Cliffs, NJ, 420 pp.

Barnett, H.J. and Morse, C. (1963). *Scarcity and Growth: The Economics of Natural Resource Availability*, Johns Hopkins Press for Resources for the Future, Inc., Baltimore, 288 pp.

Brandt, W. *et al.* (1980). *North—South. A Programme for Survival*, Pan, London, 304 pp.

Calver, E.K. (1853). *The Conservation and Improvement of Tidal Rivers Considered Principally with Reference to their Tidal and Fluvial Powers*, John Weale, London, 101 pp.

Ciriacy-Wantrup, S.V. (1952). *Resources Conservation, Economics and Policies*, University of California Press, Berkeley, 395 pp.

Clawson, M. (1969). *Economics of Outdoor Recreation*, Johns Hopkins University Press Resources for the Future, Inc., Baltimore, 328 pp.

Cole, S. (1978). The global futures debate 1965–1976. In C. Freeman and M. Jahoda, (Eds), *World Futures, The Great Debate*, Martin Robertson London, pp 9–50.

Connell, J.H. and Orias, E., (1964). The ecological regulation of species-diversity. *American Naturalist*, **98**, 399–414.

Criley, W.L. (1972). Implementing a national land-use policy: can it be done on an ecological basis? *Journal of Soil and Water Conservation*, **27**, 226–227.

Dana, S.T. (1958). Pioneers and principles. In H.C. Jarrett (Ed) *Perspectives on Conservation*, Johns Hopkins Press, Baltimore, pp. 24–33.

Dillard, A. (1976). *Pilgrim at Tinker Creek*, Picador, London, 237 pp.

Douglas, M. (1973). *Natural Symbols, Explorations in Cosmology*, Barrie and Jenkins, London, 219 pp.

Fitton, A.M.H. (1978). The reality—for whom are we actually providing? In *CRRAG Conference 1978, Countryside for All*, Countryside Commission CCP 117, pp. 38–67.

Frankel, O.H. and Soulé, M.E. (1981). *Conservation and Evolution*, Cambridge University Press, Cambridge, 327 pp.

Goldsmith, E., Allen, R., Allaby, M., Davoll, J. and Lawrence, S., (1972). Blueprint for survival. *The Ecologist*, **2**, 50 pp (also Penguin, London).

Goldsmith, F.B. (1980). An evaluation of a forest resource—a case study from Nova Scotia. *Journal of Environmental Management*, **10**, 83–100.

Goode, D. (1981). The threat to wildlife habitats. *New Scientist*, **89**, 219–223.

Gregory, R. (1971). *The Price of Amenity: Five Studies in Conservation and Government*, Macmillan, London, 319 pp.

Grove, R.H. (1981). The use of disguise in nature conservation, *Discussion Papers in Conservation No. 32*, UCL, 38 pp.

Habermas, J. (1971). Technology and science as 'ideology'. In J. Habermas (Ed) *Toward a Rational Society*, Heinemann, London, pp. 81–122.

Hall, C. (1981). Who needs a national land-use strategy. *New Scientist*, **90**, 10–12.

Harrison, C.M. and Warren, A. (1970). Conservation, stability and management. *Area*, **2**, 26–32.

Held, R.B. and Clawson, M. (1968). *Soil Conservation in Perspective*, Johns Hopkins University Press for Resources for the Future Inc., Baltimore, 344 pp.

Helliwell, D.R. (1973). Priorities and values in nature conservation. *Journal of Environmental Management*, **1**, 85–127.

Holling, C.S. (1973). Resilience and stability of ecological systems. *Annual Review of Ecology and Systematics*, **4**, 1–24.

International Union for Conservation of Nature and Natural Resources (1980), *World Conservation Strategy*, Gland, Switzerland, 30 pp.

Leach, E. (1970). *Lévi-Strauss*, Fontana, London, 126 pp.

Le Roy Ladurie, E. (1980). *Montaillou: Cathars and Catholics in a French village 1294–1324*, English Edition, Penguin Books, 381 pp.

Lévi-Strauss, C. (1966). *The Savage Mind*, Weidenfeld and Nicolson, London, 290 pp (first published 1962).

Lewis, C.S. (1943). *The Abolition of Man*, Fount Paperbacks, London.

Lowenthal, D. and Binney, M. (eds) (1981). *Our Past Before Us. Why Do We Save It?* Temple-Smith, London, 253 pp.

Mabey, R. (1980). *The Common Ground. A Place for Nature in Britain's Future*, Hutchinson, London, 280 pp.

May, R.M. (1974). *Stability and Complexity in Model Ecosystems*, Princeton University Press, Princeton, New Jersey, 235 pp.

Meadows, D.H., Meadows, D.L., Randers, J. and Behrens, W.W. III. (1972). *Limits to Growth*, Universe, New York, 205 pp.

Moss, B. (1977). Conservation problems in the Norfolk Broads and rivers of East Anglia, England—phytoplankton, boats and the causes of turbidity. *Biological Conservation*, **12**, 95–114.

Mostyn, B.J. (n.d.). Personal benefits and satisfactions derived from participation in urban wildlife projects, Nature Conservancy Council Interpretation Branch, 66 pp.

Myers, N. (1979). *The Sinking Ark. A New Look at the Problems of Disappearing Species*, Pergamon Press, Oxford, 307 pp.

Nature Conservancy Council (1977). *Nature Conservation and Agriculture*, Nature Conservancy Council, London, 40 pp.

Newby, H. (1979). *Green and Pleasant Land*, Pelican, 301 pp.

Odum, P. (1971). *Fundamentals of Ecology*, W.B. Saunders, Philadelphia, 574 pp.

O'Riordan, T. (1976). *Environmentalism*, Pion, London, 373 pp.

Page, H. and Warren, A. (1982). More about the purpose of nature conservation. *Ecos*, **3**, 27–29.

Passmore, J. (1974). *Man's Responsibility for Nature. Ecological Problems and Western Traditions*, Duckworth, London, 213 pp.

Parry, M., Bruce, A., and Harkness, C., (1981). The plight of British moorland. *New Scientist*, **90**, 550–551.

Parry, M., Bruce, A. and Harkness, C., (1982). Changes in the extent of moorland and roughland in the North York Moors, *Surveys of Moorland and Roughland Change No. 5*, University of Birmingham, 72 pp.

Popper, K.R. (1974). Scientific reduction and the essential incompleteness of all science. In F.J. Ayala and T. Dobzhansky (Eds) *Studies in the Philosophy of Biology*, Macmillan, London, pp. 259–284.

Rackham, O. (1980). *Ancient Woodlands*, Arnold, London, 402 pp.

Reed, N.P. (1981). In the matter of Mr Watt. *Sierra Club Bulletin*, **66**, 6–15.

Robinson, J. (1964). *Economic Philosophy*, Penguin, London, 104 pp.

Rose, C. and Secrett, C (1982). *Cash or crisis: The Imminent Failure of the Wildlife and Countryside Act, BANC and Friends of the Earth Special Report*, 22 pp.

Rose, F. (1974). The epiphytes of oak. In M.G. Morris and F.H. Perring (Eds), *The British Oak*, Botanical Society of the British Isles, pp. 250–273.

Sandbach, F. (1980). *Environment, Ideology, Policy*, Blackwell, Oxford, 254 pp.

Sax, J.L. (1980). *Mountains Without Handrails, Reflections on the National Parks*, University of Michigan Press, Ann Arbor, Michigan, 152 pp.

Shoard, M. (1980). *The Theft of the Countryside*, Temple-Smith, London, 273 pp.

Simon, J.L. (1981). *The Ultimate Resource*, Martin Robertson, Oxford, 415 pp.

Spearman, D. comp. (1966). *The Animal Anthology*, John Baker, London, 208 pp.

Thomas, L. (1974). *The Lives of a Cell*, Allen Lane, London, 153 pp.

Tittensor, R. (1981). A sideways look at nature conservation, *Discussion Paper in Conservation*, UCL, 29, 45 pp.

Warren, A. and Maizels, J. (1977). Ecological change and desertification. In *Desertification: Its Causes and Consequences*, UNEP, Pergamon, Oxford, pp. 169–260.

White, L. (1967). The historical roots of our ecological crisis. *Science*, **155**, (3767) 1204.

Williams, H.J. and Harding, D. (1982). Towards a land use strategy for the uplands of Wales. *Quarterly Journal of Forestry*, **76**, 7–22.

Williams, R. (1973). *The Country and the City*, Paladin, London, 399 pp.

Zimmermann, E.W. (1951). *World Resources and Industries*, Harper Bros., New York, (first published 1933), 832 pp.

PART I
THE ECOLOGICAL BASIS
OF CONSERVATION

Introduction

Nature conservation must be based on sound analyses of nature itself, and this is the rationale of the following eight chapters. All the authors of these chapters are natural scientists who show just how vital is an understanding of ecological processes to any attempt to manipulate ecosystems for some end. Our aim as editors has been to collect essays about the principal habitats in Britain and its coastline (the open sea, Chapter 3, the coast, Chapter 4, lowland heaths, Chapter 5, woodlands, Chapter 6 and upland moors, Chapter 7), although the coverage is by no means comprehensive (some other habitats are discussed in Part II). We have also included three chapters on discrete elements of ecosystems that are important in all habitats: soils (Chapter 2), animals (Chapter 8) and genetics (Chapter 9). Again, we could clearly have enlarged this list, but felt that most of the other elements such as plant population processes are covered in other chapters. One obvious and regrettable omission is hydrology.

The authors do not, however, confine themselves to academic analysis, believing this merely to be a means to an end. Three practical questions concern them. The first is the importance of selecting the best sites on which to conserve our flora and fauna. In a country like Britain, where nothing is even 'near-natural' and where almost every semi-natural site is under threat from agriculture, forestry, housing, industry or pollution, this is a critical problem. The second focus is the identification of threats to the species conserved on sites—from the destruction of their habitat by erosion or pollution, or from genetic 'drift'. And the third focus is the choice of appropriate management for the sites or populations that are chosen to be conserved: how, in the face of other pressures, should the conservationist manipulate his chosen system in the long-term interest of his charges?

These chapters are essential background for those that follow. Their theoretical and practical conclusions must underlie all practical conservation (Part II) and must also underpin the legislative, political and administrative efforts that are discussed or advocated in Part III.

CHAPTER 2

Conservation and the Land

A. WARREN

INTRODUCTION

Good naturalists are said to have a sixth sense that helps them to locate the plants, animals and soils that interest them. Ecological theories about the numbers of species in a succession or on 'islands' of different sizes are too generalized to locate specific sites or species. It is the 'lie of the land' that really leads naturalists to their rare ferns, herb-rich meadows, rich invertebrate fauna, roosts, nests, or burrows of vertebrates, old trees or ancient soils. I first want to show in this chapter how the sciences of the land—geomorphology and pedology—can help to articulate and systematize this sixth sense; second, how recent findings in these sciences relate to other ecological models; and third, how they can help in the selection and management of sites that are of special interest in nature conservation.

Nature conservationists in Britain have been busy evolving criteria for selecting sites (see Goldsmith, Chapter 14). Very few of the criteria that have emerged relate specifically to the ground itself, and yet of the ten best-known criteria (those of Ratcliffe, 1977) only 'intrinsic appeal' and 'recorded history' do not depend primarily on non-biological factors. The 'extent' of a relatively homogeneous site, for example, is controlled partly by the compartmentalization of the landscape by ridges, valleys, streams or arms of the sea, and these may also control its shape (a factor discussed by Diamond, 1975). The slopes of species—area relations, on which a large part of the value of 'extent' depends, are also controlled by the form of the land (Kilburn, 1966). The diversity of species on a site may depend as well on its inherent fertility (Grime, 1979, p. 185), and stability, and these are both dependent to a large extent on physical processes (Connell and Orias, 1964). The preservation of 'naturalness' depends partly on how vulnerable a site is to physical disturbance by accelerated erosion. Its 'typicality' and 'position in an ecological/geographical unit' depend on the geologically and geomorphologically controlled distribution of sites. Finally the 'potential' of the site is related to how quickly its physical structure can

19

recover from disturbance. If the term 'recorded history' includes history preserved in the landscape (for example as pollen), then this criterion too relates to physical protection from erosion. Even if alternative criteria are proposed, physical factors are still very important: the 'maturity' criterion of Adams and Rose (1978), for example, depends on the physical stability of sites.

The lithology of the solid rock or drift that underlies a site is very obviously an important characteristic of a site, but of equal and sometimes greater importance are its geomorphological and pedological attributes, and these are often misinterpreted or underplayed. There have been attempts to incorporate these into ecology, some by well known ecologists (such as Whittaker, 1956), but these have received scant attention from ecologists and conservationists, who have often used rather dated concepts (Pickett, 1976) or have been at a rather generalized level (see Drury and Nisbet, 1971 for animal ecology; and Wright, 1974 for wildland conservation). This lack of interest occurs in the face of evidence that species-diversity is better explained by soil type than by isolation, size of 'island' and even latitudinal position (Johnson and Simberloff, 1974) and the discovery that landscape position can explain a large amount of variance in communities (Strahler, 1978).

In the last decade there has been an explosion of both concepts and observations about the soils and landforms of temperate landscapes, much of it based on British experience, and it has several important ecological implications. Some sites are so little affected by surface erosion that their future can only be a slow decline in fertility. On other kinds of site nutrients are quite literally continuously unearthed by erosion. Some are inherently unstable at one scale and stable at another. In yet others the relationship between scales of stability and instability are quite different. In consequence, some types of site are favoured by *K*-selected species and some by *r*-selected species. Some sites are resilient to disturbance, others are not. Finally, an understanding of geomorphological processes helps to explain the distribution, shape and rarity of types of site.

TYPES OF SITE

The development of temperate landscapes is controlled by the downcutting of stream channels. The few exceptions, such as areas with a glacial inheritance or underlain by permeable limestone, have less orderly assemblages of slopes, but the elements are essentially similar to those in the monotonously repetitious patterns associated with 'fluvial' landscapes.

Many systems for describing the units of these landscapes have been developed. The one used here (Figure 2.1) is synthetic, bearing a close resemblance to those of Hack and Goodlet (1960) for the central Appalachians, Dalrymple, Blong and Conacher (1968) for New Zealand and Curtis, Doornkamp, and Gregory (1965) for England, all of which incorporated earlier experience. The descriptive units of these authors can now be justified on the

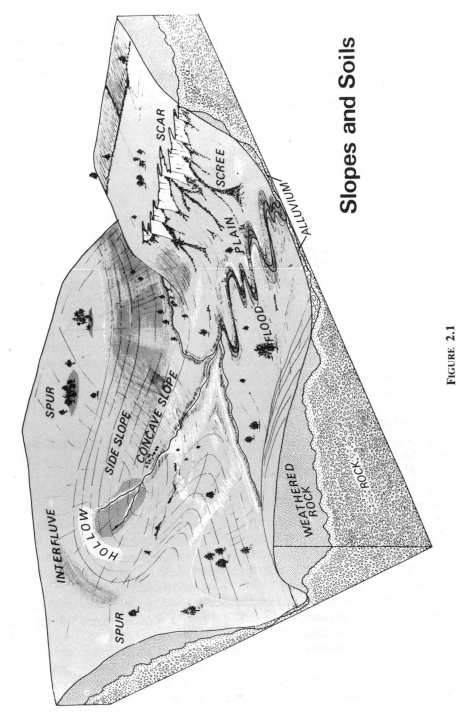

Slopes and Soils

FIGURE 2.1

theoretical grounds that they isolate distinctive combinations of processes (Kirkby, 1980). They have the further justification that since the early simple descriptions of a soil-slope system as a 'catena' (Milne, 1935), more and more analyses of soil processes are being framed in terms of their position in three-dimensional landscapes (Huggett, 1982). Moreover, the units described here are very close to the landform units used by ecologists (Whittaker, 1956), and can also be related to practical systems of land classification (Bibby and Mackney, 1969).

The simplest way to describe the differences between these types of site is in terms of a mass continuity equation (Kirkby, 1980):

$$I = O - \Delta S \tag{1}$$

I is the input of material from weathered rock, dust and organically fixed elements or from further up-slope as sediment; O is output of clastic fragments and dissolved material and ΔS is the change in the store of soil or alluvium. These elements have different values on different types of site.

It has been asserted for many parts of the landscape and for some soils that they are in 'dynamic equilibrium'; in other words, that over long periods input is equal to output and storage remains roughly the same (Hack, 1975; Yaalon, 1971). This appearance of equilibrium is, however, very dependent on the scale of enquiry (Allen, 1974). On a gentle slope soil particles are moved up and down mainly by wetting and drying cycles that occur once in about 10^4 seconds and over distances of less than 1 m. Below these scales the soil is seen as highly mobile; above them it appears to be near equilibrium. At these scales we might say that the system has low-order steady-state equilibrium (Allen, 1974)—a state probably favoured by long-lived sessile species.

In an alluvial valley in contrast, the accumulated sediment at one site is moved when the stream channel cuts its way across the point concerned. Below about $100 \, m^3$ or 100 years the system is highly mobile and, as such, a continually renewing habitat. But for long periods of time (more than 10^3 years) and over long stretches (more than 1 km) the flood plain may seem remarkably unchanged. At even larger scales $10^5 - 10^7$ years and $10^5 \, km^2$) some landscapes have been extraordinarily 'stable' (probably under dense forest canopies), losing only dissolved products to the oceans (as during the formation of the chalk in the Cretaceous), while others have been extremely unstable yielding large amounts of clastic sediment (as in the mountain-building period of the Tertiary). These two states are the '*biostasie*' and '*rhextasie*' of Erhart (1955). I will not, however, be concerned with such geological perspectives, and confine myself to scales of less than 10^2 years and 10^5 km.

Another general differentiation between types of process (and so types of site) in the physical landscape is that they have different 'relaxation times'—that is, they vary in the time they take to reach equilibrium or steady-state after being disturbed (Graf, 1977). In Britain, where a balance between slow erosion and slightly faster soil deepening on upland sites has probably not yet been reached in the 10^4 years since the end of the last glaciation, the relaxation time for upland soils is long; a stream bed, on the other hand, can reach a new equilibrium within minutes of a change in discharge.

The nature of these equilibria or steady-states also distinguishes types of site from each other. Pedologists and geomorphologists believe that as the soil is deepened by the weathering of the underlying rock, it provides more and more protection from the harsh environment of the surface (Kirkby, 1980). The rate at which the soil deepens will eventually slow down and approach the rate of erosion: in other words, an 'equilibrium' or steady-state depth of soil will be reached and maintained as the whole landscape is lowered. The character of this equilibrium varies from site to site and is of profound importance to the character of the biological part of the ecosystem.

SPURS AND INTERFLUVES

The spurs and interfluves in Figure 2.1 are the most stable of all the types of site. They usually have the deepest soils, and are potentially the least fertile. They are also the only sites that can develop 'mature' soil profiles.

The inputs here are from rainfall, dry fallout, and organically fixed carbon and nitrogen, as well as from weathering rock. The balance of inputs varies from site to site. The East Twin Brook catchment on Mendip, which is mostly covered by gentle upper slopes, actually gained more from the atmosphere than it lost over a 1-year period, and only catastropic erosion must keep the balance (Finlayson, 1978). Such natural catastrophes, however, must be very infrequent: in Maryland the June 1972 flood had a recurrence interval of 100 years and yet had virtually no effect on gentle slopes (Costa, 1974). Yet Costa believed that rare washouts must occur on these slopes, for otherwise much thicker mantles of leached quartz sand would have accumulated.

In upland Britain many spurs are covered by deep 'blanket' peat, which is clearly the result of another kind of accumulation. At Moor House in the northern Pennines where the net primary productivity was estimated at 0.034 $g cm^{-2}$ $year^{-1}$, there appeared to be an addition of 0.97 $cm year^{-1}$ to the anaerobic peat (Clymo, 1978).

On other sites, especially in the lowland east of Britain, weathering of the underlying rock or drift is the main input to sites on spurs. Hydrolysis, which is the chief weathering process, attacks some rock minerals more than others. Felspars and micas tend to break down first contributing briefly to the store of clay minerals before being entirely lost, whereas quartz, being the most resistant common mineral, accumulates in the soil (Patton, 1978). This means that the character of the soils on spurs is more distinct from that of the parent rocks than the soils of most other types of site (Walker, Hall, and Protz, 1968).

The sedimentary output from spurs is very small in lowland temperate areas, where spurs and interfluves dominate the landscape. In lowland Sussex, mechanical removal is $143\,t\,km^{-2}\,year^{-1}$ while solution removes up to 207 $t\,km^{-2}\,year^{-1}$ (Collins, 1981), and in the Maryland piedmont where mechanical erosion removes $3.2 \times 10^{-6}\,t\,km^{-2}\,year^{-1}$, it is five times less effective than solution (Cleaves, Godfrey and Bricker, 1970). This is not to say that soil particles do not move, but their movements are over small distances and in random directions (Finlayson, 1981), so that in Karcz's (1980) terms the soil system is in a high entropy state: the thermodynamic equilibrium of Chorley and Kennedy (1971, pp. 201–250). The forces that produce these movements, in Galloway soils at least, have been ranked in the following order: soil moisture > freeze-thaw > worms and other burrowing animals > temperature changes > plant roots (Kirkby, 1967).

Biological activity is at a minimum on spurs. In the Ardennes, burrowing rodents brought only $14.5\,cm^3\,year^{-1}$ to the surface on 'upper' slopes compared to $24.5\,cm^3\,year^{-1}$ on slopes near the river (Imeson, 1976). Upland soils evolve less CO_2 (de Jong 1981), and this has been interpreted as a sign of low primary productivity. There are fewer phytoliths in some 'upland' soils than in nearby side slopes, another indication of low primary productivity (Jones and Beavers, 1964).

Because soils with a full complement of horizons (zonal soils in Dokuchaiev's famous classification) can only survive on spurs, it is only spurs that retain evidence from which pedologists can deduce soil processes (Ball and Stevens, 1981). An example is the 'multiple-podzol' found only on gentle upper slopes in a wood in Northern Ireland described by Cruickshank and Cruickshank (1981). Here the two superimposed organic-rich B horizons were interpreted as evidence for two periods of woodland clearance separated by a period of woodland regeneration.

Some characteristics of the soil on spurs and interfluves may not reach equilibrium for a very long time indeed (they have long 'relaxation times'). After a period of rapid change they continue to adjust for many thousands of years, and after a peak, the trend of fertility is continuously downward. The phosphorus content of soils, being a good measure of their fertility, is a sensitive indicator of this degradation (Smeck, 1973). In a dated sequence of New Zealand soils there appeared to be a peak in phosphorus availability to plants at about 10^4 years, but after $3-6 \times 10^5$ years there is virtually no more available (Walker, 1965). In Britain many spur soils seem to have been slowly and quite naturally acidifying since the disturbance of the ice age. Even in the chalky boulder clay, decalcification has commonly extended to 70 cm and sometimes to 2 m (Catt, 1979).

The vegetation on Walker's New Zealand 'chronosequence' had changed from low early successional scrub at the start, through high, productive forest at the peak of phosphorus availability, to low sparse scrub again at the end. This picture of eventual degradation is rather different from the classic picture of 'succession' and 'climax' as described by Clements (1936). Although ecosystems do seem to develop more independence of their environment in a succession, and so escape some of the consequences of the inevitable loss of nutrients (Odum, 1969), there is probably no escape from the inexorable loss of fertility on geomorphologically stable sites. Indeed it was just such evidence that has led many to attack Clements' ideas (Drury and Nisbet, 1971; Wright, 1974).

Depending probably on the local climatic water balance over a year, some spurs in temperate climates are drier and some wetter than slopes lower down. In the Appalachians they are drier and often capable of supporting only grassy 'balds' (Hack and Goodlet, 1960; Whittaker, 1956). In the Cotswolds, spurs are wetter than steeper slopes and beech woods there support a ground-flora of blackberry (*Rubus fruticosus*) and bluebell (*Hyacynthoides non-scriptus*) (Barkham and Norris, 1970). Even in dry East Anglia, spur sites support very wet woodlands on gley soils—as at Bradfield Woods in Suffolk (Rackham, 1980). In wetter western Britain gentle spur sites are very wet indeed and only acid bog communities survive.

Some slowly degrading soils on spurs seem to reach a point of great sensitivity in their development. The brown forest soils that predated mesolithic man on the North Yorkshire Moors were being gradually and quite naturally acidified, but oak and birch scrub were partly counteracting the leaching by bringing up nutrients in their deep roots. This delicate balance was tipped towards podzolization and a new less fertile balance, when mesolithic and neolithic men cleared the trees (Dimbleby, 1974).

On the other hand spurs are the least sensitive parts of the landscape to erosion. Rain-splash and overland flow can indeed remove material on very low angle slopes, especially if the soil has lost organic matter (Faith, 1977) and where acidity discourages the formation of stable aggregates. In the North Yorkshire Moors where upland sites laid bare by intense moor-burning are only recolonized slowly, a bare mineral soil on a spur lost as much as 63.9 mm in one year, but when heather (*Calluna vulgaris*) reached 20 cm in height accumulation replaced erosion (Imeson, 1974). Nevertheless erosion is least at low slope angles (Morgan, 1981) and convex spurs tend to dispense rather than concentrate flow, so that rills and gullies tend to die out upslope (Smith and Bretherton, 1972).

If, however, all or most of the soil on a spur were to be lost, in some really catastrophic event, then it might take decades, even centuries to recover. Once eroded, the soils on spurs would probably go through several short cycles of accumulation and loss, before a long enough period without severe storms allowed the soil to deepen beyond the critical depth that would support enough vegetation to protect it from 'normal' erosion (Kirkby, 1980).

Some facts about spurs and interfluves are summarized in Table 2.1 where they are contrasted directly with other types of site.

TABLE 2.1   Characteristics of different elements of the landscape.

	Turnover	Soil storage	Fertility	Stability (return period of disturbance)
Spurs and interfluves	Low	High	Low	?1/1000?
Straight slopes	High	Low	Medium	1/100
Scars	High	Very low	Low	1/10
Hollows and concavities	Medium	Medium	High	1/2 to 1/200
Floodplains	High	High	High	1/2 to 1/300
Channels	Very high	Very low		$1/\frac{1}{2}$

SIDE SLOPES

Side (or 'straight') slopes usually separate the upland spurs and interfluves from hollows and concave slopes below. Their character is quite different to that of the spurs. Their soils are potentially richer in nutrients; they maintain form, fertility and depth of soil over long periods despite a constant turnover; and they have much more small-scale variation. The 'steady state' projected for the Hubbard Brook ecosystem by Bormann and Likens (1979, p. 183) and for ecosystems in general by Drury and Nisbet (1971) could only take place with the constant supply of nutrients on side slopes.

The constrast of side slopes with other elements in the landscape is again best seen in the input−output−storage relation (equation 1). The main 'forcer' of the side slope system is mechanical removal of sediment at the base, either by overland flow after intense showers or by actual undercutting by a stream. The slope adjusts by becoming steeper and this encourages fast mass movements of soil; in turn the thinned soil cover allows faster weathering (input) of the rock beneath (Kirkby 1980). On Appalachian side slopes soil movement is fast enough to remove all the signs of tree-throw, whereas the nearby spurs retain these mounds (Hack and Goodlet, 1960). In other words the slow 'self-cultivation' by tree-throw on the spurs is replaced by the faster self-cultivation of erosion and weathering on the side slopes.

In the Appalachians the soils of the side slope are usually less than 1 m deep compared to 4 m on the spurs (Hack and Goodlet, 1960). Even in parts of this country where parent rocks are not very base-rich, rainfall is high and temperatures low, the fast turnover of soil and flushing of nutrients from upslope keep the soils of side slopes from being podzolized; they remain as constantly rejuvenated brown earths (Glentworth and Dion, 1949). Of all the soils in a landscape (except of course those on scars) those on side slopes have the closest relationship to the underlying rock (Walker, Hall, and Protz, 1968).

These slope and soil systems are in a much lower entropy state than those uphill in Karcz's terms (1980). Movement within the soil body is still dominated by vertical motion, but much mᴗre of the horizontal movement is directed downhill (Finlayson, 1981). These are, in fact, one of the most organized parts of the landscape. A whole range of ecologically important factors such as the pH, and organic matter content of the soil, and the surface slope, are better intercorrelated here than anywhere else on a slope (Anderson and Furley, 1975). The correlation coefficients are highest when the slope is actively being undercut by a stream (Chorley and Kennedy, 1971, pp. 156–157).

'Equilibrium' in the ecology of side slopes has a very different meaning to that on spurs. In geomorphological language equilibrium here is the 'dynamic equilibrium' of Hack (1975) and in ecological language the 'steady state' of Bormann and Likens (1979) and Drury and Nisbet (1971). Slope form, soil depth, nutrient levels, perhaps even the age-structure of the woodland have a stable pattern at scales of perhaps 10^3 years and 0.5 km^2, despite constant throughput of material and energy.

Despite organization at the scale of the whole slope unit, there is often great microdiversity of habitats on side slopes. On the steep Appalachian side slopes there are often small rock outcrops and boulder-trains (Hack and Goodlet, 1960). On many steeper slopes on soft rocks in England there are 'terracettes' with different habitats on steps and rises (Anderson, 1972). In Wales, there are often small 'sheep scars' (bare patches of soil up to 30 m^2 surrounded upslope by a vertical bare soil face) from which erosion is fast (Slaymaker, 1972). In the Peak District these scars cover 0.14 per cent of some basins and their backwalls may retreat at 0.93 cm year^{-1} giving maximum rates of loss of 34.4 t ha^{-1} year^{-1} (Evans, 1971).

The soils on side slopes show many signs of higher biological activity than those on spurs. They contain more nitrogen (total N) (Aandahl, 1948), evolve more CO_2 (de Jong, 1981) contain more plant phytoliths (Jones and Beavers, 1964), and harbour more burrowing animals (Imeson, 1976).

The nutrient-supply and the range of microhabitats would seem to give side slopes the chance of greater species diversity (Grime, 1979, p. 185) than spurs but less than hollows and floodplains. In the Great Smokey Mountains, the highest diversity of species occurs in 'coves'. The woodland here has more large old and widely spaced trees than those upslope (Whittaker, 1956). Certainly some trees, such as the bristlecone pine (*Pinus aristata*) in California survive up to 4000 years on sites where the erosion rate has evidently been between 2.4 and 3.6×10^{-2} m year^{-1} for the whole of that period (La Marche, 1968). In contrast to the Appalachians British side slopes are usually drier than nearby spurs; on Cotswold side slopes the beech woods are less mature, and the trees more closely spaced than on the spurs, and they include ground flora such as sanicle (*Sanicula europaea*) and St John's wort (*Hypericum exotericum*) (Barkham and Norris, 1970).

Side slopes are much more vulnerable to disturbance than spurs. Several studies have shown how steeper slopes are more vulnerable to trampling by visitors (Quinn, Morgan, and Smith, 1980). But the 'relaxation time' on side slopes (that is, the time to re-establish a stable state) is much shorter than on spurs. Several recent measurements suggest a time to equilibrium of $10-10^2$ years at the most for such slopes (Brunsden and Thornes, 1979).

The main characteristics of straight slopes are shown in Table 2.1.

HOLLOWS AND CONCAVE SLOPES

Hollows are potentially both more valuable for conservation and more vulnerable than spurs, interfluves or side slopes. They are places where the slope is concave both in profile and in plan, so that eroded soil, nutrients and water are funnelled together. Continuous input into the storage element in the continuity equation (equation 1) builds up deep, fertile, moist soils. Whilst this is the process in 'normal' years, the store is emptied by rare catastrophic events.

Hollows are often rather small in area. A typical example is the East Twin catchment in the Mendips (Weyman, 1974). The broad concavity at the head of the valley is only the setting for several smaller narrower concavities, the 'true' hollows which are only about 5 m across and 80 m long. Bilberry (*Vaccinium myrtillus*) and purple moor grass (*Molinia caerulea*) are the dominant plants here in contrast to the *Calluna* and *Erica* of the upper slopes and *Pteridium* on the side slopes. Ecologists would call these hollows 'seepages' or 'flushes'. In our own study of ecological value in Eastnor Park in Herefordshire (Warren and Cowie, 1976) we found that these small sites contained the rarer and more interesting elements of the local flora. In their classic studies of forest vegetation in the central Appalachians, Hack and Goodlet (1960) found the tallest, most species-rich stands of hardwoods in the hollows.

The hydrology of hollows is basic to their geomorphology, pedology and ecology. In rocky, steep terrain, like the central Appalachians, the hollows are filled with coarse debris, through which water moves very quickly (Hack and Goodlet, 1960). In Britain, certainly where there are softer rocks and gentler slopes, hollows accumulate finer sediment which acts like a sponge filling up with water after a shower, and then 'spilling over' into the stream (Anderson and Burt, 1978). Water flows through those hollows in various ways. It can move simply as diffused flow through the whole mass, but more commonly it moves through soil horizons or 'percolines' that have been leached of nutrients by the flow itself (Bunting, 1961). In many parts of upland Britain runoff has completely flushed away some of the soil in the hollows to form subsurface 'pipes' in which water can be heard trickling away beneath the surface (Lewin, Cryer and Harrison, 1974).

Hollows are at a 'threshold' between different types of denudational process. In steep hollows the sediment that accumulates can become very unstable in an intense shower and may then move off suddenly as a shallow landslide, leaving a bare scar behind and a toe of debris beneath to be recolonized by vegetation.

In Japan 'new' slides that had appeared between 1946 and 1971 covered an average of 0.5 per cent of two sample areas; those that had revegetated between those dates covered 7.61 per cent, and 'old' ones that were unvegetated on both dates covered 0.49 per cent (Tanaka, 1976). In other words, there was fast early succession, but the continued creation of new scars kept the open habitat of the scars 'alive' in the area as a whole.

Small landslides are not so frequent on our gentler slopes in Britain, but they do occur. The extreme flood of 1952 on Exmoor left many scars (Gifford, 1953) and another intense shower in September 1976 left a scar at Bilsdale in North Yorkshire (Bevan, Lawson, and MacDonald, 1978). Slides are more frequent in peat, as near Moor House in the northern Pennines, where peat slides occurred in 1722, 1870, 1930 and 1963 on slopes that were no more than $14° - 17°$ at their steepest (Crisp, Rawes, and Welch, 1964.)

Another common catastrophic form of erosion in steeper hollows is the gully. On the steep, drift-covered slopes of the Lune Valley in Cumbria, gullies seem to have been forming and revegetating over many centuries perhaps in response to sequences of wetter and drier years (Harvey, 1974, 1977). Frost and heavy rainfalls move sediment off the bare sides of the gullies on some 30 occasions per year on slopes of $35° - 55°$, providing a very unstable habitat for only a very few shortlived individual plants.

One of the very few studies of the relaxation time in a geomorphological system was of a gully. Graf (1977) used tree-ring counts to find that, following a disturbance (either climatic or human), it took gullies near Denver between 15.4 and 18.7 years to grow to half their length (their 'half-life'). They then took 51 years to reach seven-eighths of their final length.

Gentler hollows are at a threshold between unconcentrated flow of water over and through the soils upslope and concentrated flow in the streams downslope (Kirkby, 1980). This threshold migrates back and forth across the hollow in response to lighter or heavier falls of rain.

When streams extend up into hollows in periods of intense rainfall, they 'clean out' the accumulated sediment. In Devon, the drainage density (length of stream per unit area) fluctuated from 1.2 to 5.7 in a moorland catchment and 1.4 to 10.5 in a catchment dominated by enclosed land (Gregory and Walling, 1978), indicating a periodic flushing out of great lengths of normally unoccupied channel. But some parts of hollows are flushed only rarely. On Mendip the East Twin hollows contain 7×10^5 kg of organic and inorganic sediment, the annual accumulation being 531 kg. In other words, this sediment probably took 1200 years to accumulate. It may still be there, for a flood with a recurrence interval of 15−40 years (that is, of moderate intensity) did not clear it out in July 1968 (Finlayson, 1978). Most of it awaits a more intense event. Many, perhaps most, hollows in Britain may be old channels that are relics of a time when the drainage net was denser, perhaps because of higher rainfall, or because of permafrost (Gregory, 1971).

Hollows provide fertile but rather unstable habitats. If trees do get away they seem to be able to survive the instability, except when it is very intense, and grow tall and old as in the Appalachian hollows (Hack and Goodlet, 1960), but many of the smaller plants survive for only the periods between the flushing.

Their position at a threshold means that hollows are habitats that are very vulnerable to disturbance. In New Zealand, steep slopes under grassland (that is, disturbed sites) were subject to landsliding by storms with recurrence intervals of only 30 years compared to the 100 years of the undisturbed forested slopes nearby (Selby, 1976). On Plynlimon (as elsewhere in Britain) summer floods, especially those that occurred over already wet ground, produce more 'slope erosion' in the form of shallow slides on grassland than on afforested land (Newson, 1980). On one occasion the peat in a grassland hollow was removed in a spectacular 'burst' 300 m long on a slope of only 12°. In North Yorkshire, moor-burning leaves bare soils that can produce fast runoff and this cuts gullies into the hollows (Imeson, 1974). In the New Forest trampling, burning and grazing can easily initiate gullies (Tuckfield, 1964) and on the sides of the Dove valley in Derbyshire even the storm drainage from an improved road initiated a channel in a hollow (Ovenden and Gregory, 1980).

Much of what has been said about hollows applies also to slopes that are concave in section though not in plan (see Figure 2.1). They do not concentrate nutrients, water and sediments to the same extent, but they do transmit nutrients and sediments and water from upslope to the stream and act as temporary stores of sediment (Moss and Walker, 1978). The ordered arrangement of features that is found on side slopes seems to disappear in the jumble of sediments on these lower slopes (Anderson and Furley, 1975). In places they are the depository for landslides from upslope and the jumble can become intense (Dalrymple, Blong and Conacher, 1968). These slopes are not as vulnerable to gullying or trenching as the hollows, but since subsurface flow can become saturated at the base of the slope, they often suffer shallow flooding and can be dissected by shallow gullies and rills (Chorley, 1978).

SCARS AND SCREES

Fascinating landforms though these are, I do not have the space to describe them in detail. On a scar, the mass continuity equation (equation 1) has little or nothing in its storage component: as soon as a rock is loosened by frost-heave or some other process, it falls to the slope below. Nevertheless, cracks do hold some soil, in which, protected from grazing and from competition many of our rarest species survive (such as individual species of *Sorbus, Hieracium* and arctic alpines).

Screes are another rather extreme environment. Many may be inactive in Britain, being inherited from harsher conditions in the last glaciation (Ball and Goodier, 1970). The ecological hazards on screes are not only the constant, if slow, 'rain' of rocks from above, but the rolling and sliding movement of the rocks themselves (Statham, 1973).

FLOODPLAINS

Floodplains have yet another input–output–storage relationship and provide yet other distinctive sets of habitat. The inputs, outputs and stores on flood plains are the sediments eroded very largely from river banks or, in smaller amounts, from flushed-out hollows. The character of this material depends, of course, on the character of the parent rock, but being derived from the less weatherable residues in the soil it is less rich in nutrients than the country rock. The river sorts the sediment according to its size and this creates four distinct types of deposit (Carey, 1969): thin layers of fine-grained (muddy) sheet accretion deposited by overbank flooding and trapped by vegetation; coarse 'point-bar' accretion, ranging from cobbles to sands, deposited in the river channel itself, the cobbles being sorted out at the upstream end of a bar parallel to a convex bank, and sands being deposited in its 'lee' downstream (as Bluck, (1971) showed on the River Endrick near Dumbarton); very fine eddy accretion in backwaters; and fine and organic backswamp accretion in sites such as channels abandoned by the river. Each of these substrates offers a different habitat by virtue of its grain size, but floodplains have other distinctive ecological attributes as well.

First among these is the transitory and cyclic nature of the habitats, created by the constant migration of the channel meanders. An example is the sequence of habitats created by the outward expansion (over 50 m) of a loop on the River Rheidol in Wales between 1845 and 1970 (Davies and Lewin, 1974). The oldest soils had organic matter contents of 7.2 per cent, free iron contents of 0.62 per cent and pH of 4.4 compared to 1.4 per cent, 0.08 per cent and 4.8 on the more recent soils.

In general, bigger rivers migrate more quickly than smaller ones as Hooke (1977, 1980) found in his comparisons of old maps of channels in Devonshire. The time taken to traverse a floodplain varied from a maximum of 101 years for a section of the River Yarty to 1176 years for a section of the Culm. Not all changes in channels are slow, for very extreme floods can produce some marked changes very suddenly, as happened on Exmoor in the 1952 floods: Farley Water shortened its course over a 4232 m stretch by 295 m, cutting across many old meander loops (Green, 1955). Nevertheless it seems to take a really extreme flood to create such a change: hurricane Agnes produced a 200-year flood that had little visible effect on the Susquehanna channel in Pennsylvania (Ritter, 1974), and such insensitivity to flooding has been found in several other floodplains.

It is only in the wilder parts of the new world that the real ecological significance of channel migrations can be seen. The little Missouri river in

North Dakota is a mobile stream; cottonwood (*Populus sargentii*) colonizes each new point bar as it is deposited, and this produces a series of even-aged stands that date back to over 250 years (Everitt, 1968). On the Beatton river in British Columbia, there is a 100–300-year succession to spruce-dominant woodland on each point bar as it appears at 5–75-year intervals (Nanson and Beach, 1977)—in other words, each stage of the succession always occurs somewhere in the system.

Because a migrating channel may not return to one particular spot for nearly a thousand years, some very large trees can grow on the deep, well-irrigated and nutrient-flushed soils of the floodplain, as Lindsey *et al.*, (1961) were able to show with old photographs of trees before their destruction by logging along the Wabash river in central Appalachia. They found a wide variety of habitats on the floodplain, differentiated largely by their frequency of inundation and distance above the water-table: aquatic, water-margin, backwater pockets, insular bars, cutbanks, the floodplain itself and floodplain depressions. On the floodplain itself each flood buries the old surface and creates a new one for a new generation of ground-layer plants; trees send out new roots into the new soil and, if they have been damaged by the flood, resprout from the broken stumps (Yanosky, 1982). The most species-rich communities in the central Appalachians are the floodplains perhaps partly because of the supply of a large variety of seeds in flood waters (Hack and Goodlet, 1960), perhaps because the soil is renewed at intervals which allow the plant assemblages to reach the peak of diversity.

Constant irrigation and seed supply as well as nutrient flushing produces many valuable habitats on flood plains. The well-known California redwoods (*Sequoia sempervirens*) are thought to survive only where streams bring constant supplies of alluvium to kill its competitors (Stone and Vasey, 1968). In Britain, alluvial herb-rich grasslands are the most valued of communities on floodplains (Kloeden, 1976; Lees, 1981). While they are vulnerable to drainage, fertilizer treatment, and river pollution, their edaphic conditions at least can be readily recreated by simply allowing the river to create its own channel again. Several studies have shown how river channelization and straightening (to improve drainage and avoid loss of land by erosion) is often thwarted when a river recreates its old meandering pattern (Blacknell, 1981), or finds its old gradient again (Emerson, 1971).

Having now descended from the heights of the landscape to the last terrestrial habitat, there is one more important implication of geomorphology for conservation.

THE DISTRIBUTION AND ABUNDANCE OF SITES

Spurs and interfluves, side slopes, scars, screes, hollows and floodplains fit together in an orderly pattern. Valleys with no tributary valleys (first-order valleys) cover about 60 per cent of most large river basins. In steep terrain spurs cover rather les than 50 per cent of these valleys and side slopes may dominate (Hack and Goodlet, 1960). In the lowlands much more of the first-order valleys (perhaps over 80 per cent) are covered by spurs, and side slopes are rare sites. In both highlands and lowlands hollows are very small, and rare. Scars cover a very small part of steep landscapes, providing the rarest of habitats.

As valleys coalesce into second, third and fourth-order systems, the landscape is dominated more and more by long spur slopes (Arnett, 1971). Side slopes become rarer, although their mean slope angle increases as slopes undercut by large streams are added to their number. The proportion of the area covered by hollows declines even further, though other concave slopes become more important. Floodplains enter as an element in valleys of about the third order, and then widen with remarkable regularity downstream, though rarely occupying more than a few per cent of the total area of a large basin (Salisbury, 1980).

This orderly progression is reflected in the distribution of soils. The Upper Elwy Valley in Denbighshire above its confluence with the Afon y Meirchion is about a fifth-order stream (on the Strahler system) draining an area of relatively high relief. The poorly drained Ynys, Powys and Hiraethog series of Ball (1960) are the soils of spurs and interfluves and cover about 25 per cent of the landscape. The better-drained brown earths of the Denbigh series on side slopes cover over 50 per cent. The Aled soils on alluvium do not appear until the second-order valleys, are unimportant until the third-order and then grow strongly as a landscape component until just above the main confluence. They never, however, cover more than a mere 5 per cent of the landscape (Warren and Cowie, 1976). Hollows are too small to be mapped at the 1/50 000 scale.

In the lowland Chilterns, in contrast, 57 per cent of the area of the Aylesbury and Berkhamsted soil map (excluding the Jurassic lowlands) is covered by the Batcombe and the Winchester series which are upperslope soils. Alluvial soils cover less than 5 per cent of the area. Since this region includes the steep slopes of the chalk scarp, we could expect the percentage of land covered by spur and interfluve soils to be even greater, perhaps nearer 75 per cent in even lower parts of eastern England.

Thus, if one were selecting sites from a natural landscape, hollows would call immediately for protection as rare, fertile sites and alluvial floodplains would come next in rarity (and might be even more ecologically favourable). In the lowlands, side slopes would be only slightly rarer than floodplains, but in the uplands they would be common. In both lowlands and uplands, spur and interfluve sites would be the most common. Scars would be absent from most lowlands—except the coast, and would be rare in most British uplands.

When we consider a cultivated landscape, the pattern of availability of these site types for conservation is quite different. Gently sloping land is the most attractive to the farmer with modern machinery, for above about 11° slopes are hard to cultivate (Bibby and Mackney, 1969). This limitation places spur sites at a premium among farmers in the lowlands and floodplains are not far behind if somebody can be persuaded to drain them and protect them from floods. As we found in Herefordshire, low angle sites become very vulnerable to 'reclamation' (Warren and Cowie, 1976). The steep side slopes are often the only ones left for the nature conservationist.

These considerations prompt the following series of rankings (Table 2.2).

TABLE 2.2   Ranking, by characteristic, of different elements of the landscape.

	Most				Least
Stability	Spurs,	side slopes,	hollows,	scars,	floodplains
Fertility	Hollows,	floodplains,	side slopes,	scars,	spurs
Small scale diversity of habitat	Floodplains,	side slopes,	hollows,	scars,	spurs
Natural rarity in the uplands	Scars,	hollows,	floodplains,	side slopes,	spurs
Natural rarity in the lowlands	Hollows,	side slopes,	floodplains,		spurs
Rarity in cultivated lowlands	Hollows,	floodplains,	spurs,		side slopes
Rarity in disturbed uplands	Hollows,	floodplains,	side slopes,		spurs
Vulnerability to disturbance other than cultivation, etc.	Hollows,	floodplains,	side slopes,	scars,	spurs

When we take geomorphology and pedology into account, we may be better able to explain the puzzling patterns in the abundance of species discovered in remnant woodlands by Kilburn (1966), Helliwell (1976), Game and Peterken (1981) and Johnson and Simberloff (1974). We may also be able to define the minimum desirable area of reserves in different areas as envisaged by Pickett and Thompson (1978), that is, 'the minimum area within a disturbance regime [that] maintains internal colonizing sources and therefore minimises extinction'. Such advances in a theory of nature conservation should help to produce a better system for selecting and managing sites.

REFERENCES

Aandahl, A.R. (1948). The characteristics of slope positions and their influence on the total nitrogen content of a few virgin soils of western Iowa. *Soil Science Society of America Proceedings*, **13**, 449.

Adams, W.M. and Rose, C.I. (1978). The selection of reserves for nature conservation. *University College London. Discussion Papers in Conservation 20*, 34 pp.

Allen, J.R.L. (1974). Reaction, relaxation and lag in natural sedimentary systems: general principles, examples and lessons. *Earth Sciences Review*, **10**, 263–342.

Anderson, E.W. (1972). Terracettes: a suggested classification. *Area*, **4**, 17–20.

Anderson, K.E. and Furley, P.A. (1975). An assessment of the relationship between the surface properties of chalk soils and slope form using principal components analysis. *Journal of Soil Science*, **26**, 130–143.

Anderson, M.G. and Burt, T.P. (1978). The role of topography in controlling throughflow generation. *Earth Surface Processes*, **3**, 331–344.

Arnett, R.R. (1971). Slope form and geomorphological process: an Australian example. *Institute of British Geographers, Special Publication*, **3**, 81–92.

Ball, D.F. (1960). *The Soils and Land-use of the District around Rhyl and Denbigh*, Memoir Soil Survey of England and Wales, HMSO, London 267 pp.

Ball, D.F. and Goodier, R. (1970). Morphology and distribution of features resulting from frost-action in Snowdonia. *Field Studies*, **3**, 193–218.

Ball, D.F. and Stevens, P.A. (1981). The role of 'ancient' woodland in conserving 'undisturbed' soils in Britain. *Biological Conservation*, **19**, 163–176.

Barkham, J.P. and Norris, J.M. (1970). Multivariate procedures in an investigation of vegetation and soil relations of two beech woods, Cotswold Hills, England. *Ecology*, **51**, 630–639.

Bevan, K., Lawson, A. and McDonald, A. (1978). A landslip/debris flow in Bilsdale, North York Moors, September 1976. *Earth Surface Processes*, **3**, 407–419.

Bibby, J.S. and Mackney, D. (1969). Land use capability classification. *Soil Survey Technical Monograph, No. 1*, 27 pp.

Blacknell, C. (1981). River erosion in an upland catchment. *Area*, **13**, 39–44.

Bluck, B.J. (1971). Sedimentation in the meandering River Endrich. *Scottish Journal of Geology*, **7**, 93–138.

Bormann, F.H. and Likens, G.E. (1979). *Pattern and Process in a Forested Ecosystem*, Springer-Verlag, New York, 253 pp.

Brunsden, D. and Thornes, J.B. (1979). Landscape sensitivity and change. *Transactions of the Institute of British Geographers*, **NS4**, 463–484.

Bunting, B.T. (1961). The role of seepage moisture in soil formation, slope development and stream initiation. *American Journal of Science*, **259**, 503–518.

Carey, W.C. (1969). Formation of floodplain lands. *Journal of the Hydraulics Division of the American Society Civil Engineers*, **95**, 981–994.

Catt, J.A. (1979). Soils and Quaternary geology in Britain. *Journal of Soil Science*, **30**, 607–642.

Chorley, R.J. (1978). The hillslope hydrological cycle. In M.J. Kirkby (Ed), *Hillslope Hydrology* Wiley, Chichester, pp. 1–42.

Chorley, R.J. and Kennedy, B.A. (1971). *Physical Geography: A Systems Approach*, Prentice-Hall, London, 370 pp.

Cleaves, E.T., Godfrey, A.E., and Bricker, O.P. (1970). Geochemical balance of a small watershed and its geomorphic implications, *Geological Society of America Bulletin*, **81**, 3015–3022.

Clements, F.E. (1936). Nature and structure of the climax. *Journal of Ecology*, **24**, 252–284.

Clymo, R.S. (1978). A model of peat bog growth. In O.W. Heal and D.F. Perkins (Eds), *Production Ecology of British Moors and Montane Grasslands* Springer-Verlag, Berlin, pp. 187–223.

Collins, M.B. (1981). Sediment yield of headwater catchments in Sussex, S.E. England. *Earth Surface Processes and Landforms*, **6**, 517–539.

Connell, J.H. and Orias, E. (1964). The ecological regulation of species-diversity. *American Naturalist*, **98**, 399–414.

Costa, J.E. (1974). Response and recovery of a piedmont watershed from tropical storm Agnes, June 1972. *Water Resources Research*, **10**, 106–112.

Crisp, D.T., Rawes, M. and Welch, B. (1964). A Pennine peat slide. *Geographical Journal*, **130**, 519–524.

Cruickshank, J.G. and Cruickshank, M.M. (1981). The development of humus—iron podsol profiles, linked by radiocarbon dating and pollen analysis to vegetation history. *Oikos*, **36**, 238–253.

Curtis, L.F., Doornkamp, J.C. and Gregory, J.K. (1965). The description of relief in field studies of soils. *Journal of Soil Science*, **16**, 16–30.

Dalrymple, J.B., Blong, R.J., and Conacher, A.J. (1968). An hypothetical nine-unit landsurface model, *Zeitschrift für Geomorphologie NF*, **12**, 60–76.

Davies, B.E. and Lewin, J. (1974). Chronosequences of alluvial soils with special reference to historic lead pollution in Cardiganshire, Wales. *Environmental Pollution*, **6**, 49–57.

Diamond, J.M. (1975). The island dilemma: lessons of modern biogeographic studies for the design of natural reserves. *Biological Conservation*, **7**, 129–146.

Dimbleby, G.W. (1974). The legacy of prehistoric man. In A. Warren and F.B. Goldsmith (Eds), *Conservation in Practice*, John Wiley, Chichester, pp. 279–287.

Drury, W.H. and Nisbet, I.C.T. (1971). Interrelationships between developmental models in geomorphology, plant ecology and animal ecology. *General Systems*, **16**, 57–68.

Emerson, J.W. (1971). Channelization—a case study. *Science*, **173**, 325–326.

Erhart, H. (1955). 'Biostasie' et 'rhexistasie', équisse d'une théorie sur le rôle de la pédogenèse en tant que phénomène geologique. *Comptes Rendus de l'Academie des Sciences*, **241**, 1218–20.

Evans, R. (1971). The need for soil conservation. *Area*, **3**, 20–23.

Everitt, B.L. (1968). Use of the cottonwood in an investigation of the recent history of a floodplain. *American Journal of Science*, **266**, 417–439.

Faith, E.B., (1977). *A quantitative analysis of soil erodibility based on physico-chemical properties of soils*, unpublished PhD thesis, University of London, 240 pp.

Finlayson, B.L. (1978). Suspended solids transport in a small experimental catchment. *Zeitschrift für Geomorphologie NF*, **22**, 192–210.

Finlayson, B.L. (1981). Field measurements of soil creep. *Earth Surface Processes*, **6**, 35–48.

Game, M. and Peterken, G.F. (1981). Nature reserve selection in central Lincolnshire woodlands, *CST Notes 30* (Chief Scientists' Team), Nature Conservancy Council, London 30 pp.

Gifford, J. (1953). Landslides on Exmoor caused by the storm of 15th August 1952. *Geography*, **38**, 9–17.

Glentworth, R. and Dion, H.G. (1949). The association or hydrologic sequence in certain soils of the podzolic zone of northeast Scotland. *Journal of Soil Science*, **1**, 35–49.

Graf, W.L. (1977). The rate law in fluvial geomorphology. *American Journal of Science*, **277**, 178–191.

Green, C.W. (1955). North Exmoor floods, August 1952. *Bulletin of Geological Survey GB*, **7**, 68–84.

Gregory, K.J. (1971). Drainage density change in South-West England. In K.J. Gregory and W. Ravenhill (Eds), *Exeter essays in geography*, University of Exeter, pp. 33–53.

Gregory, K.J. and Walling, D.E. (1978). The variation of drainage density within a catchment. *Bulletin of the International Association of Science and Hydrology*, **13**, 61–68.

Grime, P. (1979). *Plant Strategies and Vegetation Processes*, John Wiley, Chichester, 222 pp.

Hack, J.T. (1975). Dynamic equilibrium and landscape evolution. In W.N. Melhorn and R.C. Flemal (Eds), *Theories of Landform Development*, Binghampton, New York, pp. 87–102.

Hack, J.T. and Goodlet, J.C. (1960). Geomorphology and forest ecology of a mountain region in the central Appalachians, *US Geological Survey Professional Paper*, **347**, 66 pp.

Harvey, A.M. (1974). Gully erosion and sediment yield in the Howgill Fells, Westmorland. *Institute of British Geographers Special Publication*, **6**, 45–58.

Harvey, A.M. (1977). Event frequency in sediment production and channel change. In

K.J. Gregory (Ed) *River Channel Changes*, John Wiley and Sons, Chichester, pp. 301–315.

Helliwell, D.R. (1976). The effects of size and isolation on the conservation value of woodland sites in Britain. *Journal of Biogeography*, **3**, 407–416.

Hooke, J.M. (1977). The distribution and nature of changes in river channel patterns: the example of Devon. In K.J. Gregory (Ed) *River Channel Changes*, John Wiley and Sons, Chichester, pp. 265–280.

Hooke, J.M. (1980). Magnitude and distribution of rates of river bank erosion, *Earth Surface Processes*, **5**, 143–157.

Huggett, R.J. (1982). Models of spatial patterns in soils. In E.M. Bridges and D.A. Davidson (Eds) *Principles and Applications of Soil Geography*, Longman, Harlow, 132–170.

Imeson, A.C. (1974). The origin of sediment in a moorland catchment with particular reference to the role of vegetation. *Institute of British Geographers Special Publication*, **6**, 59–72.

Imeson, A.C. (1976). Some effects of burrowing animals on slope processes in the Luxembourg Ardennes. *Geografiska Annaler*, **58A**, 115–125.

Johnson, N.K. and Simberloff, D.S. (1974). Environmental determinants of island species numbers in the British Isles. *Journal of Biogeography*, **1**, 149–154.

Jones, R.L. and Beavers, A.H. (1964). Aspects of catenary and depth distribution of opal phyfoliths in Illenois soils. *Proceedings Soil Science Society of America*, **28**, 413–416.

de Jong, E. (1981). Soil aeration as affected by slope position and vegetative cover, *Soil Science*, **131**, 34–43.

Karcz, I. (1980). Thermodynamic approaches to geomorphic thresholds. In Coates, D.R. and Vitek, J.D. (Eds) *Thresholds in Geomorphology*, George Allen and Unwin, London, pp. 209–226.

Kilburn, P.D. (1966). Analysis of the species–area relation. *Ecology*, **47**, 831–843.

Kirkby, M.J. (1967). Measurement and theory of soil creep, *Journal of Geology*, **75**, 359–378.

Kirkby, M.J. (1980). The streamhead as a significant geomorphic threshold. In Coates, D.R. and Vitek, J.D. (Eds) *Thresholds in Geomorphology*, George Allen and Unwin, London, pp. 53–73.

Kloeden, J.L. (1976). An ecological assessment of alluvial grassland in the Beult River Valley, Kent. *Discussion Papers in Conservation, UCL No. 11*, 26 pp.

La Marche, V.C., Jr. (1968). Rates of slope degradation as determined from botanical evidence, White Mountains, California. *United States Geological Survey Profession Paper 352–I*, pp. 341–347.

Lees, S.J. (1981). *The management of permanent neutral grassland*. Unpublished M.Sc. Thesis University College London, 120 pp.

Lewin, J., Cryer, R., and Harrison, D.I. (1974). Sources of sediments and solutes in Mid-Wales. *Institute of British Geographers Special Publication 6*, pp 73–85.

Lindsey, A.A., Petty, R.O., Sterling, D.K., and Van Asdall, W. (1961). Vegetation and environment along the Wabash and Tippecanoe Rivers. *Ecological Monographs*, **31**, 105–156.

Milne, G. (1935). Composite units for mapping of soil associations. *Transactions 3rd International Congress Soil Science Oxford*, vol. *I*, pp. 347–354.

Morgan, R.P.C. (1981). *Soil Conservation: Problems and Prospects*, John Wiley, Chichester, 608 pp.

Moss, A.J. and Walker, P.H. (1978). Particle transport by continental water flows in relation to erosion, deposition, soils and human activities. *Sedimentary Geology*, **20**, 81–140.

Nanson, G.C. and Beach, H.F. (1977). Forest succession and sedimentation on a meandering river floodplain, northeast British Colombia, Canada. *Journal of Biogeography* **4**, 229–251.

Newson, M. (1980). The geomorphological effectiveness of floods—a contribution stimulated by two recent events in mid-Wales. *Earth Surface Processes*, **5**, 1–16.

Odum, E.P. (1969). The strategy of ecosystem development. *Science*, **164**, 262–270.

Ovenden, J.C. and Gregory, K.J. (1980). The permanence of stream networks in Britain. *Earth Surface Processes*, **5**, 47–60.

Patton, T.R., (1978). *The Formation of Soil Material*, George Allen and Unwin, London, 143 pp.

Pickett, S.T.A. (1976). Succession: an evolutionary interpretation. *American Naturalist*, **110**, 107–119.

Pickett, S.T.A. and Thompson, J.N. (1978). Patch dynamics and the design of nature reserves. *Biological Conservation*, **13**, 27–37.

Quinn, N.W., Morgan, R.P.C., and Smith, A.J. (1980). Simulation of soil erosion induced by human trampling. *Journal of Environmental Management*, **10**, 155–165.

Rackham, O. (1980). *Ancient Woodland*, Edward Arnold, London, 402 pp.

Ratcliffe, D.A. (Ed) (1977). *A Nature Conservation Review*, Cambridge University Press, Cambridge, 2 volumes, 388 and 320 pp.

Ritter, J.R. (1974). Effects of hurricane Agnes flood on channel geometry and sediment discharge of selected streams in the Susquehanna river basin, Pennsylvania. *J. Research US Geological Survey*, **2**, **6**, 753–61.

Salisbury, N.E. (1980). Thresholds and valley widths in the South River Basin, Iowa. In D.R. Coates and J.D. Vitek (Eds) *Thresholds in Geomorphology*, George Allen and Unwin, London, pp. 103–129.

Selby, M.J. (1976). Slope erosion due to extreme rainfall: a case study from New Zealand. *Geografiska Annaler*, **58A**, 131–138.

Slaymaker, H.O. (1972). Pattern of present sub-aerial erosion and landforms in mid-Wales. *Transactions of the British Institute of Geographers*, **55**, 47–68.

Smeck, N.E. (1973). Phosphorus: an indicator of pedogenetic weathering processes. *Soil Science*, **115**, 199–206.

Smith, T.R. and Bretherton, F.B. (1972). Stability and the conservation of mass in drainage basin evolution. *Water Resources Research*, **8**, 1506–1529.

Statham, I. (1973). Scree-slope development under conditions of surface particle movement. *Transactions of the British Institute of Geographers*, **59**.

Stone, E.C. and Vasey, R.B. (1968). Preservation of coast redwood on alluvial flats. *Science*, **159**, 157–161.

Strahler, A.H. (1978). Response of woody species to site factors of slope angle, rock type and topographic position in Maryland as evaluated by binary discriminant analysis. *Journal of Biogeography*, **5**, 403–423.

Tanaka, M. (1976). Rate of erosion in the Tanzawa Mountains, central Japan, *Geografiska Annaler*, **58A**, 155–163.

Tuckfield, C.G. (1964). Gully erosion in the New Forest, Hampshire. *American Journal of Science*, **262**, 795–807.

Walker, P.H., Hall, G.F., and Protz, R. (1968). Soil trends and variability across selected landscapes in Iowa. *Soil Science Society of America Proceedings*, **32**, 97–101.

Walker, T.W. (1965). The significance of phosphorus in pedogenesis. In E.G. Hallsworth and D.V. Crawford (Eds) *Experimental Pedology*, Butterworths, London, pp. 295–315.

Warren, A., and Cowie, J. (1976). The use of soil maps in education, research and planning. *Welsh Soils Discussion Group Report*, **No. 17**, 1–14.

Weyman, D.R. (1974). Runoff processes, contributing area and stream flow in a small upland catchment. *Institute British Geographers, Special Publication*, **6**, 33–43.
Whittaker, R.H. (1956). Vegetation of the Great Smoky Mountains. *Ecological Monographs*, **26**, 1–80.
Wright, H.E. (1974). Landscape development, forest fires, and wilderness management. *Science*, **186**, 487–495.
Yaalon, D.H. (1971). Soil forming processes in time and space. In D.H. Yaalon (Ed), *Paleopedology*, Israel University Press, Jerusalem, pp. 29–39.
Yanosky, T.M. (1982). Effects of flooding upon woody vegetation along parts of the Potomac River floodplain. *US Geological Survey Professional Paper*, **1206**, 21 pp.

Conservation in Perspective
Edited by A. Warren and F.B. Goldsmith
© 1983 John Wiley & Sons Ltd.

CHAPTER 3

Conservation of Nature in the Marine Environment

TOM PRITCHARD

INTRODUCTION: THE NEED FOR CONSERVATION IN THE MARINE ENVIRONMENT

Until well into the present century there was a common opinion that the oceans of the world held an inexhaustible source of food and other products, while also having an almost infinite ability to absorb the effects of industrial and domestic pollution. People today are much less sanguine about such matters. The disposal at sea of toxic wastes, including radioactive material, incidents of oil pollution, sometimes on a grand scale, and the depredations of certain short-sighted fishery practices all give concern. All these are worries in the global context. There are, however, similar anxieties about the wellbeing of our own home waters.

Interest in the ecology of coastal waters was intensified in the 1960s by the oil pollution of the sea and the beaches caused by the wrecked Torrey Canyon, a tanker which foundered in the English Channel in 1967. Other events heightened the alarm about oil pollution especially the damage it caused to bird populations, to fisheries and to shorelines that were made dirty, sticky and very unpleasant. Then in 1969, soon after the inquest on the Torrey Canyon incident, many birds in the Irish Sea died from a cause still unknown (Holdgate, 1971). Pollution was suspected as one cause. In the early 1970s, other maritime matters were publicized and scrutinized by experts: these including the environmental aspects of the extraction of North Sea oil and gas; the condition of inshore fisheries; chemical and organic pollution entering industrial estuaries and adjacent seas; the prospect that the alien seaweeds *Macrocystis* and *Sargassum muticum* might choke coastal waters; the collection by scuba divers of organisms such as the sea urchin *Echinus esculentus* and the gorgonid coral *Eunicella verrucosa*; and the conflict between powered recreation, such as water-skiing, and the other leisure activities in coastal waters.

These activities affected commercially valuable marine organisms and other species, sometimes directly as by poisoning, but often indirectly by the alteration of their environmental conditions. Before about 1970 the non-commercial species were of interest mainly to a small number of professional marine biologists, students and naturalists who lived near the sea; but scuba diving and photographic techniques for studying life in the sea became within reach of ordinary people. They found marine creatures fascinating. Television producers have enhanced their awareness by choosing wonderful marine locations for programmes filmed at home and abroad. The public now knows the sea as varied, interesting and vulnerable.

THE LEGISLATION: A MEASURE FOR STATUTORY MARINE NATURE RESERVES

Because the public, and experts, expressed growing concern about the security of the marine environment Parliament included provision for marine nature reserves in the Wildlife and Countryside Act 1981. The details of the relevant sections of the Act (namely 36 and 37) are given in the appendix to this chapter.

MOTIVATION FOR CONSERVATION

Protecting marine flora and fauna is an appealing concept for many different reasons. The motivation for some people lies in the fact that marine conservation helps certain ways of life to survive, notably those of inshore fishermen. Those who use coastal waters for leisure, such as scuba divers, favour protection because a clean sea is essential for their enjoyment. Some industries require clean seawater for their processes, and the tourist trade is strongly affected by the condition of all coastal natural resources. It is important to take each dimension of this shared concern into account when policies and practices for marine conservation are formulated. A description of the varied public interest forms part of the comprehensive review of scientific and other issues in the conservation of marine wildlife (Nature Conservancy Council and Natural Environment Research Council, 1979). The review was undertaken by the Joint Working Party on Marine Wildlife Conservation of the Nature Conservancy Council and Natural Environment Research Council which was set up in 1975.

The Joint Working Party

The terms of reference of the Working Party were as follows:

(1) To keep under review scientific information and general developments pertinent to the conservation of marine wildlife and to advise NCC and NERC on further research requirements.
(2) To identify threats to marine wildlife and to advise NCC and NERC on the urgency of the problem and the most appropriate action.

(3) To advise NCC and NERC on other matters relevant to the conservation of marine wildlife as requested.

The enquiry related primarily to the littoral and sublittoral areas (broadly equivalent to the tidal and subtidal areas) of the coast. Conservation measures had been practised for a long time to protect maritime fauna and flora occurring above the high water mark (see Tubbs, Chapter 4) and to conserve seals and seabirds. The conservation of commercial fish stocks is the statutory responsibility of fishery departments and was not considered in detail by the working party. The Working Party comprised seven scientists specializing in various aspects of marine studies, and a Professor of Law. It was supported by a technical secretariat from the Nature Conservancy and the National Environment Research Council.

The support given to the Working Party came from two sources. First, there was extensive conservation experience in Britain, and the NCC provided a focus for it. A particular strength was the use of scientific methods as a basis for conservation. Furthermore, both the NCC and NERC employed a large number of ecologists who provide a source of multidisciplinary expertise. Secondly, there was an even larger institution of research workers in the marine sciences—within NERC and in the universities, the polytechnics, the Marine Biological Association and the Field Studies Council. Some marine research stations were founded more than a hundred years ago and scientists had been gathering information around our shores for over a century, but until recently their interest in conservation, other than of commercially valuable species, had been limited. The Working Party sought to marry the terrestrial conservation experience with marine science. Their deliberations involved 15 meetings and extensive fact-finding by the secretariat who took evidence from hundreds of sources. The Report of the Working Party is a comprehensive treatment of the issues, given in seven chapters.

The most vital of all the issues was the definition of marine conservation as a concept and as a practice. This was difficult because most of the familiar conservation concepts relate to terrestrial ecosystems and cannot easily be transposed to the sea. Ecosystems on land are often very different from systems in the sea. For example, people readily understand what would happen ecologically if a peat bog were to be drained or a woodland were to be felled; unless you happen to be a scuba diver or a marine scientist the ecological consequences of threats to organisms or their habitats in the sea or on the sea bed are not so easy to appreciate.

Then there are sharp differences between the legal basis of ownership on land and in the sea and between the patterns of established rights relating to public access and usage, and so on; and the Town and Country Planning Acts do not apply below low water mark of ordinary tides.

In 1978, an Interdepartmental Working Party on Marine Conservation was established by the Department of the Environment; it consisted of representatives from all major government departments and agencies interested in the subject. Its main purpose was to consider the findings of the NCC/NERC Joint Working Party, and then to advise ministers about the position. Its deliberations led to the inclusion of the marine clauses in the Parliamentary Bill that resulted in the Wildlife and Countryside Act 1981.

MARINE WILDLIFE RESOURCES

The coastal zone is shown diagrammatically in Figure 3.1. The classification takes account of physical, biological and legal criteria. The main features that differentiate marine from terrestrial wildlife communities are as follows:

(1) Marine ecosystems are, by and large, in a natural state. Although there are many examples of localized damage, it cannot be said that the integrity of the system has been seriously threatened.

(2) Marine ecosystems appear to be dominated by animals: the important food plants of the animals are not those seen on the shore or fixed to the bed but the small ones in the water body. On land, ecosystems are characterized by their dominant higher plants, such as grassland, woodland, heathland. Marine scientists classify communities largely by the nature of their substrate, such as bedrock, sand, gravel or mud.

(3) The animals and plants that occur in terrestrial systems are usually longer-lived than marine organisms. Rock-living species are detached by storms, and substrates such as sands are highly mobile and bring about major changes to the habitat of many species during tidal and other movements. Predator-prey relationships are often fierce in the sea. Shortlived marine animals are opportunistic and occupy niches created by the mortality of others. Because of this, much more severe fluctuations occur in marine than in terrestrial animal communities.

(4) Dispersion of larvae and spores is widespread in the sea and when a community is destroyed its locality is often reoccupied with material from considerable distances. Indeed, quite different organisms from those previously present may be involved. Marine ecosystems are therefore very diverse as compared with many terrestrial systems. The bed of shallow coastal seas is occupied by organisms ranging from sedentary types to highly mobile ones, distributed in mosaics of variable stability.

These characteristics call for conservation techniques that are rather different from those applied to land or freshwater systems. The most appropriate classification available to the conservationist is based on main types of substrates (Table 3.1).

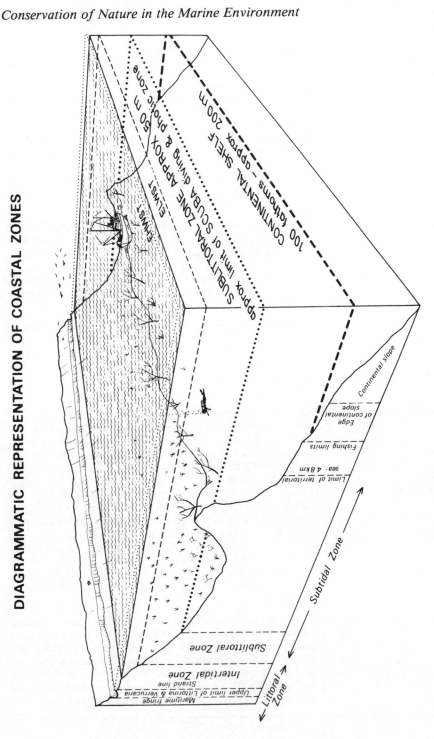

DIAGRAMMATIC REPRESENTATION OF COASTAL ZONES

FIGURE 3.1 Redrawn after Joint Nature Conservancy Council/Natural Environment Research Council Working Party Report (1979).

TABLE 3.1   Main types of habitat and associated biota.

Substrate	Biota
Mud	Characteristic of calm waters, e.g. harbours and similarly sheltered locations. Fauna similar to that of muddy littoral areas, with species tolerant of anoxic conditions. Polychaetes, certain bivalves, etc. may be locally abundant, together with certain species of crabs, shrimps and other crustaceans. Brittle stars and starfish may be locally abundant, and *Zostera* beds can provide an important habitat for many invertebrates. Muddy areas often provide important feeding grounds for fish.
Sand	Bivalves may occur in commercially important numbers in sandy areas. There are also commonly burrowing polychaetes, gastropods and crustacea and (in areas of clean sand) very large populations of heart-urchins and ophiuroids. Sandy areas inshore are feeding grounds for young fish, and in deeper offshore waters provide commercial fishing grounds.
Shell gravel	Contains many of the burrowing species found in coarse sands, but with additional specialized species. Bivalves, echinoderms and decapods are common and *Amphioxus* is characteristic of some areas. Starfish are found in the surface, together with many fish, especially bottom-dwellers such as flatfish.
Stones and boulders	As with the littoral zones, the type and abundance of organisms to be found depends upon the mobility of the substrate. The more mobile areas may be devoid of macrofauna, but the larger boulders provide a stable and rich habitat, merging into bedrock which offers the most luxuriant populations of all.
Bedrock	Kelp forests often cover extensive areas of stable bedrock, down to a depth of about 20 m. In deeper waters, algae, more tolerant of low levels of light, predominate and may continue right down to the limits for algal growth. The kelp provides a habitat for a very wide range of flora and fauna, particularly red algae, sponges, anemones, hydroids, corals and soft corals, tube worms, bryozoans, and ascidians. These in turn provide food for a very wide range of predators and grazers.
Tidal streams and rapids	These areas are characterized by suspension-feeding organisms attached to stable rock surfaces. They are usually most abundant in areas where the movement of water is sufficiently strong to prevent the settling-out of suspended matter which would cause the clogging of filter feeding mechanisms. Where these conditions extend into the littoral areas, they are much valued for teaching and study purposes.

GEOGRAPHICAL VARIATION

Because of their varied geology and topography the greatest diversity of marine habitats is found along the western shores of Great Britain. Our eastern and north-eastern shores are more uniform geologically and physiographically. Climate and hydrographic factors are also significant to the distribution of marine habitats. The British marine flora and fauna can be classified geographically into three main types:

(1) Circumboreal subarctic elements, widespread except for the south and south-west, for example, *Alaria esculenta, Balanus balanoides,* and *Sagartia troglodytes.*
(2) Celtic elements, forming the majority of the common species present throughout the British Isles, for example, *Actinia equina, Ulva lactuca* and *Corallina officinalis*; and
(3) Lusitanian-Atlantic elements, the majority of which occupy the western seaboard, for example, *Balanus perforatus, Chthalamus stellatus* and *Gibbula umbilicalis.*

THREATS TO MARINE ECOSYSTEMS

The Working Party concluded that the most damaging human impact on British coasts was the reclamation of estuarine mudflats and sandflats, and of coastal wetlands generally (see Tubbs, Chapter 4). It was also concluded that urban and industrial activities that are damaging included the disposal of organic wastes (such as sewage), and the construction of breakwaters, quays, marinas and other coastal engineering works. Generalizations were difficult because these activities could, under some circumstances, cause local increases in biological productivity and ecological diversity by altering the nature of the substratum. The example was quoted of protective structures on soft shores that offer smooth, hard surfaces which resembled bedrock. Exposed surfaces would be colonized by only a few specialized organisms; more sheltered parts would have algae, encrusting species, sea anemones, mussels and other organisms typically associated with hard surfaces. Pilings have a characteristic fauna of some interest and represented a distinct ecosystem. Each impact had to be judged on its own merits or demerits, and be evaluated as a factor for potential ecological change.

Effluent discharges and accidental discharges of other pollutants influence the environment according to the nature and volume of the discharge involved and the situation in which it was released. Oil spillage in the open sea is most damaging to auks and other seabirds, but probably has little influence on other forms of marine life.

The dumping of toxic substances is lethal to some organisms but the effects are normally very localized. Chemicals such as acids, alkalis, phenols, chromate and ammonia are actually lethal to plants and animals at the point of discharge but are usually rendered innocuous by dilution or chemical change. The extent of the lethal zone is never more than a few hundred metres radius from the source. In estuaries, where water exchange is much slower than along the open coast, the affected areas are much greater, but the effect of any one toxic substance is difficult to distinguish from those of others in a complex situation (for distribution of pollutants in an estuary, see Meyerson *et al.*, 1981).

In the absence of regulations to ensure the maintenance of sustainable yields, commercial harvesting of marine organisms is likely to lead to the loss of population of some species, at least in some areas. Development of a souvenir trade in attractive marine organisms, such as sea urchins and corals, particularly with the rapid growth of scuba diving, may have caused some depletion of populations in a few places favoured by divers. However, there was no evidence that this practice was of any significance except locally, though there are situations where such losses may be serious, for example, the Isles of Scilly.

Finally, collecting of specimens for teaching and research could sometimes be damaging, especially in much-used sites near to field stations. The Coastal Code published by NERC (Natural Environment Research Council, 1974) and the more recent NCC Code *The Seashore and You* (NCC, 1981) provide useful guidelines for those involved in field studies.

CONSERVATION PLANNING

In Britain the foundation for conservation on land is the White Paper entitled *Conservation of Nature in England and Wales* (HMSO, 1947). This paper, together with the equivalent for Scotland (Command 7235), was later embodied in the National Parks and Access to the Countryside Act 1949 which provided for the conservation of ecosystems by means of nature reserves. Various purposes were defined for terrestrial nature reserves, coming under the general headings of conservation and maintenance of the natural heritage, provision for survey and research, the establishment of reserves exclusively for experimental purposes, the provision of educational facilities, as well as other amenities where these were compatible with the protection of the natural features. The Working Party found themselves in a position analagous to that of the authors of Command 7122 and believed that the same objectives could be applied to marine nature reserves.

Until 1981, statutory conservation areas in the form of National Nature Reserves or Sites of Special Scientific Interest (SSSIs) could only be established down to low water mark of ordinary tides (LWMOT) in England and Wales and low water mark of spring tides (LWMST) in Scotland. Obviously, it was not ecologically reasonable to truncate conservation activities in this way.

Indeed, intertidal areas could be, and occasionally were, included in national nature reserves as part of maritime areas whose main interests were found above the sea shore. The 1981 Act alters this by enabling the NCC to move outward from the low water marks and to do this independently of any interest it may have landward. By expanding the principles given in Command 7122, the NCC will protect not only the examples of common marine habitats (see Table 3.1) but also unusual or rare features such as drowned valleys, sea lochs and tidal rapids. Regardless of the known or prospective threats, it is important in any case to provide such conservation status to selected parts of the marine sector of our environment.

One obvious function of marine nature reserves is to provide research facilities especially for repetitive investigations in permanent quadrats. Monitoring is an important element of marine biology and marine reserves are likely to become important sites for this kind of recording. The educational and amenity functions must also be important in some cases; scuba diving involves an ever-growing number of people who are very interested in the natural history of the sea. The scope of education is even greater in the intertidal and adjacent terrestrial areas as a large proportion of people on holiday visit the seaside. The scope for marine environmental education in schools is also considerable.

SELECTION OF SITES AS MARINE NATURE RESERVES

In the NCC, a strategy has emerged over the years for the selection of the most important sites on land. The first volume of *A Nature Conservation Review* (Ratcliffe, 1977) sets out a rationale for the selection of terrestrial sites of biological importance to nature conservation in Great Britain. The procedure involves the identification and recording of the physical and biological features of sites, and a comparative assessment of them, in accordance with a set of ten criteria, within an ecosystem classification. Dr Roger Mitchell of the NCC has tested this approach in a marine context and believes that it can be applied, in a modified form, to the selection of marine nature reserves. The method will enable the NCC to select a national series of important samples of the various types of marine system (Knight and Mitchell, 1980; Mitchell, 1977a, 1977b).

Surveys in the last decade make it possible to identify marine sites considered worthy of conservation. These include areas around small islands such as some of the Isles of Scilly, Lundy, the Monachs, St Kilda, Bardsey and Skomer; mainland coastal areas with rocks and sand, for example around Start Point in Devon, the Bembridge Ledges off the Isle of Wight, and stretches off St Abb's Head in Scotland and off the Marloes and Lleyn peninsulas in Wales; low-tide and sand-flats such as the extensive ones around Tresco and St Martins in the Isles of Scilly; sea-lochs in the Outer Hebrides; tidal rapids such as those between Linne Mhuirich and Loch Sween in Argyll, and in the Menai Strait; inlets such as the Helford and Percuil Rivers in Cornwall; and even flooded coastal quarries. Some of these areas are already non-statutory marine reserves managed by trusts and other associations of enthusiasts; others are adjacent to National Nature Reserves on land.

MANAGEMENT OF MARINE RESERVES

The establishment of statutory reserves will not be a rapid process because of the extensive consultations with other interested parties at local and national levels that will be required. The NCC proposes to consult all the organizations concerned before submitting formal proposals to the relevant secretaries of state. It will only request such byelaws as are required to complement existing byelaws such as those of Sea Fisheries Committees and it has no intention of trying to stop all current activities in the areas concerned. As with national nature reserves on land, its policy will be to permit activities that are compatible with the primary objective of conserving the flora, fauna and other features of special interest.

Though selection of the first few marine reserves will be possible on the basis of existing knowledge, further detailed survey will be required to establish the best boundaries for some areas known to be of special interest, to verify the importance of others, and to compare sites in order to select the best examples of particular habitat types.

Management will aim not only to conserve wildlife and other features but also to provide for the use of the reserve for research, education and leisure wherever they are compatible with conservation. Unlike wildlife habitats on land, which often require human control and interference, as well as protection to maintain their special interest, marine habitats, being in a more natural condition, rarely require management. Wardening will be mainly concerned with controlling potentially harmful or conflicting activities and with providing facilities for the public. Provisions for schools and other educational establishments will need to be explored with the help of the many agencies which have an interest in the maritime environment.

Nature conservation in Britain has been remarkably successful because partnerships have existed between the statutory bodies responsible for it and the non-governmental organizations of local naturalists trusts, the National Trusts, natural history societies, educational institutions and many other groups of people with an interest in the issues. The success of marine nature conservation too will depend on a continuation of such partnerships, and new perspectives are already emerging as a result of the preparatory work for the Wildlife and Countryside Act 1981; the enactment of this law will provide a fresh impetus and the necessary guidelines.

REFERENCES

HMSO (1947). *Conservation of Nature in England and Wales*, Report of the Wildlife Conservation Special Committee (England and Wales), Cmnd 7122, HMSO, London, 139 pp.

HMSO (1981). *Wildlife and Countryside Act*, HMSO, London, 128 pp.

Holdgate, M.W. (1971). *The Seabird Wreck of 1969 in the Irish Sea*, Natural Environment Research Council, London.

Knight, S. and Mitchell, R. (1980). The survey and nature conservation assessment of sublittoral epibenthic ecosystems. In J.H. Price, D.E.G. Irvine, and W.F. Farnham (Eds), *The Shore Environment, Vol. 1: Methods*, Academic Press, London, pp. 303–321.

Meyerson, A.S., Luther, G.W., Krajewski, J. and Hives, R.I. (1981). Heavy metal distribution in Newark Bay sediments. *Marine Pollution Bulletin*, **12** (7), 244–250.

Mitchell, R. (1977a). Marine wildlife conservation. *Conservation Review*, **14**, 5–6.

Mitchell, R. (1977b). Marine wildlife conservation. In K. Hiscock and A.D. Baume (Eds), *Progress in Underwater Science, 2*, Pentech Press, London, pp. 65–81.

Nature Conservancy Council (1981). *The Seashore and You*. NCC Interpretative Branch, Shrewsbury, 1p.

Nature Conservancy Council/Natural Environment Research Council (1979). *Nature Conservation in the Marine Environment*, Report of the NCC/NERC Joint Working Party on Marine Wildlife Conservation.

Natural Environment Research Council (1974). *The Coastal Code*, NERC, London, 4pp.

Ratcliffe, D.A. (Ed) (1977). *A Nature Conservation Review*, Cambridge University Press, Cambridge.

APPENDIX

The statutory basis for marine nature reserves

The Wildlife and Countryside Act 1981 (HMSO, 1981) contains sections for the establishment of marine reserves (s 36) and for their control by means of byelaws (s 37). The substance of this new measure is given in s 36 as follows:

36—(1) Where, in the case of any land covered (continuously or intermittently) by tidal waters or parts of the sea in or adjacent to Great Britain up to the seaward limits of territorial waters, it appears to the Secretary of State expedient, on an application made by the Nature Conservancy Council, that the land and waters covering it should be managed by the Council for the purpose of—

(a) conserving marine flora or fauna or geological or physiographical features of special interest in the area; or

(b) providing, under suitable conditions and control, special opportunities for the study of, and research into, matters relating to marine flora or fauna or the physical conditions in which they live, or for the study of geological and physiographical features of special interest in the area,

he may by order designate the area comprising that land and those waters as a marine nature reserve; and the Council then manage any area so designated for either or both of these purposes.

Conservation in Perspective
Edited by A. Warren and F.B. Goldsmith
© 1983 John Wiley & Sons Ltd.

CHAPTER 4

The Intertidal Zone

C.R. TUBBS

Man's attitude towards the meeting place of land and sea is dichotomous. The intertidal zone and its associated shallow waters has been valued as a fishery and as a source of wildfowl, salt, and other natural produce for 2000 years or more. At the same time the fertility of muddy embayments and estuaries has made them natural targets for agricultural reclamation whenever economic conditions have encouraged the necessary investment. It is difficult to calculate the former area of intertidal flats in western Europe but it is certain that it has been more than halved since the fifteenth century.

Embankment and drainage for agriculture, like the more direct exploitation of its natural resources, at least depends on the inherent fertility of the intertidal zone, but no such claim can be made for the widespread obliteration of the intertidal in the twentieth century for industry, port and marina development, refuse tipping and the dumping of dredged spoil, nor for such abuses as pollution from domestic and industrial effluent or the winning of sand. Most such losses are irretrievable. In this chapter I have attempted to outline the salient features of the ecology of the intertidal zone; to draw conclusions about its conservation value; and to examine the threats to its survival and the difficulties encountered in its conservation in Britain.

INTERTIDAL ECOLOGY

The physical and ecological nature of the intertidal zone is determined by the configuration of the coastline and the directions and strength of the prevailing waves, tides and currents that produce and distribute sediments. In general terms, sediment accumulates where there is shelter and erosion occurs where the coast is exposed to long wave fetch or tidal scour. In the latter case, there are few opportunities for reclamation. It is accreting coasts that are at risk and it is with these that I am essentially concerned in this chapter.

The greater the exposure to wave action or the tidal velocity, the larger the particle size of intertidal sediment. Thus, in many tidal inlets there is a gradation from firm sands around the mouth to soft mud in the sheltered inner area.

53

Though the gradation of particle size may be continuous, it is customary to regard mineral particles less than 0.05 mm in diameter as mud; between 0.05 and 2.00 mm as sand; and larger particles as shingle. The shape, size and mineralogy of the particles will depend on the parent rock from which they are derived. The transition (often a remarkably sharp one) from sand banks to mudflats is marked by the occurrence of increasing quantities of particulate organic matter.

The accretion of sediment above high water neap tide level eventually permits colonization by halophytic plants and the development of saltmarsh. The developing plant cover reduces tidal velocity and encourages the deposition and retention of yet more particles that would otherwise have been removed by the ebb tide, thus accelerating the vertical growth of the marsh. Changes in marsh level lead to a plant succession associated with an increasingly dense plant cover and a reduction in the frequency of tidal inundation. Saltmarsh development is a natural reclamation agency and various techniques have been developed to encourage the process as a precursor to final embankment and conversion to farmland. Much of the upper shore of the Dutch and German coast and much of the former area of the Wash have been reclaimed in this way, and the process is continuing (Beeftink, 1977; Ranwell, 1972).

Saltmarsh formation is of great antiquity on accretion coasts. Cord-grass *Spartina* marsh is of recent origin. Though *Spartina* has invaded high level saltmarsh, continuous monospecific swards are characteristic of somewhat lower levels in the tidal range in most of its geographical range. *Spartina anglica* appears to have arisen in the late nineteenth century in Southampton Water first from hybridization between the native *S. maritima* and the introduced *S. alterniflora*, and then by a doubling of the chromosomes to produce a fertile plant which proved to possess a phenomenal capacity for colonizing the middle and upper levels of open mudflats. The spread of *Spartina* marsh has extensively depleted the area of mudflats. In the 1960s there were an estimated 21 000–28 000 ha of *Spartina* marsh on the coasts of north-west Europe. However, in the region of the plants' origins the marshes have now ceased vertical and lateral growth, the plants are dying back and the marsh platforms are slumping and eroding. The causes remain imperfectly understood (Goodman, Braybrooks, and Lambert, 1959; Hubbard and Stebbings, 1967; Ranwell, 1967; Tubbs, 1980).

The crucial biological characteristic of the intertidal zone is that it receives a constant supply of nutrients in the form of organic material (both living and detrital) and soluble bases, derived from both marine and terrestrial sources and distributed by local currents and the ebb and flow of the tides. The role of saltmarshes in the production of nutrients within estuaries has received much attention from North American authors (Armstrong and Hinton, 1978; Teal, 1962; Welsh, 1980). Estuaries and sheltered embayments are among the world's most productive ecosystems and can yield six to eight times as much living material as a cultivated arable crop and 20 times as much as the open sea. Much of this production is exported by transient organisms such as birds and fish, and by man's harvest of fish and shellfish.

Numbers of individual invertebrates tend to increase with declining particle size and increasing organic content. Invertebrate production is especially high in mudflats, which support prodigious numbers of animals, though the numbers of species adapted to the difficult physical conditions, and the size of individuals, are relatively small. The maximum species diversity is combined with moderately high numbers and is usually achieved in sediment of intermediate texture. Table 4.1 illustrates the differences between soft mud and sands in Langstone Harbour and demonstrates the comparative richness of mudflats. The numbers of animals are commonly three or four times the average shown: the small gastropod *Hydrobia ulvae* alone can achieve densities of 80 000–100 000 per m² under the most favourable conditions.

TABLE 4.1   Invertebrate numbers in Langstone Harbour, Hampshire (from Withers and Thorp, 1978 and Withers, 1980).

	Soft mud Eastney Lake	Muddy sand Sinah Sand	Coarse sand Winner Bank
Average number of individuals/m²	24 573	670	210
Number of all species	26	45	28
Number of annelids	14	18	8
Number of crustaceans	4	21	18
Number of molluscs	5	4	0
Number of other species	3	2	2

The fauna of the three sites in Table 4.1 mainly comprise small animals. Estuaries and embayments also support dense populations of larger species, notably the bivalve molluscs, and some of these are of commercial significance. Walne (1972) gave examples of the standing crop of some commercially valuable species in various intertidal areas of north-west Europe. He recorded values as high as 1940 g m⁻² for mussels in the Menai Strait: 120 g m⁻² for cockles in the Burry Inlet; 200 g m⁻² for oysters (no locality given); and 206 g m⁻² for the hard-shell clam in Southampton Water. Yields of around 1000 g m⁻² of large molluscs should be possible in most sheltered intertidal systems in north-west Europe and cultivation can increase this many times. Walne gave examples of commercial yields which included the wholly cultivated mussel fishery of north Norfolk, where it was nearly 8000 g m⁻²—much higher than any known yield from terrestrial animals.

British estuaries support important fisheries for oysters (*Ostrea edulis*), cockles (*Cerastoderma spp.*), mussels (*Mytilis edulis*), locally winkles (*Littorina spp.*); and in Southampton Water, the naturalized hard-shell clam (*Mercenaria*

mercenaria). Though intertidal areas, at least in Britain, are less important than deeper waters for finfish, they nevertheless produce commercially significant quantities of several species, especially flatfish such as dab (*Pleuronestes limanda*) and flounder (*P. flesus*). More important, the intertidal zone functions as a nursery ground for these and other commercial species or else exports species upon which they feed. The role of the intertidal zone in these respects is difficult to quantify, but the future of some offshore fisheries may depend on the careful conservation of coastal ecosystems. Finally, estuaries are a vital link in the migration paths of salmon, sea trout and eels, which support a substantial and highly valuable fishery both in rivers and inshore waters. The estuary plays only a small part in the lifecycle of these migrants, but estuarine pollution or destruction can result in the loss of entire fisheries of these species.

Birds and fish both predate intertidal invertebrates and feed directly on algae and *Zostera* in the intertidal zone. Birds are both the more conspicuous and more easily counted, and the spectacular numbers that assemble there outside the brief breeding season are often regarded as a measure of the overall richness of the system. Fish have a greater direct commercial value, but birds are favoured with greater intrinsic, scientific, cultural and nature conservation values. Thus, the bird populations of intertidal areas of Europe have been studied intensively in recent years.

In north-west Europe, five groups of birds depend wholly or partly on the intertidal zone:

(1) waders (*Charadrii*) and the shelduck (*Tadorna tadorna*), which feed mainly on the invertebrates of intertidal sediment and very shallow water (19 species);
(2) surface-feeding ducks, which feed both on invertebrates and vegetable matter (five species);
(3) brent geese (*Branta bernicla*), which feed on green algae and *Zostera* on mudflats, on saltmarsh plants, and in recent years also on terrestrial grass and arable crops near the shore;
(4) 'sawbills', sea-ducks, grebes, cormorants and herons, which feed on fish and other organisms obtained in shallow waters (seven species), and
(5) gulls, which feed on invertebrates and fish obtained from sediment surfaces and shallows (nine species).

In the 1970s, threats to intertidal areas generated an upsurge of interest in the ecology and migration strategies of many wetland birds, waders and wildfowl in particular. It has become apparent that for most species, or for different populations of the same species, particular intertidal areas function as critical links in chains of moulting and 'refuelling' stations distributed on migration paths between breeding grounds in northern latitudes and wintering grounds in Europe and Africa. We know that the stations are not necessarily the same for different species or for different populations of the same species, nor, indeed, are they always the same in spring and autumn. Our perception of the migratory strategies of most wetland birds remains imperfect and it is essential that we learn more if we are to conserve them.

We are relatively well endowed with quantitative information about the bird populations of European and to a lesser extent of African intertidal areas. In Britain, monthly counts of ducks and geese have been made for most areas since 1952 on behalf of the Wildfowl Trust. The Birds of Estuaries Enquiry 1969–75, organized by the British Trust for Ornithology (BTO), the Royal Society for the Protection of Birds (RSPB) and the Wildfowl Trust, and financed by the Nature Conservancy Council (NCC), extended this to other species, notably waders, and since 1975 regular counts have continued at most of the important coastal sites. For a few places, notably Langstone Harbour, Hampshire, comprehensive counts at monthly or shorter intervals go back to the early 1950s and permit some conclusions about long-term trends (Tubbs, 1977).

We are fortunate that in Britain there is so large and competent a body of volunteers to make such census work possible. Elsewhere in Europe the data are less comprehensive but, nonetheless, since 1966 annual international counts of waders and wildfowl have been organized by the International Wildfowl Research Bureau (IWRB), whilst some intertidal systems, notably the Waddensee, have been studied intensively. The data show that during 1975–80 about five million individuals of the thirteen most numerous species of waders were dependent on the intertidal areas of the migratory flyway extending from the high Arctic through western Europe to north Africa. Of these, between 1.5 and 1.8 million depended on British intertidal areas, the other most important links in the chain of migratory and wintering stations being the Waddensee, the Rhine–Maas–Scheldt delta, the west coast of France and Portugal, Morocco and the Banc d'Arguin, Mauritania. In Britain, the Dee, Morecambe Bay, the Ribble, the Solway and the Wash each supported more than 100 000 waders at times; the Humber and the Severn 50 000–100 000; and another thirteen areas supported 20 000–50 000; and a further 27 areas, 10 000–20 000 (Prater, 1981). Though the larger numbers of individuals occur in the largest intertidal systems, this does not detract from the collective importance of the smaller bays and estuaries, many of which are also important for local populations and races of various species. Of the wildfowl whose whole population occurred in the intertidal zone outside the breeding season, approximately 160 000 brent geese and 130 000 shelduck wintered in north-western Europe in the late 1970s, of which 65 000 and 60 000 respectively wintered in Britain.

In the past most of the bird populations occurring in the intertidal zone were exploited as a source of protein. In particular, huge numbers of ducks were trapped in decoys, most of which were located in coastal marshlands (Payne-Gallwey, 1886). Today, most species are protected over much or all of their range or are hunted on a relatively small scale for sport. Only in Holland do a small number of commercial duck decoys survive.

THE THREAT

The preceding brief review shows that the intertidal zone is a natural resource of great inherent fertility. It is galling to the biologist that its most productive components are generally also its most accessible and, for agriculture, the most suitable for reclamation.

Until this century, most reclamation was for agriculture or was incidental to the construction of sea defences. The earliest north European embankments were probably Roman and included those which enclosed the Fens from the Wash (Darby, 1956). In much of western Europe, the first main impetus to reclamation, however, appears to have arisen from the land hunger of the eleventh, twelfth and thirteenth centuries, and to have terminated temporarily with the agricultural depression of the early fourteenth century. It is often said that until the present century most reclamation was of saltmarsh, but I think that at least on the southern shore of the North Sea, the enclosure of mudflats is medieval in origin. Some of the reclamation at this and later periods was for saltpans as well as agriculture. From the early sixteenth century to the mid-seventeenth century and again from the mid-eighteenth century to the mid-nineteenth century there were further periods of reclamation prompted by high prices for agricultural produce (Van Bath, 1963). The present century has seen a number of ambitious long-term reclamation projects, generally supported by national governments rather than by individuals or syndicates; they include such ecologically alarming enterprises as the isolation and partial reclamation of the former Zuider Zee and the Rhine–Maas barrage scheme.

In Britain one has only to consider the enormous areas of farmland protected by seawalls on the east coast from Lincolnshire to Kent to appreciate the extent of saltmarsh and perhaps mudflat reclamation. The process has enjoyed its latest boom since the 1950s. The Nature Conservancy Council has estimated that of a total area of about 38 100 ha of saltmarsh in the United Kingdom in 1950, 5630 ha or 15 per cent had been reclaimed by 1980. The greatest losses seem to have been sustained around the Wash, where about 2700 ha or 39 per cent of the marsh was embanked for farmland; 700 ha were reclaimed there in the six years between 1974 and 1980.

To the ancient practice of embankment for agriculture and flood protection, the twentieth century has added reclamation demands for ports and marinas, industry, and the tipping of refuse and fly ash. Such developments involve the irretrievable loss of the resource. In the late 1960s and early 1970s four of the largest intertidal systems in Britain, the Wash, Dee, Morecambe Bay and Solway, were the subjects of feasibility studies for fresh-water storage impoundments involving massive barrage schemes. The demand forecasts for water, however, have subsequently diminished and the threats have, at least temporarily, receded. The Severn estuary, however, is currently the subject of a government-sponsored prefeasibility study for a barrage which would permit the harnessing of the huge tidal range to generate an estimated 10 per cent of the United Kingdom's electricity demand. Several possible positions for the barrage have been examined and though there remain uncertainties about the ecological and hydrological implications of any of them, it seems certain that in any event the intertidal area would be greatly diminished. Inevitably, various associated developments have been suggested, including a road crossing, water storage impoundments, docks and an airport, but how realistic are any of these is doubtful, at least in the current economic climate. It is some comfort that the proposal to site a third London airport on the Maplin Sands, obliterating about 6200 ha of intertidal muds and sands which among other considerations are a vital early winter feeding ground for a large percentage of the dark-bellied brent geese wintering in Europe, was eventually abandoned, though ecological considerations did not figure prominently in the decision. There remains a shadow over Maplin until a third London airport is constructed elsewhere or fuel costs rise sufficiently to place air travel beyond the common means.

Langslow (1981) summarized the known threats to intertidal areas in Britain from information provided by regional staff of the Nature Conservancy Council. For fifteen important intertidal systems, schemes for agricultural reclamation were proposed or pending; another eleven areas were subject to or threatened with refuse and fly ash tipping; and a further eight were threatened by reclamation for industry, mostly associated with petrochemicals. The threat from further port development was probably most acute in what is left of the intertidal area of Southampton Water, but at least 22 estuaries and inlets were currently or had recently been proposed as locations for marinas, or yacht harbours (Figure 4.2). Innocuous though these latter facilities may appear, they too often involve massive dredge-and-fill operations linked with reclamation for 'marina villages', hotels and the like. I have known proposals which more nearly resemble housing estates with puddles in the middle than they do yacht harbours. In sum, of the 41 largest intertidal systems in Britain, in 1979–80 parts of 29 were the subject of proposals for reclamation and development other than for agriculture. In addition the outer banks of at least a dozen were being dredged for sand and there were proposals for sand extraction from the Taw–Torridge estuary and for alluvial tin extraction from Devoran Creek and the Hayle estuary in Cornwall, all proposals which would involve a reduction in the intertidal area and which would be likely to have consequences for future sedimentation patterns. Though in most localities where sand extraction takes place there is little evidence of adverse ecological effects, we remain woefully ignorant of the long-term effects, nor do we know the extent to which extraction has prevented further growth of the intertidal area. Little relevant research has been embarked upon, partly perhaps because of its inherent long-term nature.

FIGURE 4.1

Though the catalogue of threats is long, we must see it in perspective. Many of the proposals are tentative, or small in scale. Others have been rejected in the past only to surface again in different guise but with little greater chance of success. Nonetheless they collectively represent a continuous process of attrition and reflect an unabated rapaciousness towards the intertidal resource.

It is a miracle that in Britain only two estuaries have come close to obliteration—the Tees and Southampton Water. As Langslow (1981) remarked, the

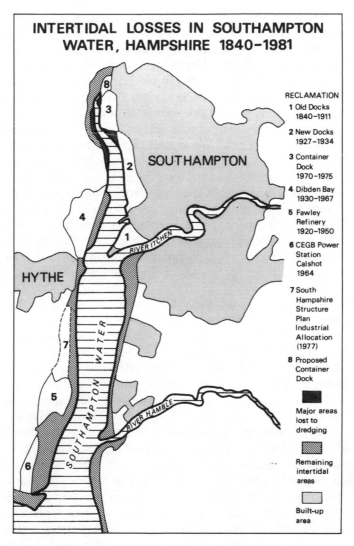

INTERTIDAL LOSSES IN SOUTHAMPTON
WATER, HAMPSHIRE 1840–1981

SOUTHAMPTON

HYTHE

RIVER ITCHEN

SOUTHAMPTON WATER

RIVER HAMBLE

RECLAMATION

1 Old Docks
1840–1911

2 New Docks
1927–1934

3 Container
Dock
1970–1975

4 Dibden Bay
1930–1967

5 Fawley
Refinery
1920–1950

6 CEGB Power
Station
Calshot
1964

7 South
Hampshire
Structure
Plan
Industrial
Allocation
(1977)

8 Proposed
Container
Dock

Major areas
lost to
dredging

Remaining
intertidal
areas

Built-up
area

FIGURE 4.2   From Coughlan (1979).

Tees could be the first British estuary to lose virtually all its intertidal zone. In the early nineteenth century the estuary comprised about 2400 ha of flats. By 1969 no more than about 240 ha remained and by 1974 this had been reduced to 140 ha, all of which is allocated for development in the local structure plan. The progressive loss of intertidal land in Southampton Water is shown in Figure 4.2: here, too, a substantial part of what remains is at present allocated for industry in the South Hampshire Structure Plan.

Where they remain undiminished in size, intertidal areas are not necessarily free of abuse. Sand extraction has already been mentioned. Pollution is often a more potent force for change. In recent decades the problems of some estuaries that have been grossly polluted since the nineteenth century have been greatly alleviated, but many still receive treated but nutrient-rich domestic effluent, heavy metals, and a variety of industrial effluents. Up to a certain threshold, nutrient enrichment may be beneficial but beyond that it can have undesirable effects. In Langstone Harbour, enrichment has stimulated the growth of the green alga *Enteromorpha* and *Ulva lactuca*, which now blanket large areas of the mudflats in summer. Many areas are anaerobic and support only an impoverished infauna, though the weed mats themselves are rich in some animals, notably *Hydrobia ulvae*. Though the algae provide an important food supply for brent geese and probably wigeon (*Anas penelope*), they appear to have depressed the carrying capacity of the harbour for some mud-probing waders. More worrying, because tidal exchange with the sea is limited, nutrient levels may continue to rise even though phosphates and some nitrates are now being stripped from the effluent before discharge (Portsmouth Polytechnic, 1977; Tubbs, 1977).

Intertidal areas receive a wide range of potential pollutants. The Wash, a comparatively clean system, receives heavy metal loads which are reflected in the seasonal loading of wader tissues (Parslow, 1973). The Firth of Forth receives distillary wastes which provide a rich source of food for some sea ducks (Milne and Campbell, 1973); and the long-suffering estuary of Southampton Water receives refinery effluent containing oil, dissolved sulphides, phenols, nitrogen and other compounds which have destroyed saltmarsh and sterilized mudflats around the discharge point. It is impossible to generalize about the effects of effluent discharges, but all modify the ecology of the receiving system in some fashion. All demand long-term monitoring and few get it.

I am unaware of any comrehensive inventory of current threats to intertidal wetlands in the western Palearctic as a whole but it is evident that the United Kingdom experience is not unique. In particular the southern North Sea coast is under constant pressure for agricultural reclamation, though plans for reclaiming the rest of the Dutch Waddensee seem to have been dropped for the moment. There remain plans for the embankment and reclamation of saltmarsh at various sites on a smaller scale eastward to the German–Danish border. Three sites currently threatened have between them supported nearly one-third of the total population of the dark-bellied brent goose in the early spring, when the birds are putting on the fat reserves which permit their migration 4820 km (3000 miles) north and east to the breeding ground on the Tamyr peninsula. We simply do not know if the population will be able to adapt its migration strategy to such a loss.

INTERTIDAL CONSERVATION

We must question the attrition of the intertidal resource, as also we must question the notion of *re*claiming it: for the prefix reflects a persistent feeling that it is a waste lost to mankind in the past and his to take back. True, some schemes for reclamation (if we must use the word) are likely to have short-term social or economic benefits but we may nonetheless question the long-term wisdom of destroying a rich natural resource whose productivity may ultimately prove of inestimable benefit to mankind. We already believe that some sea fisheries depend on intertidal nurseries. The harnessing of intertidal productivity through aquaculture is nothing new. In addition, the nature conservation and aesthetic arguments against reclamation of intertidal areas are strong even when considered separately from their value as natural resources. The power of wetland landscapes over the human mind may be difficult to quantify but it is real enough to be a valid conservation argument. We can quantify at least part of the nature conservation argument: we know, for example, approximately how many of the northern waders, ducks, geese, and other birds depend on intertidal feeding grounds outside their brief breeding season and we know that unless great care is exercised with their habitat we may ultimately deprive them of survival.

In the 1970s three international conventions to which the United Kingdom is a signatory, a Council of Europe Resolution, and the EEC Directive on the Conservation of Wild Birds, all reflect international recognition of the conservation argument.

The Convention on Wetlands of International Importance, adopted at Ramsar, Iran, in 1971, requires, among other things, that contracting states should actively promote the conservation of wetlands within their territories, and that each state should submit a list of wetlands of international importance to which they will give special protection. Criteria of international importance were finally agreed at a subsequent conference in Heiligenhafen, West Germany, in 1974. The Ramsar Convention entered into force in December 1975 and the United Kingdom ratified in January 1976 (Cmnd 6465, HMSO, 1976). From the practical point of view it is unfortunate that the criteria of international importance went beyond biological considerations to require that a site had to be administratively and physcially capable of being effectively conserved and managed; and free from threat. So far the UK government has been reluctant to commit itself to sites not already protected, mostly as nature reserves. Thus, the initial UK list was of thirteen wetlands, of which only four were intertidal and all of which were already reserves or enjoyed some other protection. Despite a few recent additions, the list fails to reflect any real political commitment to the conservation of the 40 or so estuaries and embayments of international importance with which Britain is endowed.

Council of Europe Resolution 29 (1973) expressed concern for coastal environments in unequivocal terms:

> Aware that a considerable part of Europe's coasts is in a critical condition owing to the extremely serious biological degradation and esthetic disfigurement caused by the indiscriminate siting of buildings, industry and tourist facilities in coastal areas.

It enjoined member states to take measures for the rational planning of the coast, embracing scientific criteria. The Resolution went virtually unremarked in Britain.

In September 1979 the United Kingdom signed the Council of Europe (Berne) Convention on the Conservation of European Wildlife and Natural Habitats (Cmnd 7809, HMSO, 1980a). The Convention requires contracting states to take legislative and administrative measures for the conservation of endangered natural habitats and listed species. Of particular relevance to intertidal conservation, article 4(3) requires that they:

> give special attention to the protection of areas that are of importance for the migratory species specified . . . and which are appropriately situated in relation to migration routes, as wintering, staging, feeding, breeding and moulting areas.

The intertidal areas of Britain fill precisely these roles for some species of waders, ducks, geese and other wetland birds specified in the Convention.

The EEC Directive on the Conservation of Wild Birds can be seen as amplifying the Ramsar and Berne Conventions. Article 4(2) adds the stricture that:

> Member states shall pay particular attention to the protection of wetlands, and particularly wetlands of international importance.

Finally, the (Bonn) Convention on the Conservation of Migratory Species of Wild Animals (Cmnd 7888, HMSO, 1980b) includes the migratory waders, geese and ducks in a list of species which:

> have an unfavourable conservation status and which require international agreements for their conservation and management as well as those which would significantly benefit from the international co-operation that can be achieved by an international agreement.

It is disheartening to protect a migratory species and its habitat in one country only to have its population reduced by uncontrolled shooting or the destruction of its habitat in another.

The United Kingdom's legislative response to the Conventions and the EEC Birds Directive may be seen in the Wildlife and Countryside Act 1981. In the context of intertidal conservation, its practical implications are limited to arrangements for the protection of sites of special scientific interest (SSSIs) through statutory consultative procedures in the event of threat. These are really designed to cover agricultural reclamation and invoke the concept of compensation payments. Inevitably these arrangements beg the question whether the necessary funds will be made available by government. It is arguable that the Act fails to meet the United Kingdom's international commitments in the sense that it fails to embody any system of firm statutory controls over the continued loss of habitats, and particularly of wetlands of international importance. It remains to be seen whether its consultative provisions will succeed in halting the embankment of saltmarshes for farmland.

Administratively, the Nature Conservancy Council have been responsible for government policies for nature conservation, which include the notification of SSSIs to local planning authorities, and the acquisition of national nature reserves (NNRs). However, it is essential to appreciate that SSSI notification is not in itself a protective device but invokes consultative procedures which until now have not legally embraced agricultural reclamation, though for some years consultation with the Ministry of Agriculture has occurred where grant aid was involved in reclamation schemes. Notification can, however, result in rigorous protective policies on the part of local planning authorities, though attitudes in this respect vary.

Approximately 17 500 ha of saltmarsh, mudflat and sandflat fall within NNRs, mostly in fourteen reserves. Local authorities also have statutory powers to acquire nature reserves and there are currently nine mainly intertidal local reserves (LNR) with an intertidal area of 5500 ha. In addition, naturalists' trusts and the RSPB own or control a number of important intertidal areas. However, it is doubtful if more than about 5 per cent of the intertidal area of Britain is at present protected by reserve status.

Let us glance at some of the other realities of intertidal conservation in Britain (see Figure 4.1). Agricultural reclamation is not only normally free from statutory planning control but enjoys the benefit of public subsidy from water authorities and the Ministry of Agriculture, Fisheries and Food (MAFF). Moreover, there is seldom opportunity for the public to comment on reclamation proposals. If the area is an SSSI, the Nature Conservancy Council have been consulted by the MAFF and in the absence of agreement the matter will have been decided by the Minister of Agriculture. His power to call a public inquiry to assist him in making his decision has, so far as I know never been exercised in a case of saltmarsh reclamation. However, in 1980 it proved possible for the Secretary of State for the Environment to call a public inquiry into the proposed reclamation of 80 ha of saltmarsh at Gedney Drove End on the Wash, because the embankment itself was deemed to be a matter for planning consent. At the inquiry a consortium of conservation organizations, including the NCC and RSPB, successfully opposed the application, but refusal of consent turned on the point that the marsh was periodically inundated, only irregularly grazed and therefore not agricultural land. Had it been deemed to be agricultural land the embankment would merely have constituted an agricultural improvement and would thus not have been subject to the provisions of the Town and Country Planning Acts. The lesson for would-be reclaimers is clear, though the situation can be met in the planning context by the approval of the Secretary of State for the Environment of a Direction under article 4 of the Town and Country Planning General Development Order 1977 that the reclamation of specified saltmarshes should be subject to planning consent. I am not, however, aware of an instance in which an article 4 Direction has been invoked for such a purpose.

The Wildlife and Countryside Act 1981 improves the position in that it provides a statutory framework for consultation and negotiation between the Nature Conservancy Council and owners and occupiers when land-use changes are contemplated. However, consultation is no end in itself and the Act embraces the ultimate concept of compensation or, if all else fails, compulsory purchase by the Nature Conservancy Council—thus begging the question of financial support from government for conservation. Perhaps more hopefully, the Act also requires that in considering applications for grant aid for schemes within SSSIs the Minister of Agriculture shall so far as possible further nature conservation. It will be of great interest to see in what manner this is interpreted.

The case of the Ribble saltmarshes demonstrates that government can intervene in an important conservation issue if it has the will. Stated baldly, 2226 ha of the Ribble saltmarshes were purchased by speculators who proposed to reclaim them to agriculture. The public outcry that ensued gained a large measure of support in Parliament and the government made sufficient funds available for the NCC to purchase, with the back-up threat of their compulsory purchase powers. Unfortunately the government changed before the transfer of funds was complete and a proportion of the purchase price of £1.7 million had to be found from the NCC's existing budget. Coming at a time of financial constraint this added significantly to the operational difficulties experienced by the organization and also demonstrated clearly the inadequacy of the NCC's funding in relation to modern land values. What is perhaps more fundamental is that, as the economist John Bowers (see also Chapter 21) showed in his evidence at the Gedney Drove End public inquiry, the cost/benefit ratio of such a reclamation do not survive close scrutiny. Though the individual gains from public subsidies and price support systems, the community at large gains no tangible benefit because the extra agricultural produce is not needed.

Unless land use changes are permitted development listed in schedule I of the General Development Order 1977—which includes most agricultural and silvicultural activities and also leaves open a number of other loopholes which sometimes give rise to conservation problems—they generally require consent from the local planning authority.

With the present vogue for public participation, the local authority generally consults widely both in the preparation of its structure plans and local plans and in processing individual planning applications. Thus, there should be adequate opportunities for conservation to be incorporated into planning policies and development control. In any particular case, however, decisions depend on the attitudes of individual local authorities; on the view of the Secretary of State after the examination-in-public of the local structure plan; and on the outcome of public inquiries into individual cases which have either been called in for determination by the Secretary of State or which are the subjects of appeal against the local authority's refusal of planning consent.

In this statutory planning field, the record of the NCC and other conservation organizations in securing coastal conservation and in successfully opposing individually damaging intertidal developments, has been good in the 1970s so long as what has been politically conceived as 'the national economic interest' has not been invoked. That the record was less good in the 1960s reflects the change in attitudes towards the natural environment. Since 1970, as advocate or witness, I have appeared at thirteen public inquiries into proposed intertidal developments and (in the absence of political considerations) the conservation case has prevailed in all but one. On the other hand, it is often the case that if government policies are directly involved—the desire to provide jobs in a development area, or to promote the exploitation of North Sea oil—then intertidal conservation takes a back seat, even if it can be argued that the development should take place ashore rather than at the expense of intertidal land. Classic cases are the assignment of the last 140 ha of the Tees to development, referred to earlier; and the allocation in the Highland region of a large area of Nigg Bay in the Cromarty Firth, for petrochemical and associated development. In the latter case, reclamation has already taken place for an oil production platform fabrication yard and an oil stabilization, storage and export facility, besides which outline planning permission exists for the reclamation of 270 ha of flats for a natural gas liquids fractionation plant and an ethylene plant. The outline Environmental Impact Study, prepared for the developers is, to say the least, skimpy and comes inevitably to the conclusion that 'none of the changes expected or likely to take place ... are considered to be gross, widespread, detrimental and irreversible' (O'Sullivan and Kelly, 1981). The last in particular is patently untrue: there will never again be intertidal flats on the reclaimed area.

In concluding this review, I believe it is fair to argue that the conservation of intertidal areas, which are limited natural resources of great biological richness and beauty, demand explicit recognition at both the local and the national government level. At national level we should aim for no less than the inclusion of the 40 odd intertidal systems of international importance in the UK's Ramsar list. At local level, the need for intertidal conservation demands clear recognition in structure plans prepared by local authorities and approved by the Secretary of State for the Environment—and to be fair, some structure plans have gone a long way in that direction. It follows that reclamation, however limited, should not be permitted save in demonstrably extraordinary circumstances and as a last resort. It is doubtful if concepts such as these could be enshrined in national legislation—no other country has attempted it—but it could find expression in national statements of policy, specifically, for example, through Planning Circulars issued to local government by the Department of the Environment, and also in modified procedures for dealing with proposals to reclaim intertidal areas for farmland which would expose them to more objective scrutiny than they now receive. For the future of the intertidal heritage, this is the least that can be asked.

REFERENCES

Armstrong, N.E. and Hintson, M.O. (1978). Influence of flooding and tides on nutrient exchange from a Texas marsh. In M. Wiley (Ed), *Estuarine Interactions*, Academic Press, New York.

Beeftink, W.G. (1977). Salt marshes. In R.S.K. Barnes (Ed), *The Coastline*, John Wiley, London, pp. 93–121.

Coughlan, J. (1979). Some aspects of reclamation in Southampton Water. In B. Knights and A.J. Phillips (Eds), *Estuarine and Coastal Land Reclamation and Water Storage*, Teakfield, Farnborough and the Estuarine and Brackish Water Sciences Association.

Darby, H.C. (1956). *The Draining of the Fens*, Cambridge University Press, Cambridge.

Goodman, P.J., Braybrooks, E.M., and Lambert, J.M. (1959). Investigation into dieback in *Spartina townsendii*, Agg. 1. The present status of *Spartina townsendii* in Britain. *Journal of Ecology*, 47, 651–677.

HMSO (1976). *Convention on Wetlands of International Importance especially as Waterfowl Habitat*, Cmnd. 6465, HMSO, London, 9pp.

HMSO (1980a). *Convention on the Conservation of European Wildlife and Natural Habitats*, Cmnd. 7809, HMSO, London, 20 pp.

HMSO (1980b). *Convention on the Conservation of Migratory Species of Wild Animals*, Cmnd. 7888, HMSO, London, 19 pp.

Hubbard, J.C.E. and Stebbings, R.E. (1967). Distribution, dates of origin and acreage of *Spartina townsendii* (s.1.) marshes in Great Britain. *Proceedings of the Botanic Society of the British Isles*, 7, 1–7.

Langslow, D.R. (1981). The conservation of intertidal areas in Britian. *Wader Study Group Bulletin*, 31, 18–22.

Milne, H. and Campbell, L.H. (1973). Wintering sea ducks off the east coast of Scotland. *Bird Study*, 20, 153–172.

O'Sullivan, A.J. and Kelly, P. (1981). *Outline Environmental Impact Study of Proposed Gas Related Developments at Nigg Bay*, commissioned by Dow Chemical Company Ltd, Cromarty Petroleum Ltd and the British Gas Corporation.

Parslow, J.L.F. (1973). Mercury in waders from the Wash. *Environmental Pollution*, 5, 295–304.

Payne-Gallwey, R. (1886). *The Book of Duck Decoys: Their Construction, Management and History*, Van Voorst, London.

Prater, A.J. (1981). *Estuary Birds of Britain and Ireland*, Poyser, Berkhamsted.

Portsmouth Polytechnic (1977). *Langstone Harbour Study: The Effects of Sewage Effluent on the Ecology of the Harbour*, Report commissioned by the Southern Water Authority, Portsmouth.

Ranwell, D.S. (1967). World resources of *Spartina townsendii* (sensu lato) and economic use of *Spartina* marshland. *Journal of Applied Ecology*, 4, 239–256.

Ranwell, D.S. (1972). *Ecology of Salt Marshes and Sand Dunes*, Chapman and Hall, London.

Teal, J.M. (1962). Energy flow in the salt marsh ecosystem of Georgia. *Ecology*, **43**, 614–624.

Tubbs, C.R. (1977). Waders and wildfowl in Langstone Harbour. *British Birds*, **70**, 177–199.

Tubbs, C.R. (1980). Processes and impacts in the Solent. In *The Solent Estuarine System: An Assessment of Present Knowledge*, NERC Publications Series C, **22**, 1–5.

Van Bath, B.H.S. (1963). *The Agrarian History of Western Europe*, Arnold, London.

Welsh, B.L. (1980). Comparative nutrient dynamics of a marsh–mudflat ecosystem. *Estuarine and Brackish Marine Sciences*, **10**, 143–164.

Walne, P.R. (1972). The importance of estuaries to commercial fisheries. In R.S.K. Barnes and J. Green (Eds), *The Estuarine Environment*, Applied Science Publishers, London, pp. 107–118.

Withers, R.G. and Thorp, C.H. (1978). The macrobenthos inhabiting sandbanks in Langstone Harbour, Hampshire. *Journal of Natural History*, **12**, 445–455.

Withers, R.G. (1980). The macro-invertebrate fauna of the mudflats of Eastney Lake (Langstone Harbour). *Journal of Portsmouth and District Natural History Society*, **3**, 59–67.

Conservation in Perspective
Edited by A. Warren and F.B. Goldsmith
© 1983 John Wiley & Sons Ltd.

CHAPTER 5

Lowland Heathland: The Case for Amenity Land Management

CAROLYN M. HARRISON

INTRODUCTION

At a time when most nature conservationists in Britain are despairing of their ability to ensure a rich and diverse wildlife in tomorrow's countryside, the question of how heathland might contribute to that countryside is important. In many parts of lowland England, heathlands are some of the last remaining extensive tracts of wild, open countryside, and they are valued for both their distinctive plant and animal associations and for the countryside recreational opportunities they afford to large numbers of people. Moreover, the recent fate of heathlands is salutary and symptomatic of many other seminatural areas in the lowlands. Some 4 per cent of SSSIs have been lost annually on a national basis (Goode, 1981) and lowland heaths have been among the most vulnerable, disappearing at a fast rate. All the remaining heathland areas are under constant threat from a large number of competing land uses. If these trends in losses persist (Nature Conservancy Council, 1977) the only areas of heathland in 1990 will be in nature reserves. Unlike the upland heather moors, the question about lowland heaths is not, how can nature conservation interests be integrated with productive farming in a mosaic of multiple uses?; on lowland heaths it is, how can heathlands be preserved at all? The answer to this question, I believe, lies in the reconciliation of nature conservation and recreation, in the vigilance of local, regional and national bodies who are concerned with these interests and most importantly in the recognition of the essential transient nature of the heathland landscape. The purpose of this chapter is to demonstrate why this reconciliation should take place and to demonstrate how, through it, a more positive approach to the management of lowland heathlands might be achieved.

71

GENERAL PERSPECTIVES

In contrast to upland moorland ecosystems few lowland heathlands are cropped commercially either for sheep or for game (grouse or deer) so that recreation rarely conflicts with agriculture. The notable exception is the New Forest where cattle and horses graze in considerable numbers. Outside the New Forest few commoners exercise their rights on those heathlands that are common grazings. The main conflict is therefore between recreation and wildlife conservation. A second point of contrast with upland heaths is that the reduction of the heathland area by changes to other land uses is more marked in the lowlands than the uplands; this is a function partly of the fragmented area of heathland in contrast to the more extensive tracts of heather moorland, but it is also a function of the greater demands for building, mineral extraction and agricultural reclamation in southern England. Norman Moore's study of the Dorset heaths (Moore, 1962) demonstrated this point very clearly, and the recent resurvey by Nigel Webb and Lesley Haskins (1980) reinforced his findings. Although these latter authors noted that the rate of heathland losses had slowed down recently, partly because of statutory designations of heathland areas, the general decline persisted. The fragmentation and declining acreages catalogued for Dorset are known to be representative of other parts of the lowlands such as Surrey (Harrison, 1976), the Brecklands (Sheail, 1979) and the Sussex sandlings (Armstrong and Milne, 1973). Goode (1981) reported that only 35 per cent of the rough grazing and heathland present at the turn of the century remains in lowland Devon, excluding upland Dartmoor and Exmoor and that in the period 1966−80 a 19.4 per cent reduction had taken place in the heaths of north Hampshire. In the face of competing land uses the commonland status of much heathland provides insufficient protection on its own for ensuring that reclamation will not occur because the lord of the manor can sell mineral rights and can himself initiate a change of use. Nevertheless, those heathlands such as the Purbeck heaths, which are not commonland, are even more vulnerable than the commons which are at least afforded some partial protection. Where lowland heaths are being lost at such an alarming rate, proof of longstanding and continued use by the public for recreational and amenity acts as a potentially powerful reason for maintaining heathland in tomorrow's landscape.

That people obviously enjoy heathland areas as countryside open spaces is apparent from the very large number of visitors who annually use heathlands for countryside recreation, for example, in the New Forest (Hampshire County Council, 1970), Surrey and Ashdown Forest (Countryside Commission, 1977). Many people would agree with Professor Gimingham's sentiments when he said that 'at times heathland gives an impression of barren waste and windswept monotony, but when in late summer the rich purple of heath plants in flower spreads across the country, natives and visitors respond to the beauty of the landscape' (Gimingham, 1972, p. 1). The demand for informal recreation in the south is likely to continue even if participation rates do not rise as rapidly as during the 1970s. National statistics reveal that increases in leisure time, disposable income and car ownership, are all associated with a growing demand for countryside recreation and the most recent general survey of leisure pursuits undertaken by the Countryside Commission (Countryside Commission, 1979) revealed that next to gardening, the countryside was the most attractive form of outdoor recreation. This study and others (see Countryside Commission, 1977) confirmed that millions of people, actively sought informal recreation in the countryside each weekend and that this trend was likely to continue. In the south-east as a whole this can mean over $2\frac{1}{4}$ million people on a fine Sunday afternoon.

What visitors require of open spaces in order to satisfy their recreational requirements is however, more difficult to discover. Responses to several onsite surveys (Harrison, 1981) suggested that the visitor was not particularly discriminating about the type of vegetation mosaic he or she enjoyed but was more concerned to enjoy the 'peace and quiet' afforded by an open space outside the built-up area. This attitude is as true of visitors to heathland areas as downland or woodland sites. One corollary of this is that the presence of visitors on a heathland allows the manager to plan mosaics of vegetation designed to meet a variety of recreational and other purposes. The visitor need not be regarded as a nuisance: car-borne visitors may require the manager to provide a limited number of facilities such as a carpark, toilet block, some obvious paths and access at all times of the year, where riding is concerned, the manager may need to provide and maintain bridleways, but in general, visitors do not make many other demands on management resources.

The case for managing heaths primarily for amenity is all the stronger in southern England because, away from the coast, heathlands and downlands offer the last remaining areas of open, unfarmed countryside in what has become an increasingly suburbanized landscape since the 1930s. If the heathlands were to disappear, where would these visitors, horseriders, motorcyclists and others go for their recreation? In this respect a realignment of nature conservation and recreational interests as proposed by Green (1981) is perhaps the single most important means of securing a place for heathlands in the lowland landscapes of tomorrow.

WILDLIFE CONSERVATION STATUS OF LOWLAND HEATHLAND

The distinctive plant and animal associations that are encountered on lowland heaths have long been recognized as worthy of conserving, and over 40 heathland and acid grassland areas are listed as being of high national status in Ratcliffe's *Nature Conservation Review* (Ratcliffe, 1977). These heathland communities range from the distinctive grass heaths of the Brecklands through the Surrey heaths which are dominated by *Calluna–Erica cinerea–Ulex minor* and the extensive heather tracts of the New Forest and Purbeck, to the *Erica vagans–Schoenus* communities developed on the serpentine soils of the Lizard peninsula. The bryophytes and lichen flora of the lowland heaths is very varied and these habitats also afford some of the last remaining refuges for a dwindling reptilian fauna. The southern lowland heaths harbour populations of smooth snake (*Coronella austriaca*) and sand lizard (*Lacerta gracilis*) as well as the adder. Heathland ponds, pools and bogs are important wetlands favoured by some eight different species of dragonfly; several species of butterfly and moth such as the emperor moth (*Saturnia pavonia*) are restricted to the heathlands. Among the birds, several important predatory birds hunt over the heathlands and species such as the buzzard (*Buteo buteo*), the sparrow hawk (*Accipiter nisus*) and the kestrels (*Falco tinnunculus*), and the once-abundant hobby (*Falco subbuteo*) also breed on the heaths. The Dartford warbler (*Sylvia undata*) is virtually confined to the southern heaths (Bibby and Tubbs, 1975), and several other birds with small populations feed and breed on the heaths (for example, the stonechat *Saxicola torquata*). The larger predatory birds need large areas of heathland rather than a precise composition of the plant assemblage, but for other animals, for example, the sand lizard, and the Dartford warbler (Tubbs, 1976), the structure of the vegetation mosaic is very important.

The question of how best to preserve and manage lowland heathlands for nature conservation thus involves both considerations of the size of the area and the detailed composition of the vegetation. On extensive lowland heather tracts, the controlled burning practices that have been designed to optimize the diversity of species in upland moorland can be used successfully (Tubbs, 1974). But on many of the heathlands in the south, the vigorous implementation of rotational burning is frustrated by two conflicting factors: the advanced stage of scrub and woodland colonization of many of the Surrey and Berkshire heaths; and the conflicts between the use of heathlands for recreation and their management for wildlife. It has often been said that wildlife is the victim of conflicts and only a careful analysis of their causes can uncover any possibility of reconciliation.

CONFLICTS BETWEEN RECREATION AND WILDLIFE CONSERVATION INTERESTS

The loss of species by disturbance and soil erosion

Evidence that use by visitors has led to changes in the flora or fauna of a heathland is very difficult to establish because other environmental changes are likely to have a similar effect. For example a loss of plant species is equally likely to have been the result of drainage or fire. However, we can point to particularly vulnerable species and habitats in heathlands that are more at risk from the effects of visitors than others. Habitats such as wet flushes and small valley bogs and species that are at the edge of their geographical range in southern Britain (for example, Cornish heath, the sand lizard, the Dartford warbler) are all particularly vulnerable (see discussion in *Southern Heathlands Symposium*, Tubbs, 1976).

The comparative ease with which damage to upland communities dominated by heather can be inflicted by trampling has been remarked upon by several authors, for example in the Cairngorms and in the Peak District (Bayfield, 1979). A recent study of wear by trampling on two southern heaths confirmed that these *Calluna* heathlands are equally vulnerable (Harrison, 1980–81). Areas on flat heathlands which had previously been untrampled were subject to weekly trampling throughout a summer and winter period. Plots were allowed to recover for between 6 to 7 weeks before the resurvey; 400 passages per week were applied in the summer, and 100 passages per week in winter, these levels of trampling being designed to reflect the actual number of visitors who used the sites. For both summer and winter use, the *Calluna* communities failed to regain a live vegetation cover in the intervening recovery period and on Keston Common in Kent, the site trampled in both summer and winter was still almost 100 per cent bare at the end of the recovery period, approximately 12 months after the start of the treatment. In comparison to other acidic, neutral and basic grasslands treated in a similar manner, the heathlands were approximately ten times more vulnerable to wear, whether vulnerability was measured in terms of loss of live cover, percentage bare ground or amount of recovery made after trampling had ceased. The reason for this vulnerability is partly the growth habit of *Calluna* itself, which suffered winter browning and die-back after summer and winter treatments, and partly the absence of any other species in the heathland community that are tolerant of trampling.

The lack of tolerant species may in part be explained by the peculiar ecological conditions of heathland paths. Grime (1979) identified four major types of environment based on a combination of stress and disturbance; 'stress' was defined in terms of conditions limiting plant production (water stress, nutrient deficiency) and 'disturbance' was defined as factors such as grazing or trampling that were externally imposed on the habitat. Using this analysis, heathland paths appear as high stress/high disturbance environments, to which no plant group has adapted. It is not surprising to find that, once developed, heathland paths quickly deteriorate to sandy tracks and that unlike grassland paths they have no readily recognizable 'path' flora. One corollary of this is that the manager will have to be content with bare tracks even if he is catering for low weekly numbers of visitors, and on steep slopes the bare paths may become gullied and give rise to problems of public safety.

Vehicles are potentially more damaging than human feet, and in heathland communities the location of carparks and their restriction to flat, freely drained areas away from the more botanically/zoologically interesting parts of the site would seem to be a sensible practice. Repeated use by vehicles and cars can lead to total loss of cover, and even of mor humus or peat and the exposure of what is often a highly erodible substrate of sand or pebble gravel. For example, erosion associated with vehicles and pedestrians along the shores of the Great Pond at Frensham Common in Surrey became so severe that a restoration project was implemented in order to allow the vegetation to recover (Streeter, 1976). This restoration work has demonstrated that few species are capable of establishing themselves on the highly nutrient and organically deficient siliceous sands (Streeter, 1976). Species represented in commercially available seed mixtures proved to be poor colonizers and, although over a period of 5 or more years some vegetation cover has been established, even now it is unlikely that the species which were members of adjacent heathland communities will survive in the swards in any number. The managers will no doubt have to be content with a green grass sward, and implanted shrubs and trees where heather and Ericas once grew. This is probably the only way in which these sites can support repeated and intense use by visitors. Although it would be helpful to develop better seed mixtures with native heathland species such as *Sieglingia decumbens, Agrostis setacea* and *Deschampsia flexuosa*, the inherently low growth rates of these species would prevent them from providing an effective vegetation cover even over a long time period.

Fire

One might suspect that the number of visitors to a site would be correlated with high frequency of fires, as I have argued elsewhere (Harrison, 1976), and that firing would be one of the most important ways in which recreation would conflict with the objectives of the manager who sought to conserve wildlife. On a few lowland heathlands fire is used on a rotational basis. Parts of the New Forest and the Purbeck reserves can be managed in this way, but most of the other heathlands in southern England are too small for rotational or compartmental management. Fire, when it occurs, is often started accidentally or with malicious intent; by whom is seldom discovered, and in few cases can effective steps be taken to prevent it. In most cases the manager can only hope to contain fires within the perimeter by maintaining a girdle of firebreaks—but these may themselves be costly to make and maintain and it is easy to see why some managers come to accept frequent fires as inevitable. The manager's position is made all the more difficult because heathland fires need not be attended by the local fire brigade unless property or commercial timber is threatened and even then other calls take precedence. Maltby (1980) showed that the southern area of England is a very high fire-risk area; high fire risk is associated with high population densities and with a low number of firemen (Figure 5.1).

That fires are frequent on many of the Surrey commons is demonstrated

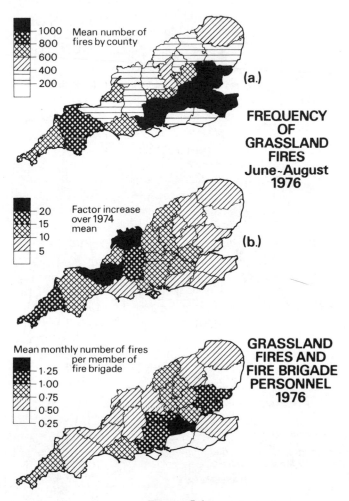

1000
800
600
400
200

Mean number of
fires by county

(a.)

FREQUENCY
OF
GRASSLAND
FIRES
June-August
1976

20
15
10
5

Factor increase
over 1974
mean

(b.)

Mean monthly number of fires
per member of
fire brigade

1·25
1·00
0·75
0·50
0·25

GRASSLAND
FIRES AND
FIRE BRIGADE
PERSONNEL
1976

FIGURE 5.1

by Figure 5.2. And while it is not possible to provide a similar analysis of visitor numbers, the diagram shows that few heathland areas escape fire in any one year, that more fires occur in the holiday periods than at other times, but that fires can occur at *any* time in the year. This is consistent with the thesis that large numbers of visitors are associated with high frequencies of fire and with the fact that all heathlands are on somebody's back doorstep and are therefore at risk at all times. This may not be the case with the more remote Purbeck heaths or the least accessible of the New Forest heaths but for many heaths in the south and east, accidental fires are common. While the frequency of fire can be determined from fire station records it is another matter to examine the extent of fires. Seldom is an attempt made to assess extent at the time of the fire and

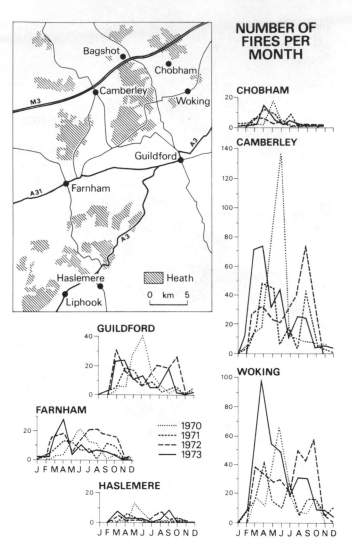

FIGURE 5.2   Fires on the Surrey heaths. From Harrison (1976); reproduced by permission of Applied Science Publishers.

while large numbers of perimeter fires need not impinge too greatly on the wildlife conservation interest of a heathland, extensive fires can be extremely damaging to wildlife and are visually highly intrusive. Few of the lowland heaths escaped without one extensive fire in 1976 and the consequence was the slow recovery of a number of sites and the loss of some wet flush communities. That many of these fires during the drought were accidental is certain, but whether or not they were associated with visitors is less clear. Sergeant (1980) reported that of 110 nature conservation sites affected by fire, some 58 were heathlands or peatlands and that the dry heaths which were most severely damaged were in those areas subject to 'excessive public pressure'. However, at times when there are many visitors, it could be argued that sufficient people are on hand to assist with the fire-fighting and therefore there is no reason to suspect a strictly linear relationship between public pressure and the number of fires. In the peculiar conditions prevailing in 1976 it proved impossible for even large numbers of adequately equipped people to contain the fires. Sergeant believed that even on the most severely damaged heathland sites, the damage caused by fire-fighting was worse than that from the fires themselves, largely because extensive firebreaks were ploughed up. Notwithstanding this, severe damage was associated with the fires, most notably in Dorset where 180 ha of *Calluna* heath was destroyed at Hartland Moor National Nature Reserve and surface and subsurface peat was burned down to a depth of 150 cm. The loss of scrub habitats favoured by the Dartford warbler and of the old heather stands favoured by a number of the reptiles (sand lizard and smooth snake) severely depleted the population of these animals. The heavy rain that brought the 1976 drought to a close led to dramatic increases in the drainage density of a number of peat and heather sites and soil erosion on the bared surface exceeded that under mature heather by a factor of 20 (Arnett, 1980). Catastrophic fires of this nature obviously have severe implications for the stability and regeneration of the heathlands. The vegetative colonization after the severe fire is being monitored on a number of sites and preliminary studies suggest that recovery of ericaceous species is delayed and extensive bare areas remain amongst the moss communities that form the early colonizers (Morgan, 1980).

The analysis of the occurrence of fires suggests that high frequencies can be associated with the use of heathlands by people. The highly suburbanized landscapes of the home counties makes the heathland manager's life here doubly difficulty in the sense that heathlands are used regularly and frequently by very many people and they are in a zone of high fire risk (Maltby, 1980). No matter how concerted the effort to educate visitors and residents of the disadvantages of frequent fires, the managers of some southern heathlands are fighting a losing battle against fire for no management practice can preclude catastrophic fire. On the other hand, Green (1976) has argued that were lowland heaths to be managed on a strictly rotational basis using fire as a management tool, then the effects of frequent and catastrophic fires would be mitigated. While many ecologists would accept the rationale for a rotational burn on lowland heaths, its implementation is not readily accepted by heathland managers. They point to the small size of many of the lowland heathland areas that makes a compartmentalized approach to management very difficult, to the labour-intensive nature of rotational management, to the advanced stage of woodland establishment on many of the lowland commons, to the high fire risk of the southern heathland zone such that even were rotational burns to be introduced accidental fires would always occur, and to the desirability of retaining very old heather stands on sites where certain reptiles are given priority. In other words, geographical location and site conditions as well as the attitudes of managers prevent many heathland areas from being managed with practices that are most consistent with nature conservation.

CONCLUSIONS

It is helpful to see the presence of visitors on heathlands as both an opportunity and a constraint. As far as the opportunity is concerned, the recreational role of heathlands in the south is real and likely to continue, particularly as the extent of the open countryside as a whole declines (Shoard, 1980). Visitors appear to require little of the manager save a well-placed, well-constructed and well-maintained carpark and other minor investment in formal facilities. In general, visitors have a permissive attitude to the management which conforms with other management objectives such as nature conservation. In terms of the constraint, visitors do erode paths and this may be a risk to public safety on some steep slopes and on easily eroded substrates. Extensive erosion with consequent loss of habitat and visual intrusion is only associated with vehicles. *Calluna* communities cannot be maintained for either frequent or intensive recreation unless they are replaced by grassland swards or by large bare areas. In highly suburbanized lowland England, heaths suffer many fires that are difficult to control and that substantially change the vegetation mosaic from one dominated by *Calluna* towards one dominated by grass or bracken. These are usually thought to be undesirable changes on sites that are valued for wildlife.

A reconciliation of the opportunities afforded by visitors to heathlands and their disadvantages leads to the following suggestions. First, all lowland heaths that are habitually used for informal recreation and that are near to urban areas should be managed first and foremost for recreation. On these sites a grass−heath−scrub mosaic could readily be sustained and this together with the careful siting of carparks, a sensitive layout of paths and the provision of a limited range of interpretative facilities would satisfy most of the regular and casual visitors. Second, on large and extensive sites that can be managed and preferably owned by a single agency, nature conservation could be promoted beside amenity, as in the New Forest (Hampshire County Council, 1970). In these circumstances, zoning could allow the most sensitive and valued wildlife habitats from damage by fire, vehicles and large numbers of pedestrians by the careful siting of amenity attractions in more accessible areas. Such conservation zones could form the nucleus of Green's Regional Conservation Areas. Third, on those heathland sites that are neither extensive, nor close to an urban area, attention should be focused on the determination of appropriate management techniques and practices which are designed to enhance both amenity and wildlife (Lowday and Wells, 1977). On many of the Surrey heaths scrub and bracken have advanced to a stage where neither amenity nor wildlife is well served. On some of these sites, especially those that are infrequently used by the general public, a strong case can be made for their role as refugia, such that rare species can be preserved through reintroductions, as for example might be the case with the sand lizard or smooth snake, and for some of the fast-disappearing plant species of the wet heath and bog communities. Equally, on other sites amenity interests could be better served through the reintroduction of grazing by domestic stock, including horses, cutting regimes based on readily available machines such as swipes and flymos, and limited, controlled burns coupled with tree felling and scrub clearance. Such an approach to heathland management deserves close attention because it would make better use of the resources available in the local community. For these latter sites the national agencies such as the Countryside Commission, National Trust, Nature Conservancy Council and the county agencies such as the county councils and naturalists trusts would serve as pump-primers and advisers, and local residents groups, parish councils or amenity groups would be entrusted with day-to-day managment. Groups such as the British Trust for Conservation Volunteers could supplement the local labour force (Earle, 1981).

I suggest an aggressive policy for the amenity management of heathlands which recognizes that the extent to which a recreational role, wildlife conservation role, or both roles can be performed by a given site depends upon its geographical location with respect to the urban edge, the peculiar biological assets and the degree of commitment of its managing body. Unless such a policy is implemented, the only heathland that will be represented in the lowland landscape of 1990 will be held in nature reserves and the once familiar landscapes described by Hardy, Belloc, Young, and Defoe will have been lost for good.

REFERENCES

Armstrong, P. and Milne, A.A. (1973). Changes in the land use of the Suffolk Sandlings: a study in the disintegration of an ecosystem. *Geography*, **58**, 1–8.

Arnett, R.S. (1980). Soil erosion and heather burning on the North York Moors. In J.C. Doornkamp, K.J. Gregory and A.S. Burns (Eds) *Atlas of Drought in Britain 1975/76*, p. 45 Institute of British Geographers, 87 pp.

Bayfield, N.G. (1979). Recovery of four montane heath communities on Cairngorm, Scotland from disturbance by trampling. *Biological Conservation*, **15**, 165–79.

Bibby, C.J. and Tubbs, C.R. (1975). Status, habitats and conservation of the Dartford warbler in England. *British Birds*, **68**, 177–195.

Countryside Commission, (1977). *Study of Informal Recreation in South East England—The Site Studies*, Countryside Commission Publication, Cheltenham.

Countryside Commission, (1979). *Digest of Countryside Recreation Statistics*, Countryside Commission Publication **86**, Cheltenham.

Earle, S. (1981). The management of heathland by the British Trust for Conservation Volunteers, unpublished MSc thesis, University College London.

Gimingham, C.H. (1972). *Ecology of Heathlands*, Chapman and Hall, London 266 pp.

Goode, D. (1981). The threat to wildlife habitats. *New Scientist*, **89**, 219–223.

Green, B.H. (1976). Heathland conservation and management. In *Southern Heathlands Symposium*, Surrey Naturalist's Trust, Godalming, Surrey, pp. 60–64.

Green, B.H. (1981). *Countryside Conservation*, Allen and Unwin, London, 249 pp.

Grime, J.P. (1979). *Plant Strategies and Vegetation Processes*, John Wiley, Chichester, 222 pp.

Harrison, C.M. (1976). Heathland management in Surrey, England. *Biological Conservation* **10**, 211–220.

Harrison, C.M. (1981). A playground for whom? Informal recreation in London's Green Belt. *Area*, **13**, 109–114.

Harrison, C.M. (1980–81). Recovery of lowland grassland and heathland in southern England from disturbance by seasonal trampling. *Biological Conservation*, **19**, 119–180.

Hampshire County Council (1970). Conservation of the New Forest. Hampshire County Council, Winchester, 62 pp.

Lowday, J.E. and Wells, T.C.E., (1977). *The Management of Grasslands and Heathlands in Country Parks*, Countryside Commission, Cheltenham.

Maltby, E. (1980). The environmental fire hazard. In J.C. Doornkamp, K.J. Gregory, and A.S. Burns (Eds), *Atlas of Drought in Britain 1975–6*, p. 60 Institute of British Geographers, 87 pp.

Moore, N.W. (1962). The heaths of Dorset and their conservation. *Journal of Ecology* **50**, 369–91.

Morgan, R.K. (1980). Post-fire plant communities on a lowland heath—Iping Common, W. Sussex, *Rogate Papers No. 3*, Kings College London, 39 pp.

Nature Conservancy Council, (1977). *Nature Conservation and Agriculture*, Nature Conservancy Council, London, 40 pp.

Ratcliffe, D.A. (1977). *A Nature Conservation Review*, Cambridge University Press, London.

Sergeant, P., (1980). Amenity restrictions. In J.C. Doornkamp, K.J. Gregory, and A.S. Burns (Eds), *Atlas of Drought in Britain 1975–76*, Institute of British Geographers, 87 pp.

Sheail, J. (1979). Documentary evidence of the changes in the use, management and appreciation of the grass-heaths of Breckland. *Journal of Biogeography*, **6**, 277–292.

Shoard, M. (1980). *The Theft of the Countryside*, Temple Smith, London, 269 pp.

Streeter, D. (1976). Problems of management of amenity grassland. In S.E. Wright (Ed), *Amenity Grassland*, pp. 8–15 Wye College, Kent.

Tubbs, C.R. (1974). Heathland management in the New Forest, Hampshire. *Biological Conservation*, **6**, 303–306.

Tubbs, C. (1976). Heathland vertebrates. In *Southern Heathlands Symposium*, pp. 53–56 Surrey Naturalists' Trust, Godalming, Surrey.

Webb, N.R. and Haskins, L.E. (1980). An ecological survey of heathlands in the Poole Basin, England in 1978. *Biological Conservation*, **15**, 281–296.

CHAPTER 6

Woodland Conservation in Britain

G.F. PETERKEN

Richard Crossman, writing in the *Times* of 22 August 1973, lamented the felling of Shirburn Wood. Lying on the Chiltern escarpment, this wood was

> known not only for its oaks in a land of beeches, but also for the bluebells under them. It was perhaps the finest and most extensive bluebell carpet you could find in the whole south of England, [However] ... one day last February the demolition squads arrived and ... a week or two later nothing was left but a morass of mud and broken timber. Nobody apparently regarded the oaks and bluebells of Shirburn Wood as a national asset worth preserving, so the oaks were felled *en masse* and the bluebell carpet was destroyed. There was no need to demolish the wood in one fell swoop ... it would have been perfectly possible to thin out the trees over the years and so preserve the bluebell carpet.

Here in one incident we see all the problems faced by a society which requires both timber and beauty from its woods. Whilst the forester will regard this as merely the harvesting of one crop to be followed by planting and the growth of another—good conservation in fact—others see it as gross destruction of a familiar and much loved scene. In this chapter we will consider these conflicting needs and develop a formula for reconciliation.

HISTORICAL FACTORS

There has long been a strong but ambivalent link between conservation and forestry. The first explicit forms of conservation in Britain were the medieval royal forests, private chases and parks, which were established mostly in well-wooded districts and served over the subsequent centuries as a brake on the rate of woodland clearance. Initially these wooded preserves provided facilities for hunting and a supply of fresh meat, but timber-growing increased in importance and was eventually enshrined in, for example, the 1543 Act for the Preservation of Woods. The importance of woodland in the landscape was recognized from the eighteenth century (Gilpin, 1789). By then the parks around county seats were being landscaped with belts and clumps of trees. Hunting never-theless remained a strong factor in the establishment and preservation of woods: indeed, many were—and still are—kept or planted only as fox coverts and pheasant preserves. Throughout the historical period most woods were treated as coppice or wood pasture (see below) in order to supply wood, timber and pasturage for mainly local consumption. Wildlife and natural features sur-vived largely as a byproduct of woodland management undertaken for other purposes.

Conservation in a more modern sense began as a reaction to the nineteenth century tide of inclosure, break-up of the traditional countryside and the expansion of urban populations. The principal aim was landscape preservation and the retention of open spaces for recreation (Eversley, 1910), a diluted form of what Americans would recognize as wilderness preservation (Nash, 1973). In this way many important woods, such as Epping Forest and Burnham Beeches were saved from extinction. At about this time traditional forestry ceased to be effective in conservation: disafforestation of Rockingham in 1796, Hainault in 1851, Wychwood in 1853 and other forests led directly to large-scale woodland clearance. Indeed, in the surviving forests, forestry in the form of expanding in-closure for plantations was seen as a threat to the natural beauty of the scenery. In the New Forest, where the foresters' plans from 1851 onwards would have placed the entire area under plantations by 1908, protests led to the 1877 Act, giving statutory protection to the ancient and ornamental woods, for which the forest is justly famous.

In the nineteenth century naturalists were still in a rapacious phase and ecology had not been born as a science, but from the early twentieth century, conservation of wildlife was increasingly emphasized, and the value of seminatural woods for ecological research became apparent (Sheail, 1976). The potential for conflict between modern forestry and nature conservation was not realized until after the establishment of the Forestry Commission in 1919, and the subsequent drive for massive afforestation and conversion of seminatural woods ('scrub' to a forester) to plantations. By the 1940s the loss of seminatural woods was so rapid that conservationists felt the need to preserve representative examples of all remaining types (HMSO, 1947). In 1946 Sir Arthur Tansley expressed a view of forestry in relation to conservation which has been con-firmed by subsequent experience:

I do not believe that the Forestry Commission, with their existing aims, policy and compulsory powers, can be depended upon to co-operate seriously in the preserva-tion of landscape beauty. Their aims are too radically different, and the position of those who hold that economic considerations must have absolute priority is in-telligible enough, even to those who do not share it. The claims of beauty, amenity and the conservation of wildlife are certain to be brushed aside when they conflict with the maximum production of wood most in demand. Persistent pressure from outside might cause the Commission slightly to modify its programme here and there, as indeed it has already done, but I am inclined to think that the only really effective course, if [The National Parks Committee's] aims are to be realised, is to acquire the legal right to enforce the absolute priority of amenity in its widest sense within certain defined areas. Attempts at detailed adjustment and compromise over wide areas can only lead to constant friction and thoroughly disappointing results. (PRO HLG/93/24, NPC 49).

The close association between landscape conservation and nature conservation was weakened after the 1940s when the emphasis in nature conservation was focused on scientific objectives, and the National Parks Commission and the Nature Conservancy were separately established. Modern forestry, like modern agriculture and unlike traditional forms of both, was widely regarded as damaging to both landscape and nature, yet it continued virtually unchecked. Protests were often outspoken (for example, the Ramblers Association, 1971). Personal ministerial intervention was required to stop the wholesale felling of broadleaf woods in the New Forest. Despite this, foresters had begun to accept the conservation arguments. In 1936 the first forest park was established in Argyllshire, and this has been followed by a deep involvement in public recreation in forests throughout Britain. Discussions about reserves for ecological research, initiated at the British Association meeting in 1938, led in 1943 to the designation of Lady Park Wood in the Wye Valley as a natural reserve. Increasingly, foresters planned their operations to harmonize with the landscape (Crowe, 1978) (though at the same time coppices were being underplanted, killed with herbicides and left standing as gaunt reminders of the mechanical mentality with has pervaded modern forestry). Furthermore, foresters have pressed their case as good conservationists, not only by defining conservation as maintaining and increasing the productive capacity of the site, but also by claiming that even conifer plantations are good for wildlife (Forestry Commission, 1977), a claim that is contested by conservationists (Ratcliffe, 1980). Today the uneasy relationship continues, but the arguments lie mainly between the urge for greater material productivity (in other words, timber) and the need for refreshment, pleasure and escape from urban artificiality which the contemplation of wild plants and animals and access to relatively natural woodland readily affords (Peterken, 1981, Chapter 12).

TYPES OF WOODLAND

Before any quantitative appraisal for woodland conservation can be attempted, various types of woodland must be defined. The most obvious distinction, which is recognized in all censuses, lies between broadleaf and coniferous woodland. The former contains trees such as oak, beech, ash and lime, whereas the latter contains pine, spruce, fir and other conifers. Virtually all the broadleaves are deciduous, and most broadleaved trees in Britain are natives. Some evergreen alien broadleaves have become naturalized, for example,

Quercus ilex. Except for larch, all the conifers are evergreen, and only pine and yew are native. Accordingly, there is some justification for the common habit of using the terms broadleaf, native and deciduous as if they are synonymous.

A second important distinction is drawn between seminatural woods and plantations. In *plantations* the composition, age structure and distribution of species is almost wholly determined by man (see Goldsmith and Wood, Chapter 17). *Seminatural* stands are those that were not planted: their characteristics are therefore determined more by natural factors such as soil conditions. No woods are completely free of human influence, and thus *natural*. Nor are plantations totally void of at least some self-sown trees. Many seminatural stands contain a few planted trees. This distinction therefore expresses in a simplified form a continuous variable (Peterken, 1981, Chapter 3).

Many of the woods we see today grew up on ground that was once used as pasture, arable, etc. Such woods are *secondary*. Other woods have existed continuously since prehistoric times when Britain was covered in woodland. These *primary* woods are the direct descendants of original natural woodland which were never cleared away. Primary woods are however almost impossible to identify, so a distinction is made for practical purposes between ancient woods, originating in the Middle Ages or before, and *recent* woods, originating since the Middle Ages. All recent woods are secondary and all primary woods are ancient, but a few woods are both ancient and secondary (Peterken, 1981; Rackham, 1980).

The silvicultural systems under which woods have been treated can broadly be classified as coppice, wood pasture and high forest (Rackham, 1976). Coppices are broadleaved woods cut periodically at ground level and allowed to sprout again from the stumps. Above the dense thicket so created there were usually scattered timber trees known as standards. *Wood pasture* combines trees with grazing animals, such as cattle or deer. The trees were usually pollarded—coppicing out of reach of the animals—and regeneration from seed was inhibited by grazing, so that the characteristic structure was an open scatter of large old trees with spreading crowns. *High forest* broadly covers all stands grown from seedlings. Numerous systems have been tried (Troup, 1928), but most high forest stands are now even-aged plantations. In continental Europe many high forests are grown from natural regeneration and some selection stands survive (Reed, 1954).

In the Middle Ages, woods consisted of native trees and were treated as coppice or wood pasture (Rackham, 1980). They were not planted and would today be regarded as seminatural. They were ancient (by definition), and most were probably primary, though there were many ancient secondary woods on the southern chalk and limestone (Dewar, *c*. 1926). The wood pastures on commons, forests and deer parks were denuded, neglected and destroyed from the late Middle Ages onwards, though some fine examples survive, such as the New Forest (Tubbs, 1968) and Woodstock Park (Bond, 1981). Planting was infrequent in the Middle Ages, but from the seventeenth century onwards (Evelyn, 1664, Sharp, 1975) numerous shelterbelts, fox coverts and amenity woods were planted on agricultural land. The Board of Agriculture reports of *c.* 1800 (Jones, 1961) distinguished clearly between 'woodland', meaning coppices surviving from the Middle Ages, and 'plantations', being recent woods established by planting. For much of the nineteenth century the two traditions of management still coexisted with little interaction, apart from an early nineteenth century phase of coppice improvement when, for example, many Kentish coppices were replanted with chestnut. Towards the end of the nineteenth century, however, woodland and plantation management declined and many coppices were no longer cut. When the revival came after 1920, forestry—profoundly influenced by German traditions inherited via the Indian Forest Service (James, 1981)—was solely concerned with plantations treated as high forest, and the coppices slipped steadily towards neglect and destruction.

THE IMPORTANCE OF ANCIENT SEMINATURAL WOODS

Ancient, seminatural woods are far more important for nature conservation than other woodland types because of the following factors.

(1) They include all primary woods. These are the lineal descendants of Britain's primeval woodland, whose wildlife communities, soils and sometimes structure have been least modified by human activities over the millenia. Their tree and shrub communities (Figures 6.1 and 6.2) preserve the natural composition of Atlantic forests (Birks, Deacon, and Peglar, 1975). Once destroyed they cannot be recreated (Peterken, 1977).

(2) Being relatively unaffected by man, they provide baselines (or 'controls' in scientific parlance) against which to measure the effects of man on, say, soils (Ball and Stevens, 1981), productivity of woodland communities, food webs, etc.

(3) Their wildlife communities are generally, but not invariably, richer than those of recent woods. The latter tend to contain only weed species or impoverished forms of heathland, meadow, bog, according to the prior land use.

(4) They contain a very high proportion of the rare and vulnerable wildlife species, in other words those most in need of protection if all species are to survive in Britain. Many of these species require the stability afforded by the continuity of suitable woodland; they cannot colonize newly created woodland, or do so only slowly (Boycott, 1934; Peterken, 1974).

(5) Where large, old trees have been present for several centuries, they provide refuges for the characteristic inhabitants of primeval woodland, such as lichens (Rose, 1976) and beetles (Buckland and Kenward, 1973; Hammond, 1974).

Acid Oak-Lime Woodlands

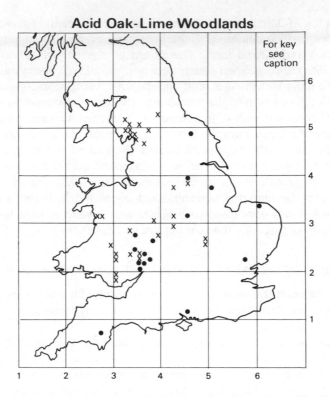

FIGURE 6.1 Distribution of acid sessile oak–lime woodland (stand type 5B of Peterken, 1981) in Britain. ● 10 km square containing at least one example confirmed by G.F. Peterken; × 10 km square containing a wood in which *Tilia cordata, Quercus petraea* and *Betula pendula*, but not *Fraxinus excelsior* coppice were recorded in a national woodland survey carried out about 1970.

(6) They contain other natural features which rarely survive in an agricultural setting, such as streams in their natural watercourses and microtopographical conditions formed under periglacial conditions (Rackham, 1980, p. 13).

(7) They are reservoirs from which the wildlife of the countryside has been maintained (and could be restored). For example, hedges and recent woods in the vicinity of ancient woods are more likely to contain woodland species than sites isolated from such woods (Peterken and Game, 1981).

(8) They have been managed by traditional methods for centuries. Not only are they ancient monuments whose value to historians and to village community consciousness is arguably as great as that of the older buildings in a parish, but, where traditional management continues or can be revived,

Barney Wood-Stand Types

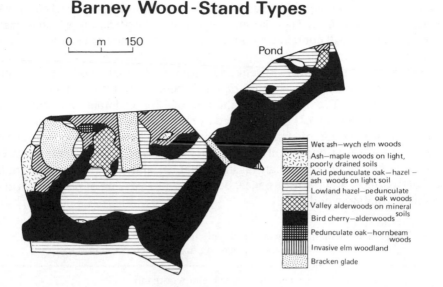

0   m   150

Pond

Wet ash—wych elm woods

Ash—maple woods on light, poorly drained soils

Acid pedunculate oak—hazel—ash woods on light soil

Lowland hazel—pedunculate oak woods

Valley alderwoods on mineral soils

Bird cherry—alderwoods

Pedunculate oak—hornbeam woods

Invasive elm woodland

Bracken glade

FIGURE 6.2   Map of stand types in Barney Wood, Norfolk (12 ha, TF(53) 989328). Mapped by Suzanne Goodfellow and G.F. Peterken. On gently undulating ground with light, acid—neutral soils, the hazel—oak and hazel—ash stands pick out the higher ground, whereas the alder stands occupy receiving sites. Descriptions of the stand types are available in Peterken (1981), Chapter 8.

they are living demonstrations of conservation in the broader sense of a stable, enduring relationship between man and nature.

With a few exceptions (for example, Holme Fen, Kirkconnel Flow), the woods of national importance listed in *A Nature Conservation Review* (Ratcliffe, 1977) are all apparently both ancient and seminatural. The same high correlation can be seen in the 124 000 ha of woodland scheduled as SSSIs in 1974 (Peterken, 1977).

CLASSIFICATION OF SEMINATURAL WOODLAND

Since many practical conservation objectives revolve around the need to perpetuate examples of all seminatural woodland types, it is important to have an agreed classification, particularly of seminatural stands in ancient woodlands. The classification developed by the founder British ecologists (Tansley, 1939) has—remarkably—continued in common use, even though it has serious gaps. Based on stand composition and site characteristics the types included, for example, damp oakwoods and calcareous beechwoods. Recently, this approach to classification has been considerably developed by Rackham

(1980), dealing mainly with East Anglia, and the author (Peterken, 1981, Chapter 8) (Table 6.1). The National Vegetation Classification, which is nearing completion, will contain a woodland section. Famine has been replaced by embarrassing plenty!

The distribution through Britain and within individual woods of the stand types recognized by Rackham and myself is believed to reflect predominantly natural factors. The natural distribution of, for example acid oak−lime woodland (see Figure 6.1) is determined by climatic, edaphic and historical factors. The distribution of stand types within Barney Wood (see Figure 6.2) can be related to variations in soil texture, base status and moisture regime. Even so, the composition of seminatural stands has certainly been affected by traditional treatment and by disturbance, but not as much as that of plantations and recent secondary woodland.

TABLE 6.1   Seminatural woodland types in ancient woods: the main groups of stand types recognized by Peterken (1981).

(1)	Ash−wych elm woodland
(2)	Ash−maple woodland
(3)	Hazel−ash woodland
(4)	Ash−lime woodland
(5)	Oak−lime woodland
(6)	Birch−oak woodland
(7)	Alder woodland
(8)	Beech woodland
(9)	Hornbeam woodland
(10)	Suckering elm woodland
(11)	Pine woodland
(12)	Birch woodland

DECLINE OF THE ANCIENT, SEMINATURAL WOODLAND

Inevitably the extent of ancient, seminatural woodland has declined over the years. More than half the 1104 kha of woodland recorded by the Board of Agriculture in 1895 was of this type, implying a total area of roughly 600 kha. In 1086, 800 years earlier, the Domesday Book records a woodland area of roughly 1650 kha in England (Rackham, 1980, Table 9.6). Adding Scotland and Wales, the eleventh-century extent of ancient, seminatural woodland may have been around 2500 kha. Between 1086 and 1895 the losses were due almost entirely to clearance for cultivation and pasture, but after 1895 a new threat emerged, the conversion of seminatural woods to plantations. Now, only about 300 kha survive, together with roughly 275 kha of ancient woodland converted to plantations (Peterken, 1977) (Table 6.2). This long decline can be seen in three East Midland study areas (Table 6.3) where, nevertheless, there were marked fluctuations in the rate from one period to another.

Since the Second World War, clearaᴜᴄe for agriculture has been common in sparsely wooded lowland districts where forestry was not worthwhile, and those ancient woods that survived mostly remained unmanaged, save for shooting (Peterken and Harding, 1975). In more densely wooded districts forestry has been more worthwhile: this has given owners an incentive to keep the woods, though many have been converted to plantations. In the uplands many seminatural woods have been engulfed in or converted to conifer plantations. Despite these regional differences in the nature of change, the losses appear to have been fairly uniform throughout the country. Some 38–44 per cent of the ancient seminatural woodland present in 1946 in three East Midland study areas had been cleared or converted to plantations by 1972, and more approximate results from elsewhere suggest a nationwide, postwar decline of at least 30–50 per cent.

TABLE 6.2   Approximate extent of major woodland types existing and foreseen in Britain about 1980 (thousands of hectares).

Kha	Woodland type	Broadleaf	Conifer
300	Ancient, seminatural	290	10
275	Ancient woodland now converted to plantations ⎫		
1425	Recent, secondary woods. Mostly plantations, but including recent seminatural stands ⎬	370	1330
1700	Future afforestation ⎭	?	1700

Current work by the NCC is revealing the present state of ancient, seminatural woods. Using recent ground survey, old Ordnance Survey maps and other sources, we are preparing an inventory of these woods for each county. Results available by October 1982 (Table 6.4) confirm that, despite substantial differences between counties, about half the ancient woods are not plantations. Ancient, seminatural woods occupy only about 0.5–2 per cent of the land surface in most areas. Of the 1924 ancient woods in the first five counties examined (Cambs, Lincs, Salop, Pembs, Central Scotland), 53 per cent (679) were less than 10 ha and only 9 per cent (115) were more than 50 ha. Forestry is clearly more active in the larger woods: just 10 per cent (12) of the woods over 50 ha contain no plantations (i.e., were completely seminatural), whereas 63 per cent (425) of the woods below 10 ha were totally seminatural. Increasingly, the lineal descendants of primeval woodland survive only as tiny woods or as fragments of larger woods.

TABLE 6.3 Decline in the extent of ancient and ancient, seminatural woodland in three East Midland study areas.

Year	Ancient woodland (ha)			Ancient Seminatural Woodland (ha)		
	Rockingham Forest	Central Lincolnshire	West Cambridgeshire	Rockingham Forest	Central Lincolnshire	West Cambridgeshire
1650	8442	?	?	8442	?	?
1796	7868	?	?	7868	?	?
1817–34	7182	3726	700	7182	3726	700
1885–87	4261	2679	682	(4200)	(2600)	(670)
1946	3971	2614	675	2819	2240	631
1972–73	3558	2558	598	1582	1169	432
1981	(3500)	2521	?	?	1132	?

Sources: Peterken and Harding (1975); Peterken (1976).
Estimated figures are bracketed.

TABLE 6.4   Extent and condition of ancient woods in sixteen 'counties', collated by the Nature Conservancy Council.

County	Number of separate ancient woods	Area (ha)	Ancient woods as proportion of county area (%)	Proportion of ancient woods still seminatural (%)
Essex	618	8618	2.34	84
Norfolk	243	2867	0.54	54
Cambridgeshire	123	2998	0.87	72
Bedfordshire	152	2591	2.10	64
Northamptonshire	197	6565	2.77	40
Lincolnshire	217	6227	1.06	46
Humberside	66	1075	0.31	71
Buckinghamshire	439	9451	5.02	72
Surrey	549	7140	4.25	66
Cornwall	434	6174	1.74	53
Shropshire	518	10517	3.01	39
North Cumbria	354	4083	1.79	62
Gwent	585	8817	6.41	37
Pembrokeshire	216	2532	1.59	49
Clwyd	662	5434	2.24	56
Central Scotland	220	5604	2.13	50

From various surveys collated by Ian Bolt, Keith Kirby, Jonathan Spencer and Graham Walker for NCC.

HOW MUCH WOODLAND?

Britain now has 2.0 m.ha of woodland (see Table 6.2) or nearly double the area at the start of the present century. Afforestation will continue until well into the next century: by 2050 the woodland area may reach 3.7 m.ha (Forestry Commission, 1977). Even this will still leave Britain poorly wooded by European standards: roughly 9 per cent of the land is now wooded, compared with 20–40 per cent in much of Europe. In the Highlands, however, forests may eventually cover 40 per cent of the ground.

We now stand part-way through a massive switch from broadleaf to conifer woodland. Over 90 per cent of our native woods were naturally broadleaved and as late as 1924 broadleaves still predominated (67 per cent according to the Forestry Commission census). Now conifers are in a clear majority (66 per cent), and by 2050 they will predominate absolutely. If all the new woodland is coniferous and there is no net loss of broadleaves, the latter will comprise around 20 per cent of all British woods. Whilst most of the shift in balance is due to afforestation, there has been an absolute decline in the extent of broadleaf woodland.

Britain's woods yielded 4.2 million m³ of timber in 1977 which, after presumably allowing for exports, supplied 8 per cent of the total demand of 39 million m³ (Centre of Agricultural Strategy, 1980). The balance of 92 per cent was imported at a cost of £2370 million in 1978. Home yield will increase as existing and future plantations come onstream perhaps to 26 million m³ each year (assuming 3.4 m.ha of productive woodland, of which 75 per cent is coniferous yielding 9 m³ ha⁻¹ year⁻¹ and 25 per cent broadleaved yielding 4 m³ ha⁻¹ year⁻¹. If by then, as some suppose, demand has risen to 75 million m³, Britain would still be only 35 per cent self-sufficient, or less if any home production is exported, or if afforestation on the scale proposed is not realized.

CONSERVATION STRATEGY

The first and most important element in a woodland conservation strategy is to keep the woods we have. The expanding upland plantations are no substitute for lost seminatural woods, nor are new field-corner plantings any compensation for mature woods cleared elsewhere. New woodland takes decades or centuries to acquire a full suite of woodland wildlife (Peterken, 1977), and the upland forests are on soils which may never support a rich woodland flora (Hill and Jones, 1978). Further woodland clearance will inevitably mean further irrecoverable loss of ancient woodland. That said, we have to decide how existing and new woods should be treated. This is a complex problem, which involves not just identifying the most important woods for conservation and determining their ideal treatment, but also involves integrating the needs of conservation with fluctuating market trends and developing forestry techniques.

Heritage woods

The close links between landscape conservation and nature conservation are fundamental to any strategy. Most of the ancient, seminatural woods are also important in the landscape. For example the mature broadleaf woods of the Lake District, Dartmoor fringes, Wye Valley, Cotswolds, Chilterns and Weald are mostly ancient, and so too are the pinewoods of Abernethy and Rothiemurchus that are essential to the attractive and distinctive landscape of Speyside. Of course, not all prominent woods are ancient or seminatural (for example, shelterbelts and plantings associated with landscape parks), nor are many of the tiny woods which relieve the landscape of intensively arable districts, but the coincidence of wildlife and amenity interest on the ground is still good. Rather than base a strategy on 300 kha of ancient, seminatural woodland which is important for nature conservation it seems better to think in terms of say 350–400 kha which is important for both nature and landscape conservation. Collectively these might be known as 'heritage woods'.

Treatment of heritage woods must take account of their great variety; it would be foolish to seek a common prescription for an ash–maple coppice in East Anglia, the beech–oak wood pastures of the New Forest, a Highland birchwood and the wooded vistas of Blenheim Park. A few should be set aside as totally untouched reserves, where they can return to an almost natural state, but this is unnecessary, impractical and even damaging in most instances, as in Epping Forest (Rackham and Ranson in Corke, 1978). A better aim, which serves a wider range of interests, would be to continue or resume cropping of native trees by traditional and near-traditional means. Not only would this generate a supply of hardwood timber and small wood, but it would also maintain the diversity of habitat that is associated with traditional management, and be a safety play (to use bridge parlance) for those species which have demonstrably coexisted with such management. Moreover, just as a trimmed hedge is more likely to survive in the long run than an unkempt hedge, so a managed wood with a management plan and a variety of customers for its products is more likely to enjoy stable treatment in the long run than a 'neglected' wood, which is more vulnerable to the changing whims of owners, planners and politicians (though admittedly the markets themselves are often even more whimsical!).

The actual silvicultural system that is preferred in each case depends on the existing state of the wood and the particular ecological characteristics which make it important for conservation. In many instances coppice treatment would be preferred, especially where coppicing continues today or did so until recent decades. In such woods, oak and other potentially valuable timber trees should be treated as standards under the free growth principle. Where high forest systems are adopted, long rotations and natural regeneration would be preferred. A small group system would be preferred where shade-bearing trees were involved—lime, beech, hornbeam—but if larger group working were required for, say, oak or ash, this would normally be acceptable. Planting should be undertaken only after natural regeneration had failed, that is, as a method of beating up, except where oaks were planted at low density into coppice. First thinning should be regarded as the decisive stage, when, if possible, a mixture of native species should be retained. In general, we should seek a low input–low output approach to investment in timber growing in most heritage woods. The timber produced should be of high quality and therefore of high value. This approach to forestry has been common in continental Europe, but rare in Britain. Its wider adoption in Britain, as recommended in the Sherfield Report (1980), would go far to reconcile conservation and timber-growing in British woods.

If conservation were to be given priority in the heritage woods and timber production were subordinate, priorities would have to be reversed in the other 1600 kha of woodland and the 1700 kha of prospective new plantations. They should be treated to produce as much timber as possible, subject to maintaining the productive potential of the ground and minimizing damaging effects outside plantations, as on water quality. The aim should be stability. A normal forest (containing a mixture of crops of all ages) should be established nationally and locally, together with an infrastructure of employment in forest management and forest-based industries. The use of pine, larch and a sprinkling of native broadleaves would diversify the plantations to the benefit of wildlife and landscape. *Nothofagus* might be used. Rotation lengths and thinning regimes would be determined by the needs of timber production, but where there was a choice, rotations should be long and thinning early and heavy. Suitable ride treatment could greatly enhance the wildlife of plantations (Steele, 1972). Special care should be given to the 275 kha of plantations on ancient woodland sites, where the need to retain broadleaf stands is greatest.

What impact will this strategy have on Britain's timber supplies? Heritage woods comprise no more than 20 per cent of existing woodland and (assuming none are lost and that afforestation proceeds as indicated) about 11 per cent of future woodland. This is equivalent to 11 per cent of our long-term potential for timber production (assuming that all land is equally productive) or no more than 3 per cent of our estimated requirements. Even if all the heritage woods were managed as productively as possible, 97 per cent of our requirements would have to be grown elsewhere. Put another way, a decision to take no more timber from heritage woods could be redeemed by a small increase in the efficiency with which timber is used. In fact, we advocate the production of some timber in most heritage woods, albeit only with native trees, and the sacrifice of timber would be significantly less than 10 per cent of home potential production. Furthermore, few if any heritage woods need be less productive than now: most are now producing only lowish yields of wood and timber from native species.

Considerable problems bar the way to this strategy, but in many cases there have been encouraging steps towards solutions. These may be summarized as follows.

(1) Markets for native hardwood timber and wood are reasonable, but fluctuating (Garthwaite, 1977; Gascoigne, 1980). Woodburning stoves have recently inceased the market for coppice wood and thinnings. Even so, hardwoods and coppice are still economically unattractive to most owners, due to small early returns and the long rotations.

(2) Most heritage woods have long been neglected, so there is a backlog need for drastic rehabilitation. Restoration of traditional treatments is difficult.

(3) Landowners and foresters usually strongly resist the introduction of planning into forestry. Those who own heritage woods would expect to be compensated for loss of profits. This is in direct contrast to the political attitude to public control and care of historic buildings. Greater public influence in heritage woods may have to be accepted, especially if

forestry is to be allowed to expand elsewhere. Woodland management orders might be more acceptable than tree preservation orders (Sherfield, 1980).

(4) The skills necessary for the management of broadleaf woodland have declined. Nevertheless the Forestry Commission, having neglected broadleaves for so long, is now taking a more positive attitude with work on the free-growth idea (Jobling and Pearce, 1977) and the use of shelters to establish oaks and other broadleaves.

(5) Heritage woods are not evenly distributed between owners: the sacrifice of income from timber already grown falls disproportionately on some private woodland owners. If compensation is paid it would have to cover up to 200 years of past growth as well as future loss of opportunity for maximum profits. Tax incentives would have to be linked to heritage woods.

(6) The new Forestry Commission grant scheme, based on a once-for-all planting grant, may be satisfactory for timber growing but is quite unsuitable for heritage woods, where planting (at least at the high density required for full rates of grant) is often inappropriate. Support should be based on management costs, not planting, and on recognition of long-term dedication to heritage woodland treatment.

(7) Unlike support for timber growing, which is channelled through the Forestry Commission, support for the treatments appropriate in heritage woods comes from a variety of sources, such as local authorities, countryside commissions, National Trust, Nature Conservancy Council, naturalist trusts, and the Woodland Trust. Many private owners carry out the appropriate treatments without support or with reduced grants. A heritage woodland dedication scheme is required into which the money earmarked for the hardwood supplement of the new Forestry Commission scheme might be paid.

(8) Before this strategy could work, ancient seminatural woods must be identified. The Nature Conservancy Council is doing this (see above). A precedent exists for special treatment of an identified group of woods in the special provisions of the old basis III dedication scheme for those native pinewoods identified by Steven and Carlisle (1959).

(9) Woods have often been established as nature reserves or amenity woods without thought to subsequent management. Whilst some must remain untreated, many are now being managed and thus producing some wood and timber. These help to point the way for heritage woods in general and, vice versa, the existence of a heritage woodland dedication scheme might encourage better management of other reserved woods.

POSTSCRIPT

Readers of Richard Crossman's article will be surprised to learn that Shirburn Wood still stands, and that it is a beech wood. The oaks, whose clear felling so outraged the politician, had been planted as a mixture with larch around 1830–40 on a small field beside the wood. The bluebells and bracken had crossed the still-visible ancient wood bank, colonized the new plantation and, as the oaks matured, had eventually formed the colourful and now famous carpet of blue. Once cleared, the land was replanted with a mixture of trees, mainly broadleaf, but the new plantation was badly neglected and only a few gean, ash, larch, spruce and self-sown birch have grown away successfully, the beech being smothered and now moribund. The ancient wood itself remains untouched, as it has for decades, a high forest beechwood of moderate growth quality originating mainly in the early to mid nineteenth century, now devoid of a shrub layer and natural regeneration, but well stocked with bluebells.

Clearly this felling was a trivial incident in the recent unhappy history of Britain's native woods. Unlike numerous changes elsewhere, the felling was on a small scale, the oakwood (being secondary) was recreatable, and some attempt was made to re-establish native broadleaves. Nevertheless, Crossman was right to make a fuss. The oaks could indeed have been heavily thinned to allow planting of successors in the shelter of the survivors, thus maintaining the appearance of mature woodland. Moreover, this treatment is urgently needed in the beechwood next door. Instead of clear felling one part of Shirburn Wood, and neglecting the rest, the whole wood should have been restored as mixed-age beech–oak–ash high forest along the lines successfully followed by Michael Reade (1957) at Checkendon. If a scheme for appropriate treatment of heritage woods had been available 10 years ago, the course of events in Shirburn Wood might have been more to the taste of Richard Crossman and all those who are sensitive to the environmental impact of forestry in British woodlands.

REFERENCES

Ball, D.F. and Stevens, P.A. (1981). The role of 'ancient' woodlands in conserving 'undisturbed' soils in Britain. *Biological Conservation*, **19**, 163–176.

Birks, H.J.B., Deacon, J. and Peglar, S. (1975). Pollen maps for the British Isles 5000 years ago. *Proceedings of the Royal Society of London B*, **189**, 87–105.

Bond, C.J. (1981). Woodstock Park under the Plantagenet kings: the exploitation and use of wood and timber in a medieval deer park. *Arboricultural Journal*, 5, 201–213.

Boycott, A.E. (1934). The habitats of land *Mollusca* in Britain. *Journal of Ecology*, **22**, 1–38.

Buckland, P.C. and Kenward, H.F. (1973). Thorne Moor: a paleoecological study of a Bronze Age site. *Nature (London)*, **241**, 405–406.

Centre of Agricultural Strategy (1980). *Strategy for the UK Forestry Industry*, Centre of Agricultural Strategy Report No. 6, Reading, 347 pp.

Corke, D. (Ed) (1978). *Epping Forest—The Natural Aspect?* Essex Field Club, London, 80 pp.

Crowe, S. (1978). *The Landscape of Forests and Woods*, Forestry Commission Booklet 44, HMSO, London.

Dewar, H.S.L. (*c.* 1926). The field archaeology of Doles. *Papers and Proceedings of the Hants Field Club and Archaeological Society*, **10**, 118–126.

Evelyn, J. (1664). *Sylva, or a Discourse of Forest-Trees* (4th edition), published, 1706, Doubleday, London, pp. 335 and 287.

Eversley, Lord (1910). *Commons, Forests and Footpaths* (revised edition), Cassell, London, 356 pp.

Forestry Commission (1977). *The Wood Production Outlook in Britain*, Forestry Commission Edinburgh, 111 pp.

Garthwaite, P.F. (1977). Management and marketing of hardwoods in S.E. England. *Quarterly Journal of Forestry*, **71**, 67–77 and 144–150.

Gascoigne, P.E. (1980). A case for coppice-with-standards—for profit and pleasure. *Quarterly Journal of Forestry*, **74**, 47–56.

Gilpin, W. (1789). *Observations on the River Wye and Several Parts of South Wales etc.*, R. Blamire, London, 152 pp.

Hammond, P.M. (1974). Changes in the British Coleopterous fauna. In D.L. Hawksworth (Ed) *The Changing Flora and Fauna of Britain*, Academic Press, London, pp. 323–369.

Hill, M.O. and Jones, E.W. (1978). Vegetation changes resulting from afforestation of rough grazings in Caeo Forest, South Wales. *Journal of Ecology*, **66**, 433–456.

HMSO (1947). *Conservation of Nature in England and Wales*, Cmnd 7122, HMSO, London, 139 pp.

James, N.D.G. (1981). *A History of English Forestry*, Blackwell, Oxford, 339 pp.

Jobling, J. and Pearce, M.L. (1977). *Free Growth of Oak*, Forestry Commission Record 113.

Jones, E.W. (1961). British forestry in 1790–1813. *Quarterly Journal of Forestry*, **55**, 36–40 and 131–138.

Nash, R. (1973). *Wilderness and the American Mind* (revised edition), Yale University Press, New Haven and London, 300 pp.

Peterken, G.F. (1974). A method for assessing woodland flora for conservation using indicator species. *Biological Conservation*, **6**, 239–245.

Peterken, G.F. (1976). Long-term changes in the woodlands of Rockingham Forest and other areas. *Journal of Ecology*, **64**, 123–146.

Peterken, G.F. (1977). Habitat conservation priorities in British and European woodlands. *Biological Conservation*, **11**, 223–236.

Peterken, G.F. (1981). *Woodland Conservation and Management*, Chapman and Hall, London and New York, 328 pp.

Peterken, G.F. and Game, M. (1981). Historical factors in the distribution of *Mercurialis perennis* L. in central Lincolnshire. *Journal of Ecology*, **69**, 781–796.

Peterken, G.F. and Harding, P.T. (1975). Woodland conservation in eastern England: comparing the effects of changes in three study areas. *Biological Conservation*, **8**, 279–298.

Rackham, O. (1976). *Trees and Woodland in the British Landscape*, Dent, London, 204 pp.

Rackham, O. (1980). *Ancient Woodland*, Arnold, London, 402 pp.

Ramblers' Association (1971). *Forestry: Time to Rethink*, Ramblers' Association, London, 16 pp.

Ratcliffe, D.A. (Ed) (1977). *A Nature Conservation Review*, Cambridge Unviersity Press, Cambridge, 388 and 320 pp.

Ratcliffe, D.A. (1980). *Forestry in relation to nature conservation*. Memorandum submitted to the Science and Technology Committee of the House of Lords, 11 June 1980, HMSO, London, 33 pp.

Reade, M.G. (1957). Sustained yield from selection forest. *Quarterly Journal of Forestry*, **51**, 51–62.

Reed, J.L. (1954). *Forests of France*, Faber, London, 296 pp.

Rose, F. (1976). Lichenological indicators of age and environmental continuity in woodlands. In D.H. Brown, D.L. Hawksworth and R.M. Bailey (Eds) *Lichenology: Progress and Problems* (Systematics Association special volume 8, Academic Press, London, pp. 279–307.

Sharp, L. (1975). Timber, science and economic reform in the seventeenth century. *Forestry*, **48**, 51–86.

Sheail, J. (1976). *Nature in Trust*, Blackie, Glasgow and London, 270 pp.

Sherfield, Lord (1980). *Scientific Aspects of Forestry*. 2nd Report of the House of Lords Select Committee on Science and Technology. HMSO, London, 59 pp.

Steele, R.C. (1972). *Wildlife Conservation in Woodlands*, Forestry Commission Booklet 29.

Steven, H.M. and Carlisle, A. (1959). *The Native Pinewoods of Scotland*, Oliver and Boyd, Edinburgh and London, 368 pp.

Tansley, A.G. (1939). *The British Islands and their Vegetation*, Cambridge University Press, Cambridge, 930 pp.

Troup, R.S. (1928). *Silvicultural Systems*, Clarendon Press, Oxford, 199 pp.

Tubbs, C.R. (1968). *The New Forest: An Ecological History*, David and Charles, Newton Abbot, 248 pp.

Conservation in Perspective
Edited by A. Warren and F.B. Goldsmith
© 1983 John Wiley & Sons Ltd.

CHAPTER 7

Heather Moorland in Northern Britain

G.R. MILLER and A. WATSON

INTRODUCTION

The northern uplands of Britain are mainly treeless and covered with vegetation less than 1m in height. Many of these moors are characterized by a predominance of heather (*Calluna vulgaris*), an ericaceous dwarf shrub. Its distribution extends from subarctic Scandinavia to the Mediterranean, and from Ireland to the Ural Mountains, where it grows in many different plant associations from the sea coast to the hill tops.

Heather flourishes particularly in eastern Britain and along the western seaboard of Europe from southern Scandinavia to northern Spain, where it may dominate the vegetation of large areas, often in association with other dwarf shrubs. On the continent, much of this land has been afforested or turned into productive farms, but this is less true of Britain where extensive heather moors remain in the east-central Scottish Highlands, north-eastern England and eastern Ireland. Here the acid and usually well-drained podzolic soils support a luxuriant growth of heather and in some places few other flowering plants occur. Heather thrives less well in the cool moist climate and waterlogged soils of western Britain where it is merely one of several main constituents of the communities on blanket peat, including purple moor grass (*Molinia caerulea*), deer sedge (*Trichophorum cespitosum*) and cotton sedge (*Eriophorum* spp.). Only locally, on the best-drained hillocks or slopes, is heather as vigorous and predominant as on the eastern moors.

Heather moorland covers roughly 0.8 million ha in Scotland and 0.4 million ha in England and Wales; there are also large tracts in eastern Ireland. Because most of the soils are freely drained and capable of economic exploitation, farmers, foresters and other land users have competed for this large area since the turn of the century. The fact that moorland makes a significant contribution to the economy of our sparsely populated uplands has stimulated much research. As a result, probably more has been written about the origin, characteristics, management and potential for development of heather moorland than about any other kind of upland vegetation. Gimingham (1972) has reviewed the literature and his book should be consulted for references to many of the original scientific papers.

ORIGIN OF HEATHER MOORLAND

Small areas of moorland have developed behind coastal sand dunes and cliffs, on drained or drying bogs and, more widely, above the climatic tree limit. These apart, ecologists now believe that most moors resulted from the destruction of woodland, both by climatic change and human interference (O'Sullivan, 1977).

Many pollen analyses suggest that moor and bog vegetation extended and woodlands declined during the cool and wet Atlantic period beginning about 7000 years ago. Nonetheless, northern Britain was still heavily forested when Neolithic man settled about 3000 BC. From this time onwards, forest destruction accelerated. People at first practised a shifting agriculture close to the sea coast and on the drier slopes of the lower hills. By deliberately burning and felling, they made open spaces for growing crops and grazing livestock, used them for a short while, and then abandoned them for fresh ground. From about 500 BC, human settlements became more permanent and forest clearance spread to poorer ground. The intensification and extension of primitive farming coincided with the onset of the cooler and wetter climate of the sub-Atlantic period. Together these factors not only destroyed woodland cover but also inhibited its regeneration.

By the end of the seventeenth century, a rapid expansion of agriculture and increasing demands for timber for buildings, boats, implements and charcoal had led to the almost complete destruction of natural woodland in England. Iron ore was taken north for smelting in Scotland but some of the more remote parts remained unscathed for a further 100 years.

Wherever soils are acid, heather was and still is abundant in open parts of pine, birch and oak woodland. Forest clearance would have allowed it to flourish, and the dwarf shrub vegetation of the scattered glades presumably spread and merged to form large stretches of heather moor. These open moors provided free-range grazing for sheep and, in Scotland, stocks of hill sheep increased rapidly in the late eighteenth and early nineteenth centuries. Grazing and burning destroyed tree seedlings and maintained heather moor wherever there were acid, freely drained soils. On many moors, viable tree seed is now so scarce that even if burning and grazing were to cease, reversion to woodland would take many decades (Gimingham, 1972).

VEGETATION AND ANIMALS

The predominance of heather and its influence on microclimate and soils restrict the diversity of plant species on moorland. Several dwarf shrubs of the family Ericaceae are common but other shrubs are few. There is, however, an abundant and varied bryophyte and lichen flora. Many of the characteristic species of heather moor also occur in woodlands.

The floristic composition of communities dominated by heather is related to regional differences in climate, soils, and management. Gimingham (1972) has described and classified several heath associations. Using wider categories, McVean and Ratcliffe (1962) recognized the following three main associations where heather predominates below the tree line in Scotland:

(1) Dry heather moor (*Callunetum vulgaris*) comprises more or less pure stands of heather with a moss layer consisting largely of *Dicranum scoparium* and *Hypnum cupressiforme*. Small patches of abundant *Erica cinerea* and *Vaccinium* spp. may occur and this association includes the *Calluna–Erica cinerea* and some of the *Calluna–Vaccinium* heaths of Gimingham. Typically, this community occurs on podzolic soils, particularly iron-humus podzols.

(2) *Arctostaphylos*-rich heather moor (*Arctostaphyleto–Callunetum*) is an exact equivalent of Gimingham's *Calluna–Arctostaphylos* heath. It is a distinct local variant of pure heather moor and contains a much greater variety of dwarf shrubs and herbs. Apart from *Arctostaphylos uva-ursi*, the herbs *Lathyrus montanus, Lotus corniculatus, Viola riviniana* and *Pyrola media* are frequent. The association usually occurs on shallow, rather stony podzolic soils which lack a deep layer of raw humus at the surface.

(3) Damp heather moor (*Vaccineto–Callunetum*) occurs on steep, north-facing slopes and belongs to the *Calluna–Vaccinium* heaths of Gimingham. *Vaccinium myrtillus* is abundant and may gain local codominance with heather. *Empetrum* spp. may also be present, and there is a conspicuous and deep layer of mosses, chiefly *Sphagnum capillaceum, Hylocomium splendens* or *Rhytidiadelphus loreus*. The association usually occurs on a deep layer of acid peaty humus—on drained or drying bog peat, on podzol rankers, or on peaty podzols.

Heather moorland supports distinctive animal communities, some of which have been described briefly by Pearsall (1950). The abundance of vertebrates on different moors is often related to the type of underlying rocks. Thus moors over base-rich rocks support larger stocks of red grouse (*Lagopus lagopus scoticus*) than moors over acid rocks (Miller, Jenkins, and Watson, 1966). This also applies to mountain hares, *Lepus timidus* (Watson, Hewson, Jenkins and Parr, 1973), and several other species of mammal and bird, both herbivorous and insectivorous (Nethersole-Thompson and Watson, 1981). Recent studies for the International Biological Programme have focused on the complex communities of moorland invertebrates in plant litter and the soil (Heal and Perkins, 1978). Nevertheless, few quantitative data on moorland animals have been published and there are many opportunities for research. For instance, no one has determined if different types of moor support different communities or densities of animals, and even the habitat selection of common and conspicuous moorland birds such as golden plover (*Pluvialis apricaria*), curlew (*Numenius arquata*) and black grouse (*Lyrurus tetrix*) is little understood.

LIFE HISTORY OF HEATHER

During their lifespan, heather plants change gradually in size, morphology and growth habit as they undergo physiological ageing and become less vigorous. This alters their ability to compete with other species and is an important consideration in deciding how best to manage heather moorland. There are four distinct phases (Gimingham, 1972).

The 'pioneer' phase is one of establishment and early growth, either from seed or from buds at the base of charred stems remaining after a fire. The contribution that other species make to the plant cover is greatest during this period, which may last for 3–10 years. Plants next pass into a 'building' phase, which may last up to the age of about 15 years. During this period, heather develops a dense continuous canopy which suppresses bryophytes, lichens and other low-growing plants. From 15 to about 25 years, heather is in its 'mature' phase, when the plants are less vigorous and the central branches begin to spread outwards, so allowing light to penetrate to the ground. Bryophytes flourish and the lichen *Parmelia physoides* may begin to colonize the lower parts of the main branches of heather. Beyond about 25 years, plants become 'degenerate'. The central branches become heavily encrusted with lichens and eventually die, so leaving a space where other species may establish. Some of the outer branches of the moribund plant may become prostrate, root adventitiously, and continue growing with renewed vigour.

The rate at which one phase succeeds another varies. Where net growth is rapid, the plant ages quickly and may become degenerate when less than 20 years old. Conversely, ageing is retarded when net growth is slow. Continual pruning, whether by clipping, grazing or wind-blast, reduces net growth and so prolongs the period when the plant is physiologically young (Grant and Hunter, 1966).

PRODUCTION AND CHEMICAL COMPOSITION
OF HEATHER SHOOTS

Because it is difficult to measure litter-fall and woody increments accurately, there are only a few estimates of total above-ground production by heather (Chapman, Hibble, and Rafarel, 1975). Most data are for the annual production of green shoots and flowers. This is easily measured and, in any case, is of greatest relevance to herbivores.

Estimates of the mean production of green shoots and flowers by heather in more or less closed communities vary from about 140 to over 300 g m^2 (Table 7.1). No doubt some of the wide variation in mean production can be attributed to differences in altitude and in the cover of heather at the various sites. Even the highest recorded production of heather, 442 g m^{-2} (Barclay-Estrup, 1970), is far less than the estimates of 1300 g m^{-2} for pine woodland (Ovington, 1957) and 800 g m^{-2} for birch woodland (Ovington and Madgwick, 1959). Although these data are from different parts of Britain and therefore are not strictly comparable, they suggest that one of the consequences of replacing woodland by moorland ecosystems might have been a decrease in the production of dry matter.

TABLE 7.1 Some estimates of the annual production of green shoots and flowers by heather growing as a more-or-less pure stand.

Site	Altitude (m)	Annual production (g m^{-2}) Mean	Range	Reference
Poole basin, Dorset	0−60	216	136−259	Chapman, Hibble, and Rafarel (1975)
Elswick House, Grampian	107	316	141−442	Barclay-Estrup (1970)
Kerloch, Grampian	140−170	249	222−282	Miller (1979)
Glensaugh, Grampian	215	139	86−205	Grant (1971)
Lochnagar, Grampian	350−410	168	130−231	Moss and Miller (1976)
Corndavon, Grampian	410−500	212	163−284	Moss and Miller (1976)
Moor House, Cumbria	515−561	196	130−240	Forrest and Smith (1975)

Some studies have considered the effects of ageing and weather on annual production. Both Chapman, Hibble, and Rafarel, (1975) and Miller (1979) found that production from stands with more or less complete cover of heather did not vary appreciably up to about 45 years after burning. Although the age of the stand may not have a major effect on production, climate does. Thus Miller and Watson (1978) showed a negative correlation between mean production and altitude at sites 100−550 m above sea level. Moreover, year-to-year variations in production at one site can be related to the weather during the growing season: warm, dry weather between April and August favours rapid growth and a high production of shoots (Miller, 1979). Flower production, on the other hand, is related to spring temperature.

The small amounts of available nitrogen and phosphorus in acid upland soils greatly limit plant growth (Munro, Davies, and Thomas, 1973). On the podzolic soils of heather moors, phosphatic fertilizer has no effect on plant production but nitrogenous fertilizer can almost double the yield of heather shoots (Miller, 1979). This suggests that the fibrous peat of heather moors has a much greater deficiency of available nitrogen than of available phosphorus. In fact, phosphatic fertilizer does stimulate heather growth on the colloidal peats of western blanket bog, where phosphorus must be very deficient (Watson and O'Hare, 1979).

In general, the concentration of major nutrients in heather shoots declines as the plant ages. Nitrogen, phosphorus and potassium all show big decreases during the first four years after burning, but there is little further change up to 27 years (Miller, 1979). Changes in the concentration of most other nutrients are less clear except for crude fat, which increases with age. Heather shoots have a high concentration of calcium, magnesium, copper and cobalt compared with other moorland plants, and Thomas (1956) has emphasized the importance of this to livestock. However, heather is a very poor source of nitrogen and phosphorus, two of the most important nutrients in animal nutrition.

Variation in Digestibility of Moorland Species

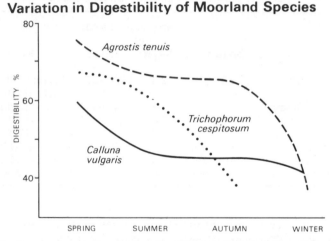

FIGURE 7.1   Seasonal variation in the digestibility of dry matter of three moorland species commonly eaten by ruminants. Data from Kay and Staines (1981).

A further disadvantage of heather as a food for livestock is its poor digestibility. Kay and Staines (1981) have summarized the relevant data which show that heather is less digestible than the common grasses and sedges found on moorland, especially in spring and summer (Figure 7.1). In winter, however, heather is a valuable and sometimes unique source of food because it remains green, whereas other species die back.

LAND USE

Traditionally, men have used heather moorland for free-range grazing by domestic livestock and for sport. In the eighteenth and early nineteenth centuries, the black cattle industry flourished in the Highlands and many store cattle were exported annually to the south. At this time sheep and goats were kept mainly for domestic use. However, the market for hill cattle eventually declined and many thousands of sheep were moved into the Highlands to replace them. By the 1830s sheep farming had spread to the Inner Hebrides and most Scottish hill grazings had been converted to sheepwalks. Free-ranging cattle are now few or absent on most hill farms. Although less profitable today than in its heyday, sheep farming still continues as one of the two main uses of heather moorland.

The other main use of heather moorland is for the shooting of red grouse, an endemic subspecies of the willow grouse (*Lagopus lagopus*). Red grouse are virtually confined to open moorland and presumably extended their range and became more abundant after the forests were destroyed. The birds had long been regarded as game, but no great importance was attached to them until about the middle of the nineteenth century. At that time many industrialists and merchants in the south were affluent, road and rail links with the north were improving rapidly, and guns became more efficient. These developments combined to make grouse shooting a highly fashionable sport that was profitable to the landowner—and it has remained so up to the present day.

The practice of stalking and shooting red deer (*Cervus elaphus*) emerged at about the same time. Although red deer originally lived in open woodland, they have adapted to a more or less treeless environment in Britain. Because of their sporting value, the Victorians deliberately fostered large stocks on deer 'forests' (though these are mostly devoid of trees) until densities far exceeded those elsewhere in Europe. In recent years, many deer have been fenced out of their wintering grounds to make way for farming, afforestation and hydro-electric schemes. Often this was done without reducing deer stocks and the ousted deer then marauded farms and plantations. In 1959, the Red Deer Commission was established and, since then, landowners have taken bigger culls. Deer forests are mostly on high ground or on westerly moors, where heather is less important in the vegetation than on the eastern grouse moors. Nonetheless, deer do seek shelter and forage on the lower heather moors, particularly in eastern Scotland in winter.

On many moors, heather is so predominant in the vegetation that there is little else for herbivores to eat. Sheep, deer, grouse, and mountain hares all depend to some extent on heather for food. Several workers have studied the diet and feeding preferences of hill sheep (Grant *et al.*, 1976) and all agree that heather can be an important food at certain times of year. In practice, the diet of sheep varies, depending on what forage is available. Where grasses are plentiful, sheep may not graze heather at all during spring and summer, but they will take heather during winter, when the supply of grass becomes exhausted. Red deer show a similar seasonal pattern of grazing on heather (Kay and Staines, 1981).

For both mountain hares (Hewson, 1962) and red grouse (Jenkins, Watson, and Miller, 1963), heather comprises nearly 100 per cent of the diet in winter and about 50 per cent in summer. Indeed, the almost total dependence of

Fate of Primary Production of Heather

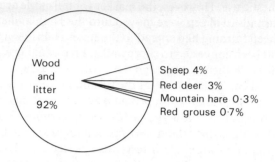

FIGURE 7.2 Notional fate of the annual primary production of heather: only about 8 per cent may be consumed by vertebrate herbivores, most of the remainder going to form wood and litter. Data from Miller and Watson (1974).

grouse on heather has been the subject of much research (Miller and Watson, 1978). It has been shown that (*a*) differences in the average number of red grouse breeding in spring on different moors are related to the quantity and quality of heather there, (*b*) year-to-year changes in breeding stocks are related to variations in the growth and dieback of heather induced by the weather, and (*c*) breeding stocks can be increased by fertilizing and by burning the moor in small patches.

In general, vertebrate herbivores are so few and the digestibility of heather is so poor (see Figure 7.1) that no more than a small fraction of the primary production is eaten (Miller and Watson, 1974). During the first year or two after burning, sheep and hares may graze heavily on the young heather. Thereafter, usually less than 10 per cent of the annual production is eaten (Figure 7.2) and utilization seldom approaches the 60 per cent thought necessary to maintain heather in a productive and nutritious state (Grant and Hunter, 1966).

Secondary production from heather moorland has received little attention. Table 7.2 provides a guide to the biomass and proportion taken as an annual crop. In fact, not all the production from sheep and deer can be validly attributed to heather moorland. In winter and spring, hill sheep are fed supplements

TABLE 7.2 Some notional figures for the cropping of herbivores from the heather moorland ecosystem in the east-central Highlands of Scotland.

Crop	Average density in spring (no. km^{-2})	Biomass (kg km^{-2})	Production of young (no. km^{-2})	Annual crop taken by man (kg km^{-2})	Main sources of data
Sheep	50	2250	40	1000	Hill Farming Research Organisation (1979)
Red deer	10	685	2	110	Mitchell, Staines, and Welch (1977)
Mountain hare	16	43	50	0	Flux (1970); Watson *et al.* (1973)
Red grouse	65	41	89	25	Jenkins, Watson, and Miller (1963, 1967)

and are allowed to 'inby' grass; and red deer seek much of their forage on high heatherless plateaux during summer and on the grassy glen bottoms in winter. Thus the sum of the annual crops taken from each animal is certainly a gross overestimate of the secondary production derived from heather. Despite this, the annual yield of about 1150 kg km² is very little compared with what can be cropped from more fertile land.

The poor production from moorland herbivores strengthens the case of those who advocate afforestation as an efficient, productive and beneficial use of moorland (see, for example, several papers in Tranter, 1978). However, in an industrialized country such as Britain the biological productivity of marginal land does not usually influence its economic value. This is determined more by the land's capacity to provide sport and recreation. For example, on many estates, red grouse give a far greater monetary return in the short term than sheep, deer or trees, although the biological production from each of these is usually greater. On a regional basis, tourism yields no biological crop, yet in some places it can produce more revenue than all the other land uses put together.

MANAGEMENT OF HEATHER

Moorland managers aim to provide a continuous supply of young heather as food for their sheep, deer and grouse stocks. It is theoretically possible to keep heather short, vigorous, physiologically young and therefore having a high nutrient content, by grazing off one half of each year's production. In practice, this is rarely achieved and production may exceed consumption by a factor of 11 to 12 (see Figure 7.2). The excess production accumulates year by year as wood and dead shoots attached to the plants and as litter on the ground. The old woody plants and litter are most easily removed by periodic burning—virtually a crude method of pruning. If it is well done, burning can rejuvenate a stand of heather and provide a crop of rapidly growing and nutritious shoots that herbivores graze preferentially. On the other hand, bad burning can do irreparable damage. The effects of fire on heather and the moorland habitat have been much studied and most of this work has been reviewed by Gimingham (1972). Two more recent publications (Muirburn Working Party, 1977; Watson and Miller, 1976) consider the practical aspects of using fire to manage heather moorland.

The practice of moor burning probably arose from the use of fire to clear the original forest cover. No doubt it was at first haphazard, as it still is in much of western Britain where burning is mainly for sheep and deer. However, in the east, many estates practise controlled rotational burning to maintain large grouse stocks as well. The aim is to burn small patches or long narrow strips of vegetation annually on a rotation such that each patch is fired once every 5 to 25 years, depending on how quickly the heather grows. Grouse should then have easy access both to short heather for feeding and to tall heather for cover when defending their small (2–5 ha) territories in late winter and spring. Although it is not always possible to achieve an ideal rotation and pattern of burning, many grouse moors are characterized by a mosaic of short and tall heather, with few patches bigger than 5 ha.

When this management began in the second half of the nineteenth century, grouse bags increased greatly. Research has shown that bags are related to the number and size of burnt areas (Picozzi, 1968) and that grouse numbers can be increased experimentally by burning small patches (Miller, Watson, and Jenkins 1970). The regular burning of small patches is therefore an essential part of grouse husbandry.

Less is known about the effects of burning on sheep and deer. Unlike grouse, these animals live in groups, with individual sheep ranging over scores of hectares and, in the case of deer, thousands of hectares. Presumably, therefore, they do not have an exacting requirement for burning to be in small patches. In fact, on most moors that are managed solely for sheep or deer, it is usual for scores of hectares, sometimes over 100 ha, to be burned with a single blaze. Grouse are unable to colonize such large areas for several years, until plants have grown tall enough to give good concealment.

Burnt heather regenerates from seeds and from the sprouting of buds at the base of stems left by the fire. Both usually contribute to the heather's recovery but vegetative regeneration is the faster. A dense stand of heather may have up to 1700 stems per m^2 and, if a large proportion of these should produce sprouts, complete cover of the ground can be restored within a couple of growing seasons. The complete regeneration of a heather sward from seeds alone can take many years. Although there may be up to 90 000 viable heather seeds buried per m^2 of moorland and an annual deposition of more than 500 000 heather seeds per m^2, seedlings establish erratically and grow very slowly during their first few years. Slow regeneration can weaken the dominance of heather, allowing the spread of species such as bracken (*Pteridium aquilinum*) or purple moor grass, which are of little or no grazing value. Regeneration from seeds is best regarded as a supplement to vegetative regeneration rather than a substitute for it.

Several factors influence the growth of new shoots from heather stems after a fire. Foremost is the physiological age of the plants at the time of burning: dense 'building' heather no taller than 30 cm and with the base of the stem no thicker than 1 cm generally regenerates much faster than taller 'mature' plants with thicker stems. In addition, the heat of the fire, the season of burning, the type of soil, the amount of grazing and the local climate all affect the success of vegetative regeneration after burning.

To sum up, the carefully controlled burning of small strips and patches according to a planned rotation will ensure a continuing supply of rapidly regenerating, nutritious heather distributed in such a way as to be of maximum benefit to all herbivores. Fire can thus be a cheap and effective management tool. In reality, however, this ideal is seldom achieved because of bad weather, shortage of labour, statutory limitation of the burning season and, above all, a lack of planning and determination on the part of many moor managers.

ECOLOGICAL CONSEQUENCES OF BURNING

There seems little doubt that the moorland flora can be impoverished by repeated burning. Such management eliminates all trees and shrubs as well as the many herbaceous species with overwintering buds above ground level. A few fire-resistant species are encouraged to spread and become dominant. Heather, with its copious production of seeds and its ability to regenerate rapidly from buds at the base of the stem, is particularly well-adapted to fire. Its dense, finely branched canopy can suppress most competitors, so that great tracts of hill ground become completely blanketed. Other fire-resistant species with rhizomes (for example, bracken) or dense tussocks (for example, purple moor grass) can also become dominant, sometimes replacing heather.

Most plant associations of heather moors contain few species compared with those in other habitats (McVean and Ratcliffe, 1962). Gimingham (1964) ranked five main heath communities from the 'species-rich', averaging 41 species per 4 m^2, to 'pure stands' of heather, averaging 13 species. High floristic diversity was associated with fertile soils but the impoverished flora of pure heather moor was attributed to burning as well as to poor soils.

Whereas burning in the relatively dry climate and poor soils of eastern Britain tends to result in a monoculture of heather, similar management elsewhere may eliminate it. Frequent burning along with heavy grazing may convert heather moor into grassland, dominated by bents (*Agrostis* spp.) and fescues (*Festuca* spp.) on the better brown-earth soils, and by mat-grass (*Nardus stricta*) on poorer, gleyed soils. On the poorly drained soils of western Britain and Ireland, frequent burning favours purple moor grass, deer sedge and cotton sedge at the expense of heather. Purple moor grass is especially well adapted to resist fire and, once established, is difficult to eliminate (Grant, Hunter, and Cross, 1963). It is a deciduous species and therefore—unlike the evergreen heather—valueless as winter grazing.

In the 1960s several ecologists studied the possible depletion of nutrients from the moorland ecosystem as a direct result of burning. It had been argued that serious losses in smoke and in solution from ash might accrue over a long series of fires. The pathways of such hypothetical losses and of the possible gains to the system are outlined in Figure 7.3.

The volatilization of nutrients in burning vegetation depends on the heat of the fire, but well over 50 per cent of the nitrogen, sulphur and carbon can commonly be lost as smoke (Evans and Allen, 1971). Potassium salts are readily

Heather Moorland-Nutrient Pathways

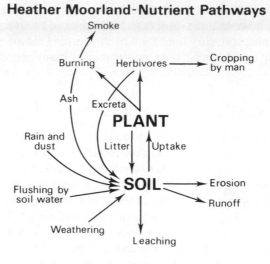

FIGURE 7.3   Main pathways of gains and losses of nutrients within the heather
moorland ecosystem.

dissolved from ash, calcium and magnesium less so, whereas phosphorus compounds are relatively insoluble (Allen, Evans, and Grimshaw, 1969). Losses due to leaching by rainwater are reduced due to the absorption of nutrients in the organic matter at the surface of the soil. The degree of retention appears to be related to the thickness of the layer, and some leaching may occur in sandy soils with a thin A_0 horizon. These findings emphasize the need to ensure that combustion is not so fierce as to burn off the layer of humus, a point that is frequently overlooked.

Several attempts have been made to draw up a balance sheet of nutrients within the heather moorland ecosystem, setting the total losses that might result from burning and from the removal of animal crops against the total income from rain and atmospheric dust (Allen, Evans, and Grimshaw, 1969). The Muirburn Working Party (1977) provide up-to-date information, including the effects of different intensities of sheep production. The possible losses of calcium and potassium seem to be only a small fraction of the income from rainfall during a 10-year interval between successive fires in the one area. Slightly more nitrogen may be deposited than is lost but there is clear evidence of a steady depletion of phosphorus.

More detailed studies of the amount, distribution and fate of nutrients deposited in ash are needed. It is not known how quickly and to what extent these nutrients are recycled within the ecosystem. Nor, for that matter, is there any information on the fate of the nutrients deposited by rain. This is a particularly important point because the case for largely discounting the effects of burning and grazing rests on the assumption that these nutrients are wholly absorbed. On the contrary, it is likely that a proportion of the income from rainfall will be lost by leaching and by surface runoff. Moreover, little attention has been paid to the possible losses of nutrients from west Highland moors, where fires are bigger and the same area is burned much more frequently than in the east.

Burning may have undesirable effects on the physical properties of the soil. Peat that is freqently burnt develops a tough, rubbery skin which possibly reduces the penetration of moisture. The skin is due partly to physical changes in the peat and partly to the growth of algae and lichens. Soil erosion has also been associated with frequent burning (Kinako and Gimingham, 1980). On flat or gently sloping ground, small patches burned on a rotation of 10 to 15 years should result in only a redistribution of soil particles from one site to another on the moor. Large, frequent fires on steep slopes could, however, result in serious erosion of the soil and its loss in streamwater. Potentially, this is a pathway for a major wastage of nitrogen and phosphorus from moorland in the west of Scotland.

One of the most important consequences of management by fire concerns the effects of the vegetation itself on the soil. Many of the species that are maintained by periodic burning produce acid, fibrous litter which does not decompose easily. Deposition of litter by heather increases steadily with the age of the plant. Beyond the 'building' stage, litter is shed faster than it is incorporated into the soil, and so large amounts accumulate at the soil surface. The thick layer of black peaty humus which is thus formed encourages acidification of the soil. This acidification has been attributed to inhibition of earthworms, shallow rooting, accelerated leaching and, most important, the removal of bases, particularly calcium (Grubb and Suter, 1971). Soil acidification promotes leaching and podzolization, reduces the availability of bases, and inhibits nitrifying bacteria. Moreover, since heather and many of its associates are shallow-rooted species, leached nutrients deposited in the B horizon of podzols can no longer circulate within the ecosystem. Thus the monoculturing of heather, besides causing floristic impoverishment, also contributes to soil impoverishment.

THE FUTURE

The development and use of the British uplands in general, and of heather moorland in particular, has been much discussed for three decades. Some changes in land use have occurred recently and others may be imminent. Two recent publications (Institute of Terrestrial Ecology, 1978; Tranter, 1978) give much of the background and describe current trends.

The traditional system of ranching hill sheep on large tracts of hill ground is relatively unproductive and uneconomic in so far as it is heavily subsidized by the state. Hence many people consider that sheep farming in the uplands will decline. It would be quite feasible to increase the output from hill sheep by fertilizing the existing vegetation, by creating more productive kinds of vegetation, or by closely controlling the grazing sheep (as outlined in Hill Farming Research Organization, 1979). It is sometimes suggested that restocking with cattle might so improve upland vegetation that not only sheep but other animals would benefit. But there are great economic and practical difficulties in doing any of these things over wide areas of hill land. Some experts urge that livestock husbandry should be intensified only on the most productive hill ground and that the poorer hills of the north and west should be abandoned to forestry, sport, recreation or, possibly, to deer farming.

Recently, planted woodland has encroached increasingly on to heather moors as foresters have overcome the technical problems of planting and growing trees on acid, infertile and sometimes waterlogged soils. Some welcome this as a return to a type of plant cover that better expresses the natural potential of this environment. However, afforestation almost invariably involves the substitution of one kind of monoculture for another: at present the exotic lodgepole pine (*Pinus contorta*) and sitka spruce (*Picea sitchensis*) are widely planted, although little is known about their impact on vegetation, on animal life or, most important, on soils (see Goldsmith and Wood, Chapter 17).

Forest plantations are now widely used to help satisfy the rising demand for outdoor recreation. Naturalists, walkers, orienteers, skiers, pony trekkers, horse-riders, cyclists and even motor rally drivers use the forest rides and roads; the shooting of roe deer (*Capreolus capreolus*) or capercaillie (*Tetrao urogallus*) can command big fees. However, forests can never satisfy all recreational needs. Many people like open spaces with unobstructed views and freedom to wander at will. This demand will always be served above the tree limit, but even at lower elevations open spaces are unlikely to disappear completely. Affluence in western Europe and North America will probably ensure that grouse shooting and deer-stalking will continue to prosper, at least in the short term, and that many open heather moors will survive. The game 'industry' earns foreign currency and contributes to the rates, local employment and viable human settlements in the uplands. On some estates, the income from the letting of the shootings exceeds the rentals from the hill farms (Airlie, 1971). Many other kinds of recreation are becoming popular in the uplands. Year by year more and more people go there to camp, walk, ski, watch birds and so on; and the purple heather moors of late summer are undoubtedly a big attraction for tourists. All this is verified by the political row that arose over the conversion of heather moor into sown grassland by Devon and Somerset farmers (see McEwen and McEwen, Chapter 22). Indeed, many landowners are worried that an influx of visitors from the cities will damage their shooting interests and so reduce rents. These new conflicts in land use are adding to the antagonisms that already exist between foresters and hill farmers.

The creation of large tracts of heather moorland was fortuitous in that they were a byproduct of man's exploitation of a 'natural' woodland ecosystem. No matter how profitable rough grazing and sport might be, clearly they cannot be justified either in respect of biological production or, because of the effects of heather on the soil, of environmental conservation, although the northern heather moors are not as vulnerable to changing land uses as are the heaths of the south (see Harrison, Chapter 5). There is now a greater possibility of changes occurring than in the recent past. The development of new concepts in managing hill sheep and a growing demand that Britain should produce more home-grown timber is already leading to the afforestation of large areas of heather moor. There is even a possibility that some moors may revert to natural scrub or woodland following a withdrawal of free-ranging sheep to the most fertile ground. Nevertheless, open moorland will survive in northern Britain wherever tree-planting is uneconomic or where grouse-shooting and deer-stalking remain profitable.

REFERENCES

Airlie, Earl of (1971). Making full use of an upland estate. *Landowning in Scotland*, **142**, 3−6.

Allen, S.E., Evans, C.C., and Grimshaw, H.M. (1969). The distribution of mineral nutrients in soil after heather burning. *Oikos*, **20**, 16−25.

Barclay-Estrup, P. (1970). The description and interpretation of cyclical processes in a heath community. II: Changes in the biomass and shoot production during the *Calluna* cycle. *Journal of Ecology*, **58**, 243−249.

Chapman, S.B., Hibble, J. and Rafarel, C.R. (1975). Net aerial production by *Calluna vulgaris* on lowland heath in Britain. *Journal of Ecology*, **63**, 233−258.

Evans, C.C. and Allen, S.E. (1971). Nutrient losses in smoke produced during heather burning, *Oikos*, **22**, 149−154.

Flux, J.E.C. (1970). Life history of the mountain hare (*Lepus timidus scoticus*) in North-East Scotland. *Journal of Zoology (London)*, **161**, 75−123.

Forrest, G.I. and Smith, R.A.H. (1975). The productivity of a range of blanket bog vegetation types in the northern Pennines. *Journal of Ecology*, **63**, 173−202.

Gimingham, C.H. (1964). The composition of the vegetation and its balance with environment. Land Use in the Scottish Highlands. *Advancement of Science (London)*, **21**, 148−53.

Gimingham, C.H. (1972). *Ecology of Heathlands*, Chapman and Hall, London, 265 pp.

Grant, S.A. (1971). Interactions of grazing and burning on heather moors. 2. Effects on primary production and level of utilization. *Journal of British Grassland Society*, **26**, 173−181.

Grant, S.A. and Hunter, R.F. (1966). The effects of frequency and season of clipping on the morphology, productivity and chemical composition of *Calluna vulgaris* (L.) Hull. *New Phytology*, **65**, 125−133.

Grant, S.A., Hunter, R.F., and Cross, C. (1963). The effects of muirburning *Molinia*-dominant communities. *Journal of the British Grassland Society*, **18**, 249−257.

Grant, S.A., Lamb, W.I.C., Kerr, C.D. and Bolton, G.R. (1976). The utilization of blanket bog vegetation by grazing sheep. *Journal of Applied Ecology*, **13**, 857−869.

Grubb, P.J. and Suter, M.B. (1971). The mechanism of acidification of soil by *Calluna* and *Ulex* and the significance for conservation. In E. Duffy and A.S. Watt (Eds) *The Scientific Management of Animal and Plant Communities for Conservation.* Blackwell, Oxford, pp. 115–133.

Heal, O.W. and Perkins, D.F. (Eds) (1978). *Production Ecology of British Moors and Montane Grasslands*, Springer-Verlag, Berlin, 426 pp.

Hewson, R. (1962). Food and feeding habits of the mountain hare *Lepus timidus scoticus* Hilzheimer. *Proceedings of the Zoological Society of London*, **139**, 515–526.

Hill Farming Research Organization (1979). *Science and Hill Farming*, Hill Farming Research Organization, Penicuik, 184 pp.

Institute of Terrestrial Ecology (1978). *Upland Land Use in England and Wales*, Countryside Commission, Cheltenham, 140 pp.

Jenkins, D., Watson, A. and Miller, G.R. (1963). Population studies on red grouse, *Lagopus lagopus scoticus* (Lath.) in north-east Scotland. *Journal of Animal Ecology*, **3**, 313–326.

Jenkins, D., Watson, A., and Miller, G.R. (1967). Population fluctuations in the red grouse, *Lagopus lagopus scoticus*. *Journal of Animal Ecology*, **36**, 97–122.

Kay, R.N.B. and Staines, B.W. (1981). The nutrition of red deer (*Cervus elaphus*). *Nutrition Abstracts Review, B*, **51**, 601–622.

Kinako, P.D.S. and Gimingham, C.H. (1980). Heather burning and soil erosion on upland heaths in Scotland. *Journal of Environmental Management*, **10**, 277–284.

McVean, D.N. and Ratcliffe, D.A. (1962). *Plant Communities of the Scottish Highlands*, HMSO, London, 445 pp.

Miller, G.R. (1979). Quantity and quality of the annual production of shoots and flowers by *Calluna vulgaris* in north-east Scotland. *Journal of Ecology*, **67**, 109–129.

Miller, G.R., Jenkins, D., and Watson, A. (1966). Heather performance and red grouse populations, I. Visual estimates of heather performance, *Journal of Applied Ecology*, **3**, 313–326.

Miller, G.R. and Watson, A. (1974). Some effects of fire on vertebrate herbivores in the Scottish Highlands. *Proceedings of the Annual Tall Timbers Fire Ecology Conference*, **13**, 39–64.

Miller, G.R. and Watson, A. (1978). Heather productivity and its relevance to the regulation of red grouse populations. In O.W. Heal and D.F. Perkins (Eds), *Production Ecology of British Moors and Montane Grasslands*, Springer-Verlag, Berlin, pp. 277–285.

Miller, G.R., Watson, A., and Jenkins, D. (1970). Responses of red grouse populations to experimental improvement of their food. In A. Watson, (Ed.) *Animal Populations in Relation to their Food Resources*, Blackwell, Oxford and Edinburgh, pp. 323–335.

Mitchell, B., Staines, B.W. and Welch, D. (1977). *Ecology of Red Deer*, Institute of Terrestrial Ecology, Cambridge, 74 pp.

Moss, R. and Miller, G.R. (1976). Production, dieback and grazing of heather (*Calluna vulgaris*) in relation to numbers of red grouse (*Lagopus lagopus scoticus*) and mountain hares (*Lepus timidus*) in north-east Scotland. *Journal of Applied Ecology*, **13**, 369–377.

Muirburn Working Party (1977). *A Guide to Good Muirburn Practice*. HMSO, Edinburgh, 44 pp.

Munro, J.M.M., Davies, D.A. and Thomas, T.A. (1973). Potential pasture production in the uplands of Wales. 3. Soil nutrient resources and limitations. *Journal of the British Grassland Society*, **28**, 247–255.

Nethersole-Thompson, D. and Watson, A. (1981). *The Cairngorms*, 2nd edition, Melven Press, Perth, 317 pp.

O'Sullivan, P.E. (1977). Vegetation history and the native pinewoods. In R.G.H. Bunce and J.N.R. Jeffers (Eds), *Native Pinewoods of Scotland*, Institute of Terrestrial Ecology, Cambridge, 120 pp.

Ovington, J.D. (1957). Dry-matter production by *Pinus sylvestris* L. *Annals of Botany, (London)* NS., **21**, 287–314.

Ovington, J.D. and Madgwick, H.A.I. (1959). The growth and composition of natural stands of birch. I. Dry-matter production. *Plant and Soil*, **10**, 271–288.

Pearsall, W.H. (1950). *Mountains and Moorlands*, Collins, London, 312 pp.

Picozzi, N. (1968). Grouse bags in relation to the management and geology of heather moors. *Journal of Applied Ecology*, **5**, 483–488.

Thomas, B. (1956). Heather (*Calluna vulgaris*) as a food for livestock. *Herbage Abstracts*, **26**, 1–7.

Tranter, R.B. (Ed.) (1978). *The Future of Upland Britain*, 2 volumes, Centre for Agricultural Strategy, Reading, 724 pp.

Watson, A. and Miller, G.R. (1976). *Grouse Management*, 2nd edition, Game Conservancy Booklet No. 12, Fordingbridge, 78 pp.

Watson, A., Hewson, R., Jenkins, D., and Parr, R. (1973). Population densitities of mountain hares compared with red grouse on Scottish heather moors. *Oikos*, **24**, 225–230.

Watson, A. and O'Hare, P.J. (1979). Red grouse populations on experimentally treated and untreated Irish bog. *Journal of Applied Ecology*, **16**, 433–452.

Conservation in Perspective
Edited by A. Warren and F.B. Goldsmith
© 1983 John Wiley & Sons Ltd.

CHAPTER 8

The Conservation and Management of Animal Populations

J. BRIAN WOOD

Populations of wild animals are influenced by mankind in several different ways. Most commonly, the influence is fortuitously as a consequence of the use of the land. The gross change in habitats that is produced by this type of management often has dramatic consequences for the animal populations concerned, and may lead to their complete eradication. Occasionally, elimination may be deliberately planned: the draining of swamps to eradicate malaria-carrying mosquitoes is one example of this.

The direct manipulation of animal populations is normally practised only if they have a significant value to man, and so warrant some expenditure. These manipulations may be one of three kinds.

(1) The number of animals in the populations may be reduced, either because it is considered to be an undesirable or pest species, or because it may provide a utilizable resource which is taken without due thought to continuing exploitation.
(2) Species whose numbers are considered to be too small or whose geographical range is considered insufficiently extensive may be manipulated so as to become more numerous or widespread; generally these are species which may provide cultural benefits to man, varying from aesthetic enjoyment to research.
(3) There are those species which may provide material resources of considerable value, which are managed so as to stabilize their population at some optimum level and thus provide a sustainable yield.

Many species in the last category have become domesticated (Jewell, 1969; Short, 1976) for it is normally only then that they can be sufficiently closely controlled and sufficiently understood to maintain their status as a truly sustainable resource. Both the enhancement and the stabilization of the size of a population fall within the realms of conservation management, although

119

because adequate resources for the successful management of wild animals often only become available when they are endangered, conservationists are most usually involved in management that is aimed at increasing numbers.

APPROACHES TO CONSERVATION MANAGEMENT

If we had a perfect knowledge of an animal population and completely understood the ways in which it might respond to manipulative changes in numbers, then management, for whatever purpose, would be a relatively simple matter. In this circumstance we would also require to have freedom of action and adequate resources with which to carry out our plan of management. Unfortunately, our knowledge is generally far from perfect and is often least about the ways in which species interact with each other and with their environment. We may nevertheless often attempt to manipulate these complex interactions through conservation management (Figure 8.1).

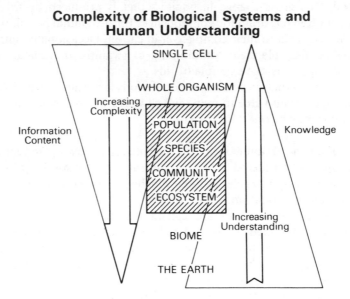

Complexity of Biological Systems and Human Understanding

FIGURE 8.1   The relationship between biological complexity and human understanding. The shaded box indicates the levels of organization to which conservation management is usually applied.

In many instances we can do little better than rely on the results of trial-and-error manipulation. Unfortunately, the speed at which agricultural, urban and industrial development has proceeded is so great that, such a clumsy and time-consuming method of attaining a working understanding of an animal population is now woefully inadequate if we are to act quickly enough to save more than a fraction of the earth's animal species from extinction. In Britain in recent years there has been at least one spectacular failure of this type of approach to conservation management. Despite the considerable knowledge of the ecology of the species, trial-and-error manipulation was unable to prevent the Large Blue butterfly (*Maculinea arion*) from becoming extinct at its last-known site in England (Thomas, 1980).

Although events such as the local extinction of the large blue butterfly may be considered as disasters for conservation, they should not be viewed as entirely negative in their consequences. Experiences of this sort are sometimes invaluable for the insights they provide into the ecology of an animal species, but the knowledge thus gained needs to be absorbed and built upon, so that it can be used to advantage in the event of similar circumstances in the future. For the conservationist who attempts to manage animal populations, time is undoubtedly the worst enemy. Rarely is adequate time available to carry out the surveys and analyses that would provide the data for a sensible management programme. If the status of an animal population is changing rapidly there may be insufficient time to assess the change before the species either dwindles to extinction or becomes so numerous that hasty and sometimes ill-advised courses of action are forced upon the manager, with potentially damaging consequences for the image of conservation, if not for the species itself. Furthermore, as greater and greater proportions of the earth's resources are channelled through the ever-increasing human population, leaving less and less to be utilized by wildlife, so demands increase daily for those reserves we are able to set aside to also serve purposes in addition to the conservation of a species. Even in well-planned developments, the conservationist is thus faced with increasingly complex questions about the way in which animal populations might respond to interference.

Remedial management

Most management of animal populations for conservation purposes is applied in an attempt to counteract the adverse effects of human activities and is instigated when the undesirable factors are clearly apparent. Legislation is the weapon most commonly applied in seeking the resolution of this type of issue, although this creates its own problems of enforcement. In Britain the Royal Society for the Protection of Birds (RSPB) annually spends thousands of pounds on the surveillance of the nest sites of birds of prey, notably the golden eagle (*Aquila chrysaëtos*), peregrine falcon (*Falco peregrinus*), and osprey (*Pandion haliaetus*). By doing so, they effectively reduce human predation in the guise of egg-collectors and falconers seeking young birds to train. Because most raptor species tend to use traditional nest sites and as their populations are often rather small, this remedy is the most effective one possible. It is necessitated by the difficulty of enforcing protective legislation, and the often inadequate fines imposed on lawbreakers. As a consequence of the activities of the RSPB, several of our populations of birds of prey are the most robust and productive in Europe.

Remedial management is also frequently applied in reserves mostly to counteract adverse factors emanating from outside the boundaries of the reserve. Particularly when managing small areas of land which may be surrounded by extensive cropped areas, the bulk of the work undertaken on a reserve may be to compensate for the adverse influences of its surroundings. Not infrequently, the potential damage caused by externalities can have extremely complex consequences for the ecology of the reserve and the animal populations it may support which would require detailed consideration if effective countermeasures were to be taken. One example will serve to illustrate this point.

Garaet el Ichkeul is a shallow, euryhaline lake of about 8700 ha in extent, situated in northern Tunisia. It is a National Park, listed in Tunisia's ratification of the Ramsar Convention and designated a Biosphere Reserve. In winter it supports internationally important concentrations of at least five species of waterfowl and may occasionally support up to 200 000 ducks at a time. The water level of the lake is currently subject to marked fluctuations. Four major rivers flow into it and it has one outlet via the Lake of Bizerte to the Mediterranean. After substantial rainfall, the level of the lake may rise by up to 2 m, causing extensive flooding and flushing salts from the lake to leave it almost entirely fresh. In the absence of rain, the lake level falls to, or just below, that of sea level, allowing a reserved movement of saline water up the outflow channel. This, together with the effects of evapotranspiration from the lake surface and adjacent marshes causes major parts of the lake to become quite saline. As a consequence, a complex zonation of vegetation occurs, from the permanently flooded margins of the lake to the upper marshes subjected to inundation only during the highest floods. In turn, the vegetation zonation influences the distribution and abundance of animals, including the internationally important waterfowl species.

Because the Tunisian economy is heavily dependent upon tourism and the products of irrigated agriculture, both creating a considerable demand for water, there are plans to dam most of the rivers flowing into Garaet el Ichkeul and divert water away from the lake. The consequences of the reduced water inflow that will ensue have been modelled by computer (Holli., 1981; Warren *et al.*, 1979); this indicates a future pattern of less frequent flooding and particularly less extensive floods. Nevertheless, the marked changes in water level that occur at present will persist, albeit in somewhat attenuated form.

It is often difficult for the conservationist to attempt to counteract the effects of major developments such as those at Ichkeul and, as the modifications in water-level that dams would produce appear to be slight, there is a strong temptation to accept the changes and hope that consequences for wildlife of the national park will also be slight. However, a closer examination of the ecology of the waterfowl inhabiting Garaet el Ichkeul reveals that reduced flooding could have very drastic effects on their numbers. A marked zonation of feeding ducks occurs in winter (Figure 8.2), but most species are heavily dependent upon the least saline end of the spectrum of the available habitats. In particular, a few species of salt-intolerant marsh plants form major food resources for many of the wildfowl; these would be particularly likely to decline as a consequence of less frequent inundation and a general rise in salinity levels produced by smaller throughputs of freshwater. Projected increases in salinity have also been modelled by computer (Hollis, 1977), and are much more serious than the predicted changes in the water levels that would follow the construction of dams.

The zonation of the principal waterfowl at Ichkeul and their food supplies

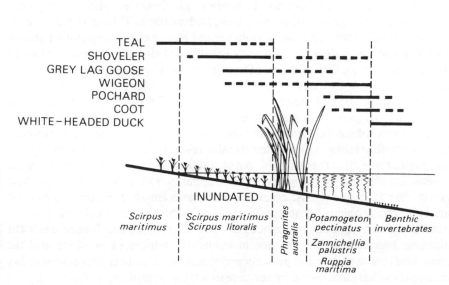

FIGURE 8.2 The zonation of the principal waterfowl at Ichkeul and their food supplies.

The solution recommended in this case was the construction of a sluice on the outflow from the lake (Hollis, 1977). By closing this sluice at appropriate times an artificial rise in lake-level can be produced so as to inundate the higher marshes. But a rapid release of this water, to allow a flushing of salts out of the basin, plus some compensatory waterflow down incoming rivers, will also be necessary to maintain the suitability of the freshwater marshes for the wildfowl species that presently occur. This remedial management, although based upon a consideration of many interacting factors, is nevertheless more readily prescribed because it is merely designed to mimic the present processes that are thought to favour wildlife. If the natural lake level were static, few managers would foresee the beneficial consequences of, or be willing to advocate, the management recommended above.

Management based upon characteristics of populations and ecosystems

It is more difficult to develop appropriate techniques for the management of animal populations when there are no examples of the desired state, which may be compared with the populations to be manipulated. Equally, restorative management is problematical if the former state of an animal population was not well studied. In these circumstances it becomes necessary to rely on a knowledge of how populations and ecosystems in general respond to management.

A major aim of wildlife conservation is to set aside examples of functioning natural ecosystems as instructive models of processes and to engender enlightened comparisons with more modified systems. Thus it is particularly worrying that in many reserves populations of large animal species have shown a tendency to fluctuate quite dramatically following their protection from hunting and disturbance (Buechner and Dawkins, 1961; Glover and Sheldrick, 1974; Laws, 1970; Wyk and Fairall, 1969). In some instances this has led to considerable degradation of the rest of the ecosystem. The development of tourism within national parks that were created for the purpose of conserving wildlife can provide a substantial income and source of employment and this in turn can help justify the setting aside of the extensive reserves that are needed to support big populations of large animals. An instability in the numbers of the more spectacular elements of the fauna, and the apparent imbalance that this implies, can detract from the tourist potential of the area and at the same time create management problems in an ecosystem that many visitors would judge to be natural and consequently not in need of direct manipulation. Consequently the dilemma becomes whether to accept dramatic changes in numbers and the associated loss of tourist and possibly ecological value, or to manage for stability in the belief that such would be the case in a truly natural situation.

A second phenomenon which is possibly of even greater concern is the tendency for the species complement of reserves gradually to decline in the years following their creation. This loss of species is not confined to the bigger animals alone, so the two problems may not be directly interconnected. However, both detract from the value of the reserve and it therefore becomes imperative to seek solutions based on an understanding of the functioning of ecosystems so as to be able to advocate appropriate management of animal populations.

A large area of land is likely to support more species than a small area of a similar type of land (Moore and Hooper, 1975). Indeed, when this observation is considered in relation to the number of species found on islands of different sizes (Diamond and May, 1976, and references therein), it becomes apparent that, except when comparing areas showing very different degrees of modification from a natural state, a constant relationship can be found between the size of an area and its species complement. The correlation is not a linear one, but takes the form of a power-function best described by the equation

$$S = CA^z \tag{1}$$

where S equals the number of species present, A equals the area, C is the species richness of unit area and z is a dimensionless constant related to diversity and the degree of isolation of the areas concerned. For isolated areas such as true islands, z has a value of around 0.27, but for sample areas delineated within a continuous more extensive ecosystem its value is often much lower, commonly betwen 0.11 and 0.18 (MacArthur and Wilson, 1967; May, 1975; Preston, 1962). Increasing isolation amplifies the degree of dependence of species complement upon area. When reserves are first designated they are often chosen because they represent the least degraded part of a more extensive ecosystem. At this stage their degree of isolation is relatively slight. However, with time, the land surrounding reserves tends to become more and more modified, until the reserve becomes an island of seminatural vegetation within a surrounding sea of very different land uses, and through which rather few wildlife species are able to immigrate successfully. The situation becomes analogous to that of an offshore island, but with a species complement equivalent to a similarly sized section of a continuous mainland habitat. The natural tendency is for the artificial island to lose species through extinction and emigration until it reaches a new equilibrium level appropriate to an island situation, with its much-reduced levels of immigration by new species. Indeed, this is exactly what seems to be happening to some extensive wildlife reserves in East Africa, which, since their intitial creation, have become isolated by agricultural development in the surrounding areas (Miller and Harris, 1977; Miller, 1978; Soulé, Wilcox and Holtby, 1979). Thus, the important message is that the exact species complement of any area is not static, but a dynamic consequence of a balance between losses and gains, and if isolation produces a drastic reduction in the rate of incoming species then the overall tendency will be for the species complement to decline. The most dramatic losses will be incurred by species whose populations are small or those whose distributions are normally highly mobile (Dempster, 1977), whilst, for largely sessile species the losses will be almost unnoticeable (Gilpin and Diamond, 1980). The resolution of this problem by management will at least partly depend upon the known or inferred incidence functions of individual species (Diamond, 1975) and especially on a consideration of patch dynamics (Pickett and Thompson, 1978). This is therefore a problem largely peculiar to animals and of much lesser importance in the conservation of plant species (Higgs and Usher, 1980). Ultimately, the problem may only be resolved by the translocation and deliberate reintroduction of threatened species, thus mimicking the natural process of immigration that would prevail if the reserve were part of a continuous natural ecosystem.

If species diversity were considered by conservationists to be more important than naturalness or, as is usually the case, if they can only afford to set aside relatively small areas as reserves that must serve to ensure the survival of the full species complement of formerly extensive natural areas, management will have to focus on the prevention of the loss of any species from reserves. Attention will then be concentrated on the rarest species, since these are the ones most likely to be lost, and management may be aimed at achieving a more equitable population size for all the species present. This would be greatly assisted if it were possible initially to identify areas with a relatively equitable division of individuals amongst species and choose these as reserves. Some ecologists have argued intuitively that diverse communities are more stable than simple ones, either because they would have more food chains through which energy is processed (Elton, 1958) or because a multiplicity of links in a trophic web will generate interactions that tend to be compensating rather than additive (MacArthur, 1955). If this is genuinely the case, then diverse ecological systems should be the ones which lose species at the lowest rate. There is a link between stable environments and species diversity because global comparisons have shown us that species diversity is highest in tropical areas and declines progressively towards the poles (Fischer, 1960; Richards, 1969; Simpson, 1964) and

Distribution of frequency of breeding pairs within the bird community

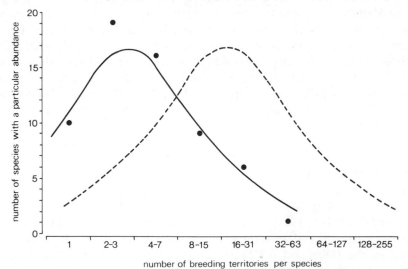

FIGURE 8.3 Distribution of frequency of breeding pairs within the bird community of a valley of 173 ha extent in the New Forest (dots) with fitted log-normal curve (solid line). The broken line shows the expected distribution of breeding pairs in the community if a larger area of 10 km² of similar habitat had been surveyed. Data from Glue (1973).

this has been equated directly with certain measures of climatic stability (MacArthur, 1975). However, so far there is no evidence that a more diverse species-assemblage produces trophic stability, and attempts to model ecosystems using mathematical formulae actually indicate that more complex systems are inherently more unstable (May, 1971). Indeed, if we examine the apportionment of individuals between species in almost any natural assembly of organisms we are likely to find a log-normal distribution (Figure 8.3). This, of course, is to be expected from knowledge of the species-area relationship found to be common for most natural communities, as this relationship equates with a log-normal pattern when z is equal to about 0.25 (May, 1975). An increase in the area sampled is thus merely equivalent to shifting the log-normal plot further to the right, revealing a hitherto hidden tail of rare species. Such distributions can result from the product of random interactions between numerous elements (Diamond, 1979; May, 1976) which suggests that communities are not themselves structured entities but merely chance assemblages of species. In consequence, more equitable distributions are unlikely to be found in nature, but, if they were to be required to ensure the survival of the rarer species, they could be produced only by management designed to serve this specific purpose. Since no structural rules for communities appear to exist, this management will need to be based upon knowledge of the ecology of the individual species in the community.

Management based upon the characteristics of individual species

All animals, by the law of natural selection, produce as many surviving offspring as possible during the course of their lifetime. However, the strategies they adopt to attain this end vary dramatically according to the environment of each animal. Some species live where their resources may be extremely abundant, if only for rather short periods of time; the best strategy for them to adopt is to reproduce as prolifically as they can during these periods, and to move elsewhere, or adopt a dormant phase, when resources disappear. Animals of this type are characteristically small, short-lived, mature quickly, produce large numbers of offspring each time they breed, and often reproduce only once in their lifetime. They have been termed 'r-selected' species (MacArthur and Wilson, 1967) and are characteristically found in early successional stages of ecosystem development or in environments that are subjected to frequent disturbance. They are thus commonly associated with short-term agricultural crops and other disturbed systems. At the other extreme are animals which live in situations where competition for resources is almost always severe. These species characteristically adopt strategies to maximize competitive ability; they are usually large, long-lived, breed many times in their lifetime but produce few offspring each time, have delayed maturity and their populations often remain very stable over long periods of time. They are termed 'K-selected' species and are characteristically found in stable climax ecosystems. Indeed, it has been suggested that K-selected species are likely to predominate in tropical areas and r-selected species in polar regions (Krebs, 1978), or that all terrestrial vertebrates are K-selected and invertebrates r-selected (Pianka, 1970). However, all natural ecosystems are likely to contain some animals showing attributes of each of these extreme types, whilst perhaps the majority of species will be rather intermediate in lifestyle; otherwise it would be unlikely that tropical rain forest would be so rich in insect life nor that polar bears (*Thalassarctos maritimus*) would exist.

Conservationists are mainly concerned to protect self-perpetuating systems which are likely to contain a high proportion of K-selected animal species. Furthermore, there is often a tendency for management to be aimed at producing a higher level of resource-predictability than may often be the case in entirely natural systems (see Wood Chapter 14) and thus further favour K-selected species. Indeed, it is this type of species that conservationists are advised to protect (Southwood, 1976), as more r-selected types are seen as having a greater capability of tolerating or escaping temporary adverse perturbations and having populations which may rapidly bounce back to a comfortably high level following their depletion. If this is genuinely so, then the polarization of lifestyles envisaged by the concept of r- and K-selection should enable conservation management to be focused upon that end of the spectrum where it is most needed, and hence enable a greater efficiency of the application of limited time and other resources. Consequently, it would become imperative to recognize the more K-selected species in any ecosystem and devote most attention to their welfare.

One very clearly discernible attribute that is associated with the $r-K$ axis of variation is an animal's size. The preferential management to favour large species, which is indicated by this correlation, also has added utility to conservationists. Large animals are often the most spectacular species that occur in any ecosystem and are thus culturally favoured, whilst they are often also species at the end of food-chains, so that their preservation may entail the preservation of adequate numbers of the lesser elements in the chain. However, size is also associated intimately with a plethora of other lifestyle characteristics, many of which are of direct concern when attempting to manage animal populations.

Because they are subjected to the universal laws of physics and therefore constrained by the ratio of body-size to surface-area (Kleiber, 1961), the metabolic rate of homeothermic animals is directly proportional to their body size. Similar relationships exist for heterotherms and unicellular animals, in which metabolic rate is proportional to metabolic size (which itself is usually positively related to body size). As it is also directly governed by metabolic rate, constant relationships can also be found between body size and the intrinsic rate of natural increase (r_m) for each of these groups of organisms. This is of direct consequence for the management of animal populations, since it indicates the maximum rate at which a population will grow in numbers when resources are freely available. Of almost equal importance is the evidence that lifespan (Sacher, 1959) and growth rate (Millar, 1977) are also correlated with size. Moreover, in a recent review, Western (1979) has shown that a host of ecologically important parameters in mammals are allometrically scaled on body-size; these include gestation time, age at first reproduction, litter weight and net reproductive rate, in addition to the factors already mentioned. In each case the relationship takes the form

$$P = a W^x \tag{2}$$

where P is the parameter being considered, W is body weight, x is an exponent common to all orders of mammals and a is a constant which varies between orders but is specific to each. Thus, at least for mammals, and almost certainly also for other classes of organisms, an animal's life-history is order-specific but with its rate of action governed by body-size. Accordingly, carnivores have a shorter gestation period than artiodactyls and primates are considerably longer-lived than similar sized animals in either of these two orders, but within each order the timespan for these events is controlled by body-size. Between-order variation seems to be related principally to relative brain size, as the slow growth of neural tissue imposes an upper limit on the rate of all other processes (Sacher and Staffeldt, 1974).

What is particularly important to the consideration of a species' lifestyle in relation to its conservation management is the revelation that, for the range of mammalian data available for comparison, Western (1979) also found there to be a constant relationship between body-size and birth-rate (Table 8.1 and Figure 8.4). His examples are all drawn from East African ecosystems which are largely natural and relatively stable, and so could be considered to be K-selective. Nevertheless, it is salutary to realize that there is no clear distinction in the way that a 10 g mouse or a 2500 kg elephant adjusts its life-history characteristics to the problems posed by the same type of environment. This is perhaps to be expected, since Caughley (1966) has shown that the pattern of mortality suffered by natural populations of mammal species is essentially the same, despite enormous differences in the body-size of the species investigated.

All animals maximize their individual fitness by producing as many surviving offspring as they can during their lifetime. The best they could hope to achieve in this respect would be equal to the product of intrinsic rate of natural increase and lifespan ($r_m \times \lambda$). For all orders of mammals studied (not just in East Africa) the exponents of the relationships between these two parameters and body-size are -0.28 and 0.20 respectively (Fenchel, 1974; Sacher, 1959). Their product is thus -0.08, indicating very little dependence on body-size of the

TABLE 8.1 Some parameters that are allometrically related to the body weight of animals

Parameter	Group of animals	Exponent	Authority
Physiological and life-history parameters			
Metabolic rate	Homeotherms	0.75	Kleiber, 1961
Lifespan	Mammals	0.20	Sacher, 1959
Lifespan	Artiodactyls	0.22	Western, 1979
Lifespan	Primates	0.24	Western, 1979
Lifespan	Carnivores	0.17	Western, 1979
Life expectation (e_0)	Artiodactyls	0.20	Western, 1979
Intrinsic rate of natural increase (r_m)	All animals	−0.275	Fenchel, 1974
Birth rate	East African mammals	−0.33	Western, 1979
Gestation time	Mammals	0.16−0.20	Kihlström, 1972
Gestation time	Artiodactyls	0.16	Western, 1979
Gestation time	Primates	0.14	Western, 1979
Gestation time	Carnivores	0.12	Western, 1979
Age at first reproduction	All animals	0.30−0.35	Fenchel, 1974
Age at first reproduction	Artiodactyls	0.27	Western, 1979
Age at first reproduction	Primates	0.32	Western, 1979
Age at first reproduction	Carnivores	0.32	Western, 1979
Litter weight	Mammals	0.83	Leitch, Hytten, and Billewicz, 1959
Litter weight	Artiodactyls	0.72	Western, 1979
Litter weight	Primates	0.65	Western, 1979
Litter weight	Carnivores	0.67	Western, 1979
Growth rate	Mammals	0.69−0.73	Millar, 1977
Area and distance parameters			
Foraging radius	Herbivorous mammals	0.40	Pennycuick, 1979
Territory size	Predatory birds	1.31	Schoener, 1968
Territory size	Herbivorous birds	0.70	Schoener, 1968
Home range area	Predatory mammals	1.41	Schoener, 1968
Home range area	Mammal 'croppers'	0.69	McNab, 1963
Home range area	Mammal 'hunters'	0.71	McNab, 1963
Number of species per unit area	Continents	0.11−0.18	Preston, 1962 MacArthur and Wilson, 1967
Number of species per unit area	Islands	0.20−0.40	May, 1975

maximum number of offspring an animal may produce in its lifetime. This is also of considerable significance for conservation management since, with only limited numbers of any species protected within reserves, the continuing survival of a species may depend upon each individual's lifetime production of offspring. It also serves to emphasize that what may be of most significance in determining survivorship and fecundity is the length of time in relation to its lifespan that conditions remain favourable to a species (Southwood, 1976, 1977). The same disruptive event, if harmful to all species, would have much more significant consequences for a small than a large animal, since it would occupy a significantly greater proportion of its lifetime. Perhaps, as a consequence, larger animals tend to occupy environments where significant perturbations are

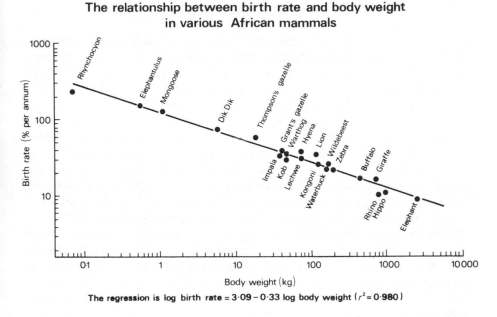

The relationship between birth rate and body weight in various African mammals

The regression is log birth rate = 3·09 − 0·33 log body weight ($r^2 = 0.980$)

FIGURE 8.4 From Western (1979), p. 194, reproduced by permission of Blackwell Scientific Publications Ltd.

typically of much longer duration than in environments utilized by small animals; the collapse and decay of a forest tree may take much longer than the entire life of a mouse, but would scarcely be of long-term significance to an elephant inhabiting the same forest (compare Gould, 1979; Warren, Chapter 2). Furthermore, our own perception of temporal stability is coloured by the length of our lives, and such dramatic interactions as those between elephant populations and savanna woodland can come to be viewed in a new light if this is fully appreciated (Caughley, 1976).

However, consideration of the rules of allometric scaling does not entirely eliminate the idea that animals may follow life-histories that are *r*- or *K*-selected. Deviations from the standard regression-line linking reproductive rate and size may occur, indicating the nature of an individual species' strategy which may tend towards one or other of the two extremes (Western, 1979). By this means it becomes possible to make realistic comparison between species of markedly different body-size. But elimination of that proportion of the variation due to allometry alone makes it clear that undue consideration of body-size may be misleading, and that the *r*−*K* axis of variation may be of rather less importance in determining a conservation strategy than other ecological parameters.

AREA REQUIREMENTS

In the search for ecological rules which may provide guidance in seeking to optimize management, one further parameter deserves particular consideration: the size of the area required by a species in order to survive and fulfil its basic daily functions. This too is related to size (see Table 8.1). McNab (1963) and Schoener (1968) examined the area of home-range or feeding territory for North American mammals and birds respectively. Both found rather precise relationships between the area utilized and body-weight, although, in each case, the area required by predators was considerably greater than for herbivores of equivalent size. This is explainable in terms of the lower food-density per unit area for predatory species, which are exploiting higher trophic levels (Schoener, 1968). But, as a consequence it is also vitally important to consider an area's overall productivity when assessing how big a particular species' territory is likely to be. Indeed, Newton *et al.* (1977) found for the sparrowhawk (*Accipiter nisus*), whose nesting territories are arranged with very regular spacing, that the average separation of adjacent pairs was positively correlated with altitude and inversely correlated with a measure of land productivity. Thus, an animal's body size is insufficient alone to predict the size of area it may occupy, although if the relationship between body-size and area can be characterized by reference to a few species and the productivity of the environment can be taken into consideration, then fairly accurate estimations may be possible for most species.

When attempting to conserve animal populations, particularly if they are of species which may be harassed once they move outside the protection afforded by a reserve, it may be inadequate merely to consider the typical size of territory or home range as a sufficient area to ensure a reasonable chance of survival. Evidently, the ability to exploit space is not merely a consequence of body-size and food availability alone, but is also dramatically affected by an animal's mobility. The most highly migratory species present very special problems, perhaps warranting a series of reserves within different parts of their annual range. However, for non-migratory mobile species, it may be more realistic to consider the maximum area likely to be covered in a day as a basis for the estimation of the required size of a reserve. Pennycuick (1979), in a perceptive analysis, has examined this parameter in some detail, using East African savanna mammals and birds as examples. The distance that an animal is able to travel during the course of its search for food is constrained by the rate at which it consumes energy during locomotion and also by the amount of energy it can store within its body. The first of these factors can be further subdivided into energy used to power locomotion and energy used to fuel basal metabolism whilst on the move. Both of these factors and the maximum size of energy store are related to body-weight by power functions. However, for a wide range of values, energy expenditure on locomotion is independent of the speed of travel (Taylor, Schmidt-Nielsen, and Raab, 1970). Thus, the faster an animal travels, the lower will be its overall rate of consumption of energy per unit distance covered, and the bigger an animal is, the further it will be able to go on a full load of fuel. Absolute limits are set, however, by the inability of very large animals to utilize the most rapid forms of locomotion; elephants cannot canter and birds above a body-mass of about 12 kg are unable to remain airborne (Pennycuick, 1972, 1979). Consequently, the ability to exploit a large range is favoured by the selection of large body-size and the utilization of energy-rich food sources. Because animal protein is a better source of energy than plant tissue, a lion is able to carry a much greater store of energy in its gut and hence travel further than a grazing ungulate of equivalent body weight. Schoener (1968), in his investigation of the size of territory in relation to body weight, found different exponents of scaling for carnivores and herbivores. He explained the different exponents by invoking differences in the scale of patchiness of environments experienced by animals feeding at different trophic levels. However, Pennycuick's (1979) predictions of foraging radius based on energetics, and the different energetic value of different foods, are themselves sufficient to explain the disparate sizes of range utilized by herbivores and carnivores of equivalent body-size, without any need to consider the concept of patchiness. Indeed, the energetic density of food sources seems more likely to relate directly to population density of the animals using these foods (Coe, Cummings, and Phillipson, 1976); carnivores tend to adopt exclusive feeding territories, whilst herbivores more commonly occupy extensively overlapping home ranges (Jewell, 1966). The determination of foraging radius thus provides us with values for the maximum area that an individual animal is able to exploit from a fixed home-base. This is likely to be a real constraint for most species when breeding, and also for species dependent upon a particular fixed resource, such as a waterhole. Thus, we can calculate that an elephant may exploit an area of about 5000 km², a wildebeest (*Connochaetes taurinus*) an area of 800 km² when walking or up to 1960 km² if it canters, whilst a mouse weighing 20 g could utilize an area of 0.55 km² if walking, and up to 2.0 km² at the fastest possible speed of travel. These figures may provide a useful rule-of-thumb for the smallest size of reserve we should consider adequate for any particular species.

One further important consideration can be derived from Pennycuick's (1979) analysis. Since area is proportional to the square of distance, the allometric relationship between exploitable area and body size will have an exponent of 0.80. This value is rather close to that observed to link body-size with territory size in herbivorous birds (Schoener, 1968) and for home-range area in mammals (McNab, 1963), although considerably smaller than the exponent for predatory birds and mammals found by Schoener. Despite any minor discrepanices, the area-requirement of individual animals is clearly very considerably affected by their body-size, much more substantially so than is the number of species dependent on area, for either continental or island situations (see Table 8.1). Thus big animals will always tend to be less numerous and, unless reserves are very large, disproportionably high numbers of the larger animal species are likely to be lost from reserves during the course of their isolation from surrounding areas (Wilson and Willis, 1975).

SUSTAINABLE-YIELD HARVESTING AND STABILIZATION OF POPULATION SIZE

In the absence of immigration and emigration, the population size that can be attained by any animal species is affected solely by two factors, birth-rate and mortality-rate. If it is considered important to stabilize the size of a population it would be necessary to adjust one or both of these factors until they were equal. Obviously, when numbers are increasing, mortality rate is usually the easiest of the two to manipulate, since additional mortality can be produced by trapping or shooting a suitable proportion of the population. To do this effectively, however, it is necessary to identify the stage in the life-cycle at which the maximum rate of natural mortality occurs: key-factor analysis is a useful tool in this respect (Podoler and Rogers, 1975; Varley and Gradwell, 1960). Only if an additional mortality is imposed on a population subsequent to the operation of a key-factor mortality will this be effective in raising overall generation mortality. The failure to observe this simple fact has been the reason for the negligible effect of many schemes that have attempted to control populations of pest species of animals (Murton, Westwood, and Isaacson, 1974). Conversely, if it were wished to crop a species to obtain a usable product, the crop should be taken before the operation of the key mortality factor, for the yield obtained can thus be much greater, without any risk of causing the population to decline.

An inadequate birth-rate is rarely the principal cause of the demise of an animal population, except in extreme cases such as when the habitat has become unsuitable or when sublethal doses of poisons may cause an impairment of reproduction (Newton and Bogan, 1978), and it is usually very obvious when the birth-rate of an animal population is insufficient to maintain its present numbers. Whilst it may be unwise to extrapolate between environments, the allometric scaling of birth-rate on body-size found by Western (1979) for mammals in a stable East African savanna system, would enable an estimate to be made of the minimum birth-rate likely to be adequate. Accordingly, a 10 g mammal should have a birth-rate of at least 600 per cent per annum and a 100 kg mammal a rate of not less than 27.5 per cent per annum if populations of each were to maintain themselves. If the birth-rate of a protected population were substantially below that predicted by this scaling-function, it would be necessary to investigate the reasons for such a low reproductive success in some detail.

Allometric scaling laws can also provide us with a crude estimate of the maximum cropping-rate that any animal population might be able to sustain without suffering an overall decline in numbers. Since the logistic equation is a fairly adequate description of the rate of change of size of many real animal populations, and, in the simplest of cases, the maximum sustainable yield from a population is obtainable when its numbers are reduced to about half those of the carrying capacity of the environment, the highest mortality rate that any population will sustain without declining to extinction will be equal to $r_m/2$ (Beddington, 1979). Since Fenchel (1974) has shown that the intrinsic rate of natural increase is proportional to body-weight, we can calculate this value for all animal species: for an elephant population $r_m/2$ equals 12.5 per cent per annum and for a 10 g animal it equals 385 per cent per annum. To approach these rates, mortality would need to be evenly spread throughout the year, but clearly any population with overall mortality approaching the values predicted in this way is in danger of suffering an irreversible decline. If rates of this order were detected in a population, conservation management would be needed urgently to increase survivorship.

In practice, models for harvested populations need to be much more complex than this (Beddington, 1979) since few animal populations fit a logistic curve exactly. In particular, populations of large animals tend to show maximum-sustainable yield curves that are skewed to the right, thus demanding maximum mortality rates of less than $r_m/2$ for stability. However, mortality due to hunting is rarely evenly spread throughout the year, and cull-rates may be greater than predicted by the logistic equation if animals are removed from a population at the right time of year and sex ratios are adjusted for maximum productivity. Thus a theoretical value or $r_m/2$ for red deer, of 30.7 per cent per annum, is rather less than may be attainable under ideal management (McVean and Lockie, 1969). Similarly, natural mortality is rarely evenly spread, and if conservation management is to be effective in reducing overall mortality we need to recognize this fact. In practice a maximum sustainable cropping regime is only adequately attainable if detailed studies of the population dynamics of the managed species are available, and suitable life-tables have been prepared and analysed (Caughley, 1977). With data of this type and the help of appropriate calculations (the use of Leslie matrices being particularly useful in this instance) it is possible not only to deduce sensible cropping-rates for a harvestable population, but also the effects on population stability of mortality acting on different sections of the population. Calculations of this form for grey seals (*Halichoerus grypus*) which breed in British waters, have been published recently (Harwood, 1978; Harwood and Prime, 1978; Summers, 1978) and are an excellent example of methodology. However, they also serve as a reminder that a management plan, even if based on sound scientific criteria, may be inoperable if its basic philosophy is founded on a political decision which proves to be unacceptable to a large segment of the interested human population. Such is the nature of conservation.

Just as plans such as that for the British population of grey seals seek to stabilize numbers for commercial reasons, so conservationists will need to seek to increase the stability of numbers of the species they wish to conserve. In doing so, the body-size of animal species must be a factor of prime importance since it is an all-pervading constraint on a host of factors which determine the way in which animals respond to their environment. There is some justification for focusing management on the larger species in a community. Their considerable area requirements make it necessary to set aside very large reserves if they are to stand any chance of survival (Kitchener, Chapman, and Muir, 1980) and the high exponent relating body-size to home-range indicates that large species are always likely to be disproportionately represented amongst the rarer animals on any reserve, making them more prone to extinctions. Furthermore, size-related life-history characteristics mean that big species respond only slowly to manipulations of their habitat, so they need particularly expedient and well-planned management.

Nevertheless, we cannot afford to be complacent about the fate of smaller species. Whilst many of these may succeed in surviving outside reserves, because of their small area requirements, the eventual isolation of protected areas could lead to losses of species of all sizes, unless positive management is continually applied to maintain an unnatural equitability of numbers. As the number of species inhabiting any area is the result of a state of dynamic equilibrium, conservation management would be most effective if it were to combine some control on the population size of the most numerous species with continual reintroduction of rare species as they disappear from a reserve. To achieve this will require several reserves in every type of environment, thereby reducing the chance of simultaneous extinction of the rarer animals over extensive areas. Without this type of management, we must face a future with a substantially depauperate fauna, in particular lacking the more dramatic species of highest cultural value, which are, for most of the human population, the very species most synonymous with the ethic of conservation.

REFERENCES

Beddington, J.R. (1979). Harvesting and population dynamics. In R.M. Anderson, B.D. Turner and L.R. Taylor (Eds), *Population Dynamics*, 20th Symposium of the British Ecological Society, Blackwells, London, pp 307–320.

Buechner, H.K. and Dawkins, H.C. (1961). Vegetation change induced by elephants and fire in Murchison Falls National Park, Uganda. *Ecology*, 42, 752–766.

Caughley, G. (1966). Mortality patterns in mammals. *Ecology*, 47, 906–918.

Caughley, G. (1976). The elephant problem—an alternative hypothesis. *East African Wildlife Journal*, 14, 265–283.

Caughley, G. (1977). *Analysis of Vertebrate Populations*, John Wiley, Chichester, 234 pp.

Coe, M.J., Cummings, D.H., and Phillipson, J. (1976). Biomass and production of large African herbivores in relation to rainfall and primary production. *Oecologia*, 22, 341–354.

Dempster, J.P. (1977). The scientific basis of practical conservation: factors limiting the persistence of populations and communities of animals and plants. *Proceedings of the Royal Society of London B*, **197**, 69–76.

Diamond, J.M. (1975). Assembly of species communities. In M.L. Cody and J.M. Diamond (Eds), *Ecology and Evolution of Communities*, Belknap Press, Harvard, pp. 342–444.

Diamond, J.M. (1979). Community structure: is it random, or is it shaped by species differences and competition? In R.M. Anderson, B.D. Turner and L.R. Taylor (Eds) *Population Dynamics* 20th Symposium of the British Ecology Society, Blackwells, London, pp. 165–182.

Diamond, J.M. and May, R.M. (1976). Island biogeography and the design of natural reserves. In R.M. May (Ed) *Theoretical Ecology: Principals and Applications*, Blackwell Scientific Publications, Oxford, pp. 163–186.

Elton, C.S. (1958). *The Ecology of Invasions by Animals and Plants*, Methuen, London, 181 pp.

Fenchel, T. (1974). Intrinsic rate of natural increase: the relationship with body size. *Oecologia (Berlin)*, **14**, 317–26.

Fischer, A.G. (1960). Latitudinal variations in organic diversity, *Evolution*, **14**, 64–81.

Gilpin, M.E. and Diamond, J.M. (1980). Subdivision of nature reserves and the maintenance of species diversity. *Nature (London)*, **285**, 567–568.

Glover, P.E. and Sheldrick, D. (1974). An urgent research problem on the elephant and rhino populations of the Tsavo National Park in Kenya, *Bulletin of Epizootic Diseases Africa*, **12**, 33–38.

Glue, D.E. (1973). The breeding birds of a New Forest valley. *British Birds*, **66**, 461–471.

Gould, S.J. (1979). One standard lifespan. *New Scientist*, **81**, 388–389.

Harwood, J. (1978). The effect of management policies on the stability and resilience of British Grey Seal populations. *J.Applied Ecology*, **15**, 413–421.

Harwood, J. and Prime, J.H. (1978). Some factors affecting the size of British grey seal populations. *Journal of Applied Ecology*, **15**, 401–411.

Higgs, A.J. and Usher, M.B. (1980). Should nature reserves be large or small? *Nature (London)*, **285**, 568–569.

Hollis, G.E. (ed.) (1977). A management plan for the proposed Parc National de l'Ichkeul, Tunisia, *Conservation Reports No. 10*, University College, London, 240 pp.

Hollis, G.E. (1981). [In Russian] The effect of a water diversion scheme on the hydrology and ecology of Garaet el Ichkeul, Tunisia, *Proceedings of the International Symposium on New Methods of Computation for Water Resources Projects*, Leningrad, 1979, UNESCO, pp. 451–459.

Jewell, P.A. (1966). The concept of home range in mammals. *Symposium of the Zoological Society London*, **18**, 85–109.

Jewell, P.A. (1969). Wild animals and their potential for new domestication. In P. Ucko and G.W. Dimbleby, *The Domestication and Exploitation of Plants and Animals*, Duckworth, London, pp. 101–109.

Kihlström, J.E. (1972). Period of gestation and body weight in some placental mammals. *Comparative Biochemistry and Physiology*, **43A**, 673–680.

Kitchener, D.J., Chapman, A. and Muir, B.G. (1980). The conservation value for mammals of reserves in the western Australian wheatbelt. *Biological Conservation*, **18**, 179–207.

Kleiber, M. (1961). *The Fire of Life*, John Wiley New York.

Krebs, C.J. (1978). *Ecology: The Experimental Analysis of Distribution and Abundance* (2nd edition), Harper and Row, London, 678 pp.

Kushlan, J.A. (1979). Design and management of continental wildlife reserves: lessons from the everglades. *Biological Conservation*, 15, 281–9P.

Laws, R.M. (1970). Elephants as agents of habitat and landscape change in East Africa. *Oikos*, 21, 1–15.

Leitch, I., Hytten, F.E., and Billewicz, W.Z. (1959). Maternal and neonatal weights of some mammalia. *Proceedings of the Zoological Society of London*, 133. 11–28.

MacArthur, R.H. (1955). Fluctuations of animal populations, and a measure of community stability. *Ecology*, 36, 533–536.

MacArthur, R.H. and Wilson, E.O. (1967). *The Theory of Island Biogeography*, Princeton University Press, Princeton, NJ.

MacArthur, J.W. (1975). Environmental fluctuations and species diversity. In M.L. Cody and J.M. Diamond (Eds) *Ecology and Evolution of Communities*, Belknap Press, Harvard, pp. 74–80.

McNab, B.K. (1963). Bioenergetics and the determination of home range size. *American Naturalist*, 47, 133–140.

McVean, D.N. and Lockie, J.D. (1969). *Ecology and Land-use in Upland Scotland*, Edinburgh University Press, Edinburgh, 134 pp.

May, R.M. (1971). Stability in model ecosystems. *Proceedings of the Ecological Society of Australia*, 6, 18–56.

May, R.M. (1975). Patterns of species abundance and diversity. In M.L. Cody and J.M. Diamond (Eds), *Ecology and Evolution of Communities*, Belknap Press, Harvard, pp. 81–120.

May, R.M. (1976). Patterns in multi-species communities. In R.M. May (Ed.) *Theoretical Ecology: Principles and Applications*, Blackwell, Oxford, pp. 142–162.

Millar, J.S. (1977). Adaptive features of mammalian reproduction. *Evolution*, 31, 370–86.

Miller, R.I. (1978). Applying island biogeographic theory to an East African reserve. *Environmental Conservation*, 5, 191–195.

Miller, R.I. and Harris, L.D. (1977). Isolation and extirpations in wildlife reserves. *Biological Conservation*, 12, 311–315.

Moore, N.W. and Hooper, M.D. (1975). On the number of bird species in British woods. *Biological Conservation*, 8, 239–250.

Murton, R.K., Westwood, N.J. and Isaacson, A.J. (1974). A study of wood-pigeon shooting: the exploitation of a natural animal population. *Journal of Applied Ecology*, 11, 61–81.

Newton, I., Marquiss, M., Weir, D.N., and Moss, D. (1977). Spacing of sparrowhawk nesting territories. *Journal of Animal Ecology*, 46, 425–441.

Newton, I. and Bogan, J. (1978). The role of different organo-chlorine compounds in the breeding of British sparrowhawks. *Journal of Applied Ecology*, 15, 105–116.

Pennycuick, C.J. (1972). *Animal Flight, Studies in Biology No. 33*, Edward Arnold, London, 68 pp.

Pennycuick, C.J. (1979). Energy costs of locomotion and the concept of 'foraging radius'. In A.R.E. Sinclair and M. Norton-Griffiths (Eds) *Serengeti: Dynamics of an Ecosystem*, University of Chicago Press, London, pp. 164–184.

Pianka, E.R. (1970). On r- and K- selection. *American Naturalist*, 104, 592–597.

Pickett, S.T.A. and Thompson, J.N. (1978). Patch dynamics and the design of nature reserves. *Biological Conservation*, 13, 27–37.

Podoler, H. and Rogers, D. (1975). A new method for the identification of key factors from life-table data. *Journal of Animal Ecology*, 44, 85–115.

Preston, F.W. (1962). The canonical distribution of commonness and rarity. *Ecology*, 43, 185–215, 410–432.

Richards, P.W. (1969). Speciation in the tropical rain forest and the concept of the niche. *Biological Journal of the Linnean Society of London*, **1**, 149–153.

Sacher, G.A. (1959). Relation of lifespan to brain weight and body weight in mammals. In *CIBA Foundation colloquia on ageing, 5, The Life Span of Animals*, Churchill, London.

Sacher, G.A. and Staffeldt, E.F. (1974). Relation of gestation time and brain weight of placental mammals: implications for the theory of vertebrate growth. *American Naturalist*, **105**, 593–615.

Schoener, T.W. (1968). Sizes of feeding territories among birds. *Ecology*, **49**, 123–141.

Short, R.V. (1976). The introduction of new species of animals for the purpose of domestication. *Symposia of the Zoological Society of London*, **40**, 1–13.

Simpson, G.G. (1964). Species density of North American recent mammals. *Systematic Zoology*, **13**, 57–73.

Soulé, M.E., Wilcox, B.A. and Holtby, C. (1979). Benign neglect: a model of faunal collapse in the game reserves of East Africa. *Biological Conservation*, **15**, 259–272.

Southwood, T.R.E. (1976). Bionomic strategies and population parameters. In R.M. May (Ed.) *Theoretical Ecology: Principals and Applications*, Blackwell, Oxford, pp 26–48.

Southwood, T.R.E. (1977). Habitat, the templet for ecological strategies? *Journal of Animal Ecology*, **46**, 337–365.

Summers, C.F. (1978). Trends in the size of British Grey Seal populations, *Journal of Applied Ecology*, **15**, 395–400.

Taylor, C.R., Schmidt-Nielsen, K., and Raab, J.L. (1970). Scaling of energetic cost of running to body size in mammals. *Americal Journal of Physiology*, **219**, 1104–1107.

Thomas, J.A. (1980). The extinction of the Large Blue and the conservation of the Black Hairstreak butterflies (a contrast of failure and success). *Institute of Terrestrial Ecology Annual Report for 1979*, 19–23.

Varley, G.C. and Gradwell, G.R. (1960). Key factors in population studies. *Journal of Animal Ecology*, **29**, 399–401.

Warren, A., Hollis, G.E., Wood, J.B., Hooper, M.D., and Fisher, R.C. (1979). Ichkeul, the problems of a wet park in a dry country. *Parks*, **4**, 6–10.

Western, D. (1979). Size, life-history and ecology in mammals. *African Journal of Ecology*, **17**, 185–204.

Wilson, E.O. and Willis, E.O. (1975). Applied biogeography. In M.L. Cody and J.M. Diamond *Ecology and Evolution of Communities*, Belknap Press, Harvard, pp. 522–534.

Wyk, P. van and Fairall, N. (1969). The influence of the African elephant on the vegetation of the Kruger National Park. *Koedoe*, **12**, 57–89.

Conservation in Perspective
Edited by A. Warren and F.B. Goldsmith
© 1983 John Wiley & Sons Ltd.

CHAPTER 9

Genetics and Conservation

R.J. BERRY

Not many years ago, any suggestion that conservationists ought to know something about genetics would have been met with amused scepticism. Inbreeding depression might cause problems for zoos and possibly therefore for animals and plants reduced to near-extinction, but these were special cases; general conservation management, so it was assumed, had no genetical implications.

This simple belief is now completely untenable for four reasons:

(1) As plant (and to a lesser extent, animal) breeders have improved and distributed more and more efficient domesticates, the genes in the old land-races from which the domesticated forms came have become an increasingly necessary resource for future development. This was recognized long ago by such pioneers as Otto Frankel, and the collection of seeds, etc. from land-races was an important element in the International Biological Programme (Frankel and Bennett,1970). One of the three primary objectives of the World Conservation Strategy is the preservation of genetic diversity as both an insurance and an investment: it is 'necessary to sustain and improve agricultural, forestry, and fisheries production, to keep open future options, as a buffer against harmful environmental change, and as the raw material for much scientific and industrial innovation'.

(2) Parallel to the recognition of international genetic impoverishment has come the scientific realization that genetical variation in natural populations is several orders of magnitude greater than previously believed, and that this variation is maintained in part by adaptation, producing rapid and precise genetical adjustment (Berry, 1977; Nevo, 1978). The classical idea that all members of a species are virtually identical has had to be replaced by one that all individuals are effectively unique and, as we shall see, subject to change by apparently innocuous management practices.

(3) The increased environmental awareness which has been the stimulus for conservation concern has brought also a questioning of man's relationship to nature, and a recognition that we have a moral responsibility for posterity and nature, as well as the strictly utilitarian involvement expressed by the World Conservation Strategy (Black, 1970; Moss, 1982). Although we are managers and apart from nature, we are also a part of nature and inextricably bound up with it. 'There is a growing feeling that to end all evolution that is not induced by our own species— except that of organisms we are as yet unable to control—is an arrogant, if not a fatal, step for man to undertake, and that as biologists we bear a special responsibility' (Frankel and Soulé, 1981). To put this into a genetical context: if, as has been claimd, the motive for all conservation is to prevent extinction, then there is a need for a 'genetics of scarcity' to make conservation realistic.

(4) Biological conservation depends upon an understanding of biological variability in both time and space; only when we understand why ecosystems differ from each other or change with time will we be able to devise rational management plans. It is the business of ecologists to investigate this variability. One reason for their relative failure in explaining it is the lack of attention hitherto paid to inherited differences between individuals in a population. A realistic approach to unravelling the key problem of survival over successive generations is Southwood's (1977) dissection of the environmental variables important for an organism's reproductive success into favourableness, probability of being in a particular place at a particular time, and uncertainty. The reaction of an organism to each of these is dependent on genes: reproductive success depends on a complicated series of interactions between a variable genome and a variable environment (Berry, 1979). Only when these interactions are accepted and understood will significant advance take place in the science of ecology, and its applied offshoot, conservation management.

MAN-MADE GENETIC CHANGES

A sceptic may still argue that genetical considerations are trivial for conservationists; that the major problems in conservation can continue to be solved by ignoring the inconvenient vagaries of inherited variations. After all, genetical change is effectively synonymous with evolution, which concerns us only in terms of geological and not ecological time. This assumption is wrong: many genetical changes are now known to have resulted as a direct or indirect result of man's activities (Bishop and Cook, 1981).

Mutation

Ironically, the most widely known genetic effect produced by man is also the least important. It is an unfortunate fact that for many, genetics is historically and emotionally entangled inextricably with the hazards of ionizing radiation, either from nuclear weapons or from the industrial uses of atomic energy. These hazards are almost always overstated. Although there is apparently no threshold below which the amount of radiation received does not increase mutation rate (including somatic mutations producing cancers and ageing) and most of the mutations induced will be deleterious, nevertheless less than one 10 000 th of 'spontaneous' mutation is produced by normal background radiation (cosmic rays, medical radiology, etc.) (Berry, 1972, 1982). Moreover, the usual fate of a new mutation is to be repaired by intracellular processes. Low-level irradiation (or exposure to environmental mutagens, such as nitrosamines, etc.) has a smaller genetic effect than predicted by extrapolation from acute exposures. 'Permissible' radiation doses were laid down in the 1950s on the basis of laboratory experiments involving acute doses. Although it is obviously sensible to keep mutation rates as low as possible, it must be recognized that variation produced by new mutation represents an extremely small part of naturally occurring genetic damage, that some mutations are advantageous (induced mutation is an important tool of plant breeders, albeit not as useful as once seemed likely), and continuing mutation is the basis for evolutionary change and survival.

Intentional environmental activity

Inherited resistance to specific pesticides has occurred in every species exposed to them for more than a short time: by 1975, 364 animal species were known to be resistant to one or more poisons; 225 of these were agricultural pests and 139 disease vectors (including 110 dipteran species) (Wood, 1981). In Denmark, the housefly (*Musca domestica*) is said to survive every insecticide which can safely be used, except pyrethrum in a few places. In the same way, pathogens (both microorganisms and parasites) have become resistant to drugs directed against them. Recognizing and avoiding resistance is now a major and chronic problem in agriculture and medicine.

Incidental genetical results

Melanism in a range of invertebrate groups has been a sequel to industrialization in both Europe and North America. The best known examples are the Lepidoptera, where over 200 species of trunk-sitting moths in Britain alone have lost their camouflage against bird predation as vegetative lichens have been destroyed by air pollution; in these species, melanic forms have replaced the formerly cryptic ones (Lees, 1981). In the twin-spot ladybird (*Adalia bipunctata*) (and perhaps some spiders) black forms occur at high frequences in areas where the sun is commonly obscured by smoke; the melanic advantage here is the speed with which pigmented animals absorb heat when compared with the non-pigmented form.

A comparable change in selective processes results when soils toxic to plant growth are produced as the waste-products of human activity. Some species have developed tolerance to such conditions. For example, 21 species of higher plant have been recorded growing on mine spoil-heaps containing high concentrations of lead, zinc or copper (Bradshaw and McNeilly, 1981). Even tolerance in algae to antifouling paint on the hulls of ships is beginning to occur.

Man's long-term activities

An unknown number of other human activities are liable to produce genetical change. In principal these effects are the same as described in the previous section, but it is as well to recognize that human 'improvement' of land not normally regarded as pollution also changes the adaptive pressures on animals and plants. For example, shell colour and banding in the common snail *Cepaea nemoralis* are subject to change by the selective predation of thrushes (*Turdus ericetorum*). In woodland, the cryptic advantage changes from yellow in the spring to brown in the autumn. A similar but more permanent change will obviously occur if woods, hedges, etc. are removed.

Introductions

The most drastic genetical changes produced by man are due to the founder effect (Berry, 1977), which is likely to result when a small number of individuals colonize a previously empty habitat. Man has been an extremely important transporting agent for such colonizing events—both unintentionally, as in the case of commensals like rodents, and intentionally, through the introduction of exotics like rabbits to Australia and red deer to New Zealand.

The members of a founding group are unlikely to carry the same alleles (that is, different forms of the gene at any locus on a chromosome) in the same frequencies as in the ancestral population. Consequently the new population will differ genetically from the ancestral one at a large number of gene-loci, often leading to 'instant sub-speciation'. Such founder effects are being increasingly recognized as important in evolution: one of the best investigated cases is that of *Drosophila* species in the Hawaiian islands, where more than 650 species are recognized (out of around 1000 species worldwide), all but seventeen being endemic. Studies of chromosomal inversions, supported by behavioural, biochemical and ecological investigations have allowed the tracing of many of the relationships between these endemics (Williamson, 1981). The common pattern seems to be that a population new to an island is founded by a very small number of immigrants from a neighbouring island, necessarily involving abrupt, non-adaptive changes; comparable differentiation from the parent stock has occurred in other Hawaiian groups, in house sparrows (*Passer domesticus*) deliberately introduced into North America from Europe in 1851, and in field mice (*Apodemus sylvaticus*) inadvertently taken to the islands of the north Atlantic by the Vikings (Berry, 1969, 1977; Mueller-Dombois, Bridges, and Carson, 1981). The rapid change of the founder effect, incidentally, provides a genetic basis for the apparently sudden appearance of new taxa in the fossil record, described by palaeontologists as 'punctuations'.

Domestication

Finally, we have the situation already noted in which farm animals and crops are increasingly drawn from a small number of 'improved' strains. This would not matter in the long run, were it not for the progressive destruction of the wild relatives of agriculturally important forms, a significant process of 'genetic erosion' (Frankel and Bennett, 1970).

THE GENETICAL CONSEQUENCES OF CONSERVATION PRACTICE

Genes can be cared for in 'gene banks' (that is, seed or sperm stores, as frozen embryos, etc.), in zoological or botanic gardens, or in nature reserves of one sort or another. Only the last involve anything approaching natural populations managed by conservationists, so it is relevant to spend more time discussing the genetic effects of reserve establishment than the more artificial systems.

The influences which determine the genetical constitution of any population are summarized in Figure 9.1. Two agents (mutation and immigration from a genetically different population) increase variation, and two (genetic drift acting in small populations and natural selection) decrease it. Of these four main influences, migration, selection, and drift are dependent on the environment as we normally understand it (mutation also may be affected by environmental factors, but this is a rather special case).

Conservation commonly involves the establishment of nature reserves or sanctuaries. These are perhaps more a feature of conservation management in Britain than in countries with a lower density of human population. It should be recognized that the very existence of nature reserves may produce genetical pressures, because the outbreeding plants and less mobile animals in the reserve will be increasingly isolated from their relatives in other parts of their range, as habitats change around the reserve. There will thus be a reduction of gene-flow into the population (in other words, immigration will fall), and hence a reduction in the number of individuals able to breed together. In other words, two of

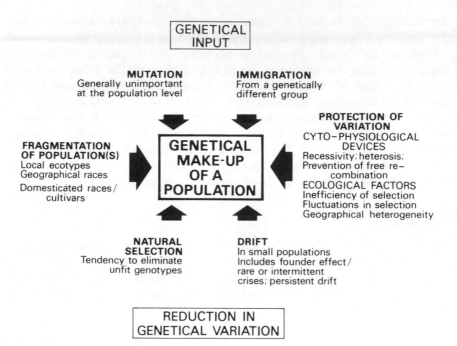

FIGURE 9.1 Factors contributing to the genetic make-up of a population or group of populations. Note that all the factors affecting genetical constitution are influenced by the environment.

the four factors affecting genetic variation (migration and drift) could be directly affected by reserve establishment.

But to what extent are these genetical pressures deleterious? How strong are they, and what is their likely effect? It is unfortunately not possible to give definitive answers to these key questions. Frankel and Soulé (1981, p. 84) condemn me (Berry, 1971) for 'phyletic optimism', in other words for concluding that the genetical results of conservation practice may not be damaging; they believe that 'nothing but the incisive action by *this* generation can save a large proportion of now-living species from extinction within the next few decades' (see also Ehrlich and Ehrlich, 1981).

Why the difference between these two views? Can they be reconciled? Who is right? Clearly extinctions will inevitably follow the elimination of particular habitats, and the number of these extinctions will be high if major habitats (such as wetlands or forest) are destroyed. But the question for the conservationist concerns the fate of species in his nature reserves, and the question can be refined in the following way: does variation ever become limiting, and restrict response to present environmental fluctuations or future environmental changes? Frankel and Soulé believe that it may; I interpret the evidence to show

TABLE 9.1 Genetic variation as measured by electrophoresis.

		Average number of loci scored per species	Mean per cent of loci	
			Polymorphic per population	Heterozygous per individual
Insects	*Drosophila*	24	52.9	15.0
	other insects	18	53.1	15.1
Molluscs	Land snails	18	43.7	15.0
	Marine snails	17	17.5	8.3
INVERTEBRATES			39.7	14.6
Fish		21	30.6	7.8
Amphibians		22	33.6	8.2
Reptiles		21	23.1	4.7
Birds		19	14.5	4.2
Mammals	Rodents	26	20.2	5.4
	Large mammals	40	23.3	3.7
VERTEBRATES			17.3	5.0
PLANTS			46.4	17.0

that it is unlikely to be, although in particular cases (such as the larger mammals, which concern Frankel and Soulé to a great extent) problems might arise.

To understand the situation, it is necessary to describe the interplay of genetical forces in more detail.

We have already seen that there is very much more inherited variation in populations than was previously thought. Indeed, theoretical calculations on 'genetic load' and 'the cost of selection' previously set upper limits to the amount of variation which could be tolerated by a population (Haldane, 1957; Müller, 1950). Great upset was caused by estimates from electrophoretic studies of proteins and enzymes which showed that actual levels were very much higher than these. In round terms, most individuals have more than one allele at about 10 per cent of their loci and approximately one-quarter of all loci in a population are segregating (in other words, are 'polymorphic') (Table 9.1).

There is still no complete agreement about the mechanisms which maintain these high levels of variation, but it is now generally accepted that they are important for their carriers; in other words, they are more than random chemical 'noise' (Berry, 1977, 1979). In particular, every example where polymorphic variation has been studied in depth has shown either different biochemical characteristics of the primary allele products of the locus concerned or the operation of selective forces. We are faced then with enormous possibilities of genetical adjustment, far more than under the traditional assumption of population structure, in which individuals were believed to possess 'wild-type' alleles at all except a few loci and genetical change was largely dependent on new mutations (Berry, 1978).

TABLE 9.2   Intensities of selection in natural populations.

Selection for	Per cent strength of selection
A. *Directed selection (i.e. extreme phenotype affected)*	
Heavy metal tolerance of grasses on mine spoilheaps	46−65
Non-banded snail (*Cepaea nemoralis*) in woodlands	19
Melanic (*Carbonaria*) form of peppered moth (*Biston betularia*) in various regions of Great Britain	5−35
Spotted form in overwintered leopard frog (*Rana pipiens*) (*versus* unspotted)	23−38
Unbanded water snakes (*Natrix sipedon*) (*versus* heavily banded)	77
B. *Stabilizing selection (i.e. intermediate phenotype favoured)*	
Coiling in snails (*Clausilia laminata*)	8
Size and hatchability in duck eggs	10
Birth-weight and survival in human babies	2−7
Inversion heterokaryosis in *Drosophila pseudoobscura*	Up to 50
Tooth variability (i.e. *de*stabilizing selection) in the house mouse *Mus musculus*	21−26
Shell variability in dog whelk (*Nucella lapillus*)	0−91
Colour morphs in the isopod *Sphaeroma rugicauda*	50+

There has been another advance in knowledge about population structure which has implications not often realized: selection intensities in nature are often 5, 10, or more per cent (Table 9.2). Fifty years ago R.A. Fisher, J.B.S. Haldane, and others carried out calculations on the effects of intensities of 0.1−1.0 per cent, and showed that genetical change could follow. However, these selection strengths have turned out to be too low in virtually every situation that has been studied, beginning with the demonstration by Haldane himself as early as 1924 that the initial spread of the dark form of the peppered moth (*Biston betularia*) in the Manchester area between 1848 and 1895 could only be accounted for by a 30 per cent disadvantage of the typical pale form when compared to the melanic (Haldane, 1924).

If we take together the amount of variation and the strengths of selection commonly found in populations, it is not surprising that stabilizing selection (in other words, the elimination of phenotypic extremes) is apparently universal in nature. For example, 4.5 per cent of all babies born in London between 1935 and 1946 were stillborn or died before 28 days, but only 1.8 per cent of average weight babies died (Karn and Penrose, 1951). Since about half the variance of birth-weight is inherited, this differential survival almost certainly results in selection for the mean at the expense of the extremes.

The common dog whelk (*Nucella lapillus*) experiences stabilizing selection of up to 90 per cent during life on wave-exposed shores, although whelks living on

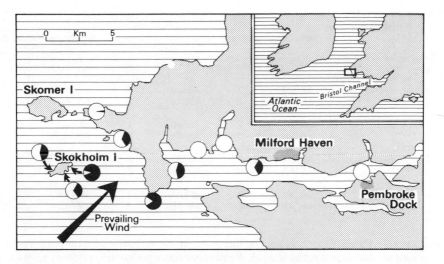

FIGURE 9.2 Stabilizing selection in the common dog whelk (*Nucella lapillus*). The proportion of black in each circle represents the difference in variance in a shell character between young and old members of the same population (= amount of selection). Based on Berry and Crothers (1968).

sheltered shore undergo no selection of this nature (Figure 9.2). There is a high correlation ($r \approx 0.7$) between the amount of environmental stress (wave action) and the intensity of selection (Berry and Crothers, 1968).

The easiest way to achieve the situation of advantage of the mean or average over the extremes is for a heterozygote to produce a phenotype intermediate between two more extreme homozygotes. In practice, most characters are controlled by alleles acting to increase the expression of a character and others acting to decrease it. In such a case, Fisher (1930) showed that selection will favour linkage between responsible loci, and also the accumulation of 'balanced' chromosomes with 'positive' and 'negative' alleles. The simplest situation to consider (Figure 9.3) involves two segregating loci *A, a* and *B, b* on the same chromosome, with *A, B* acting in one direction and *a, b* in the other (that is, normal additive genes). The intermediate type will be either the attraction or the repulsion heterozygote, respectively *AB/ab* or *Ab/aB*. Fisher pointed out that the repulsion heterozygote (*Ab/aB*) will be more favoured by selection, since it will be less likely than the attraction heterozygote (*AB/ab*) to produce zygotes giving the adaptively disadvantageous extreme phenotypes (dependent on *AB/AB* or *ab/ab* genotypes). Any mechanism bringing about tighter linkage between the loci concerned (such as chromosomal inversion) will tend to spread. This theoretical argument has been subjected to experimental test (mainly through selection experiments in *Drosophila*) by a number of workers and shown to be basically correct (Mather, 1974).

PARENTS		OFFSPRING					
		FROM NON-RECOMBINANT GAMETES		WITH RECOMBINATION IN ONE PARENT		WITH RECOMBINATION IN BOTH PARENTS	
Chromosomes	Phenotypic score	Chromosomes	Phenotypic score	Chromosomes	Phenotypic score	Chromosomes	Phenotypic score
REPULSION A b / a B	2	Ab/aB	2	AB/aB	3	AB/ab	2
		Ab/Ab	2	AB/Ab	3	AB/AB	4
A b / a B	2	aB/aB	2	ab/aB	1	ab/ab	0
		aB/Ab	2	ab/Ab	1	AB/ab	2
ATTRACTION A B / a b	2	AB/ab	2	AB/Ab	3	Ab/Ab	2
		AB/AB	4	AB/aB	3	Ab/aB	2
A B / a b	2	ab/ab	0	ab/Ab	1	aB/Ab	2
		ab/AB	2	ab/aB	1	aB/aB	2

FIGURE 9.3 Effect of linkage on the incidence of the extremes of manifestation of a continuously distributed variable. If *A, B* each produces a phenotypic effect of 1, and *a, b* an effect of 0 each, then matings between two repulsion heterozygotes are most likely to produce non-extreme phenotypes.

Beanbag genetics and genetical architecture

We have seen that there are four agents capable of changing the genetical constitution of a population (see Figure 9.1). This simple fact can easily be deduced from the Hardy–Weinberg principle (defining the relationship between allele and genotype frequencies), which is nothing more than a particular statement of the binomial theorem. There are no problems about deriving precise expressions for the effects of disrupting agents under a variety of conditions (see Crow and Kimura, 1970, among others). However, such expressions are too simplistic. They treat genes as largely independent entities subject to sampling laws, rather like beans in a bag. With such assumptions, allele loss is directly related to effective population size; as population numbers fall, so alleles will disappear from the population.

Genetic erosion in small populations is the main reason for the pessimism of Frankel and Soulé (1981). They define conservation genetics as 'the genetics of scarcity', where the scarcity is that of population numbers.

'Whether our concern is the wild relatives of cultivated plants or wild animals, the conservationist is faced with the ultimate sampling problem—how to preserve genetic variability and evolutionary flexibility in the face of diminishing space and with very limited economic resources. Inevitably we are concerned with the genetics and evolution of small populations, and with establishing practical guidelines for the practising conservation biologist.'

Now it is obviously sensible to base any guidelines on valid genetical criteria, but it is clear that Frankel and Soulé adopt an extreme 'beanbag' genetics point of view, and ignore the effects of what may be termed 'genetic architecture' in retarding allele loss in small populations. They are, of course, aware of the problem. They quote Mayr (1970): 'It is a limited number of highly successful epigenetic systems and homeostatic devices which is responsible for the severe restraints on genetic and phenotypic change displayed by every species', but they do not pursue the point. Their argument about genetic erosion is based entirely on the two 'beanbag' conclusions that the input of new variation by immigration will be negligible ('Gene flow is in most cases irrelevant when dealing with the capacity of an endangered species to respond to novel selection pressures. The reason is that endangered species are usually isolated remnants of hitherto widely distributed forms. Isolates, by definition, receive no immigration') and that genetic drift will result in a significant loss of variation ('The conservation biologist is perforce operating with relatively small populations; he courts disaster when pretending that such populations have the same evolutionary plasticity and potential as do larger ones').

The power of genetical architecture to protect variation cannot be measured at the present moment, but it is undeniably strong. There is abundant evidence from laboratory populations that inherited variants persist to a theoretically surprising extent under either intense selection or close inbreeding, and natural populations which have been through a bottleneck in numbers do not seem to be adversely affected in any way (Nei, Maruyama, and Chakraborty, 1975). The palaeontological observations on 'punctuated equilibria' are also pertinent: many species remain apparently unchanged for millions of years, showing the operation of strong conservative forces (Jones, 1981). There is little doubt that the next major advance in evolutionary biology will be an understanding of genetic architecture; such a better understanding should not be omitted from conservation genetics. Frankel and Soulé seem to have fallen into the error of assuming that things that cannot be currently measured must be unimportant.

The right-hand side of Figure 9.1 shows the more important devices which act as buffering forces when a population is subjected to action which would otherwise cause genetical change (see also Grant, 1964 and Chapters 9 and 10 in Mayr, 1963).

The fine dynamic equilibrium between environmental change and genetical adaptation is not the whole of the story; a balanced assessment of the importance and value of genetics in conservation must include both 'beanbag' results and conservatism coming from genetical architecture.

GENETICAL RULES FOR CONSERVATION PRACTICE

An environmental change may be temporary or permanent—due to normal climatic fluctuations or to longer-term factors (including habitat destruction). For example, an increase in numbers of the meadow brown butterfly (*Maniola jurtina*) in southern England in 1956 led to an increase in wing-spot variability, but by 1958 the previous (pre-1956) pattern had been regained; a similar temporary population increase in a colony of the marsh fritillary (*Euphydras aurinia*) led to a new wing phenotype even after numbers returned to normal (Ford and Ford, 1930). Without knowing the modes of action and interaction of all the components of environment, phenotype, and genotype, it is virtually impossible to forecast the genetical result of any environmental alteration. This leads to a generalization: *any environmental change may produce a genetical change*. We must qualify this immediately to take account of the fact that selection is much more powerful than the other agents shown in Figure 9.1. This leads to a second generalisation: *adaptation is rapid and precise*. It is, however, self-evident that adaptation is dependent on available variation. For example, the rosy minor moth (*Procus literosa*) became extinct in Sheffield in its typical form early in the Industrial Revolution, and only recolonized the area many years later when a melanic mutation occurred and spread. Differences in genetical content are very clear-cut when isolated races of a species are each founded by a small number of colonizers. Table 9.3 shows the frequencies of some alleles in samples of house mice (*Mus domesticus*) caught on different north Atlantic islands. Mice on all the islands are derived from the same ancestral population within the last few hundred years; they have spread as commensals with man.

TABLE 9.3   Genetic differences between house mouse (*Mus domesticus*) populations founded on north Atlantic islands.

	$Mod-1^b$	$Pgm-1^b$	Hbb^d	$Got-2^b$	$Es-2^a$
Caithness	57.6*	4.1	25.8	46.4	35.4
Orkney					
Mainland	45.2	3.9	0	0	15.3
Sanday	30.0	3.5	14.0	0	11.8
Shetland					
Dunrossness	20.8	3.0	14.9	46.9	0
Yell	—	0	85.0	—	0
Faroe					
Fugløy	0	0	100.0	100.0	0
Sandøy	35.7	0	3.8	90.0	0

Based on Berry and Peters (1977).
* The figures are percentage frequencies of alleles at loci scored by electrophoresis

It is the question of the availability of genetical variation (or rather its loss, genetic erosion) that worries Frankel and Soulé. They claim, albeit with little evidence, that 'any loss of variability, whether due to natural causes (small population size, bottlenecks, directional selection) or to artificial inbreeding, is going to reduce the chances of survival in the wild'. It is the same concern that led the authors of the World Conservation Strategy to include 'the preservation of genetic diversity' as one of their main objectives. Consequently it is encouraging to find that: *individual variation is large and resistant to loss*.

Some of the reasons for this are summarized on the right-hand side of Figure 9.1. As we have noted, in general they are not well-understood, and it is possible that incorrect weight is given to some of the pieces of evidence quoted. For example, the observation that forms exposed to a variable environment at the edge of their range have less variation than at the more equable centre could mean danger, or may represent nothing more than the adaptive loss of markers which reduce crossing-over and hence phenotypic variation (Carson, 1967; Zohary and Imber, 1963). However, it is extremely unlikely that any normal management practice could significantly affect the amount of variation in a local population to the extent of making that population unable to respond to environmental change.

CONSERVATION RULES?

The only certain principle in conservation genetics is to use every available stratagem to protect variation. Frankel and Soulé extend this to assert 'the basic rule ... [is] there is no safe amount of inbreeding for normally outbred organisms'. They apply this by taking an assumption of animal breeders that natural selection for performance and fertility can balance inbreeding depression if the rate of fitness loss per generation is no more than 1 per cent, which will result (using a formula due to Sewall Wright) if the effective breeding size of the population is less than 50. They then argue that genetic drift may be important up to a population size of around 500 (which assumes, wrongly, that natural selection is weak), and conclude that 'a genetically effective size of 500 is a satisfactory first approximation of the minimum size for the accommodation of continuing evolution'. This conclusion certainly errs on the safe side, but it is the correct side on which to err.

Population numbers are obviously related to the area of available habitat; an associated practical question is whether the maintenance of variation is affected by nature reserve design (in other words, the size, shape and distribution of patches of usable habitat). Various criteria for this have been derived, using implications from MacArthur and Wilson's theory of island biogeography (Diamond and May, 1981). It is not worth spending much space on these here, since the theory itself is certainly oversimplified, and is anyway concerned with species diversity rather than species evolution (Williamson, 1981). Suffice it to say that the main conclusions are that a large reserve is better than several small ones, since some species depend on large areas for their success; that if a large reserve is unattainable, the small ones that can be declared should be close together to allow migration between them; and a reserve should be circular to minimize dispersal distances and reduce extinctions in 'peninsulars', which are vulnerable to all sorts of pressures and hence reduce the reserve's effective area.

Simberloff and Abele (1976) have argued completely differently. They believe it would be better to establish several small reserves because different groups of species (or genomes) will survive in the different reserves, and thus the total diversity surviving in the group of reserves will be greater than if there was only one large reserve. Furthermore, catastrophes (disease, storms, etc.) are unlikely to affect every element in a system of reserves.

Such opposing views suggest that it is premature (at least) to formulate firm conservation rules. The only sure guide at the moment is to take reasonable care to protect all available variation, and probably every system will have to be approached from first principles.

The genetical constitution of any population is a summation of its past history and its past environments. Any management activity is likely to produce genetical changes, but on the whole these will be on a fairly small scale. The worst genetical problems are likely to come through the elimination of forms through habitat destruction. It is too soon to predict the effect of genetical adjustment through less drastic alterations in habitat. All we can do is re-emphasize the importance of variation for nature itself and for man's use of nature, and to reiterate:

(1) Any environmental change may produce a genetical change.
(2) Adaptation is rapid and precise.
(3) Individual variation is large and resistant to loss.

REFERENCES

Berry, R.J. (1969). History in the evolution of *Apodemus sylvaticus* (Mammalia) at one edge of its range. *Journal of Zoology, London,* **159**, 311–328.

Berry, R.J., (1971). Conservation aspects of the genetical constitution of populations. In E. Duffy and A.S. Watt (Eds). *The Scientific Management of Animal and Plant Communities for Conservation,* Blackwell, Oxford, pp.177–206.

Berry, R.J. (1972). Genetical effects of radiation on populations. *Atomic Energy Review,* **10**, 67–100.

Berry, R.J. (1977). *Inheritance and Natural History,* Collins New Naturalist, London. 350 p.

Berry, R.J. (1978). Genetic variation in wild house mice: where natural selection and history meet. *American Scientist,* **66**, 52–60.

Berry, R.J. (1979). Genetical factors in animal population dynamics. In R.M. Anderson, B.D. Turner and L.R. Taylor (Eds). *Population Dynamics,* Blackwell, Oxford, pp. 53–80.

Berry, R.J. (1982). Atomic bombs and genetic damage. *British Medical Journal,* **284**, 366–367.

Berry, R.J. and Crothers, J.H. (1968). Stabilizing selection in the dog whelk (*Nucella lapillus*). *Journal of Zoology,* **155**, 5–17.

Berry, R.J. and Peters, J. (1977). Heterogeneous heterozygozities in *Mus musculus* populations. *Proceedings of the Royal Society of London B,* **197**, 485–503.

Bishop, J.A. and Cook, L.M. (Eds) (1981). *Genetic Consequences of Man-Made Change*, Academic Press, London, 409 p.

Black, J. (1970). *The Dominion of Man*, Edinburgh University Press, Edinburgh, 169 p.

Bradshaw, A.D. and McNeilly, T (1981). *Evolution and Pollution*, Edward Arnold, London, 76 p.

Carson, H.L. (1967). Permanent heterozygosity. *Evolutionary Biology*, 1, 143–168.

Crow, J.F. and Kimura, M. (1970). *An Introduction to Population Genetics Theory*, Evanston, New York, and Harper and Row, London, 591 p.

Diamond, J.M. and May, R.M. (1981). Island biogeography and the design of nature reserves. In R.M. May (Ed.). *Theoretical Ecology*, 2nd edition, Blackwell, Oxford, pp. 228–252.

Ehrlich, P. and Ehrlich, A. (1981). *Extinction: the Causes and Consequences of the Disappearance of Species*. Random House, New York, 305 p.

Fisher, R.A. (1930). *Genetical Theory of Natural Selection*, Clarendon, Oxford, 272 pp.

Ford, H.D. and Ford, E.B. (1930). Fluctuation in numbers and its influence on variation in *Melitaea aurinia*. *Transactions of the Royal Entomological Society of London*, 78, 345–351.

Frankel, O.H. and Bennett, E. (1970). *Genetic Resources in Plants. IBP Handbook*, no. 11. Blackwell, Oxford, 554 p.

Frankel, O.H. and Soulé, M.E. (1981). *Conservation and Evolution*, Cambridge University Press, Cambridge, 327 p.

Grant, V.M. (1964). *The Architecture of the Germplasm*, Wiley, New York, 236 p.

Haldane, J.B.S. (1924). A mathematical theory of natural and artificial selection. *Transactions of the Cambridge Philosophical Society*, 23, 19–40.

Haldane, J.B.S. (1957). The cost of natural selection. *Journal of Genetics*, 55, 511–524.

Jones, J.S. (1981). An uncensored page of fossil history. *Nature, London,* 293, 427–428.

Karn, M.N. and Pensose, L.S. (1951). Birth weight and gestation time in relation to maternal age, parity and infant survival. *Annals of Eugenics*, 16, 147–164.

Lees, D.R. (1981). Industrial melanism: genetic adaptation of animals to air pollution. In J.A. Bishop and L.M. Cook, (Eds). *Genetic Consequences of Man-Made Change*, Academic Press, London, pp. 129–176.

Mather, K. (1974). *Genetical Structure of Populations*, Chapman and Hall, London, 197 p.

Mayr, E. (1963). *Animal Species and Evolution*, Harvard University Press, London and Cambridge, Mass., 797 p.

Mayr, E. (1970). *Populations, Species and Evolution*. Harvard University Press, Cambridge, Massachusetts, 453 p.

Moss, R. (1982). *The Earth in Our Hands*, Inter-Varsity Press, Leicester, 125 p.

Mueller-Dombois, D., Bridges, K.W., and Carson, H.L. (Eds) (1981). *Island Ecosystems*, Hutchinson Ross, Stroudsburg, Penn., 583 p.

Müller, H.J. (1950). Our load of mutations. *American Journal of Human Genetics*, 2, 111–176.

Nei, M., Maruyama, T. and Chakraborty, R. (1975). The bottleneck effect and genetic variability in populations. *Evolution*, 29, 1–10.

Nevo, E. (1978). Genetic variation in natural populations: pattern and theory. *Theoretical Population Biology*, 13, 121–177.

Simberloff, D.S. and Abele, L.G. (1976). Island biogeography: theory and conservation practice. *Science, New York,* 154, 285–286.

Southwood, T.R.E. (1977). Habitat, the templet for ecological strategies? *Journal of Animal Ecology*, 46, 337–365.

Williamson, M. (1981). *Island Populations*, Oxford University Press, Oxford, 286 p.

Wood, R.J. (1981). Insecticide resistance: genes and mechanisms. In J.A. Bishop and L.M. Cook (Eds). *Genetic consequences of man-made change*, Academic Press, London, p. 53–96.

Zohary, D. and Imber, D. (1963). Genetic dimorphism in fruit types in *Aegilops speltoides*. *Heredity*, **18**, 223–231.

PART II

ECOLOGY AND CONSERVATION
IN PRACTICE

Introduction

Nature conservationists have, over the last two decades, moved beyond passively accepting the remnants of our natural heritage and towards active selection, practical management, involvement with control and even the creation of new habitats. These interests underlie the chapters in this section. Here, as in the rest of our collection, we have selected only a few examples from a wide spectrum of practical involvement.

Practical conservation is involved with much more than nature itself (the starting point of the chapters in Part I). Practical projects involve techniques of crop and soil manipulation, administration, politics, aesthetic and even moral principles. Part II in this way illustrates projects at the 'sharp end' of conservation and is a bridge between the science of the chapters in Part I and the social concerns of those in Part III.

The section begins with two potentially very serious threats—pesticides and derelict land. Pesticides are valuable tools in the production of food and fibre, but by their very nature pose a major threat to wildlife. In Britain, voluntary agreements rather than legislation appear to be checking the escalation of ecological damage. Derelict land is more an aesthetic than a nature conservation problem but is being tackled from an ecological basis. The restoration of these areas depends on our knowledge of the ecology of the plant species involved.

Informal recreation and its impacts have also become the subject of restoration projects (Chapter 12), and Chapter 13 is an account of how we can use native species to enhance the ecological interest of otherwise low-value areas such as roadside verges, and rough grass in urban areas and parks.

The need for, and means of producing, management plans for nature reserves (Chapter 15) is topical and important and obviously involves an assessment of the relative values of the different areas under study (Chapter 14). These two chapters summarize recent work carried out by the Ecology and Conservation Unit at University College London.

Finally we turn to two topical conservation issues—a summary of the nature conservation interest, the potential and the enhancement of urban areas (Chapter 16) and the thorny problem of how extensively and intensively our uplands should be planted with conifers (Chapter 17).

Conservation in Perspective
Edited by A. Warren and F.B. Goldsmith
© 1983 John Wiley & Sons Ltd

CHAPTER 10

Ecological Effects of Pesticides

N.W. MOORE

PESTICIDES—A NEW ECOLOGICAL FACTOR

Pesticides are chemical substances used by man to control those living organisms which are inimical, or are believed to be inimical, to his interests. Most are poisons insofar as the target species are concerned: but the term pesticide is also used to cover chemosterilants and growth inhibitors. In Britain we associate pesticides principally with agriculture and horticulture, although they are also used widely in food storage and for the protection of wood and wool and other fibres. In many countries they are also applied very extensively to control vectors of disease and in forestry. Today pesticides are used throughout the world in a wide range of habitats—mainly in agricultural and urban environments, but also in natural or seminatural ones such as forests, rangelands, inland and coastal marshes, lakes and even deserts. The more persistent compounds disperse into all environments, including those which are never sprayed, notably the oceans and polar regions. This means that most life on the earth comes into contact with pesticides, and therefore nowadays they must be taken into account in all ecological studies; although there are still many ecosystems where their effects are likely to be negligible.

Pesticides have been used on a small scale for centuries, but they have only become an integral part of agriculture, preventive medicine and forestry during the lifetime of many of the readers of this book. Most modern pesticides are entirely novel organic substances, and so until recently no species has had previous experience of them in its evolutionary history. Their particular interest lies in the fact that they are a new environmental phenomenon.

Research on the ecological effects of pesticides concerns many applied biologists working on agricultural and conservation problems, and it should also be of great interest to a wide range of workers concerned with theoretical aspects of ecological and evolutionary biology.

THE EFFECTS OF PESTICIDES AND THEIR STUDY

Usually the biocidal effects of pesticides have been discovered empirically—by screening a vast number of compounds which happen to be available to industry. No pesticide has been especially designed to deal with a particular pest, and therefore it is not surprising that a list of pesticides covers a wide range of heterogeneous substances, both inorganic and organic, naturally occurring and synthetic, and that none are specific to target species. The most commonly used pesticides today are synthetic organic substances, and many of these are fairly selective. For example most herbicides have low toxicities to animals, several kill monocotyledonous weeds but not dicotyledonous ones and *vice versa*, and many insecticides have very low toxicities to mammals and birds. Of course all pesticides are toxic to some species or they would not be used.

Studies of pesticide effects emphasize the importance of studying cause and effect at different levels of integration and the need to relate the results at different levels to each other. Most studies have concerned physiology and autecology, but an increasing number are being carried out at lower and higher levels—on the cell and the ecosystem. Some, but not enough, work has been done to integrate toxicology and population dynamics.

Much information exists on the acute oral and dermal toxicity and on the chronic toxicity of pesticides to laboratory mammals, because this is required for registration schemes designed to protect the human operator and consumer and also livestock. Pesticides frequently have multiple pathological effects; in some cases the primary mechanism is quite well known, as in the large family of organophosphorus insecticides which act by inhibiting acetyl cholinesterase. In others, such as DDT, the mechanism is not known, but knowledge of biochemical mechanisms is not usually necessary for understanding population effects. On the other hand, studies on the toxicology of pesticides are essential for this purpose, because if we are to determine the effects of a pesticide in the field we must understand the toxicological significance of the doses which organisms in the population are likely to receive. The single dose administered orally which kills 50 per cent of the experimental population of laboratory animals (the acute oral LD_{50}) is the most widely used yardstick, but since organisms in the field are subject to stresses additional to those provided by the pesticide, the investigator into field effects must also have information about sublethal toxicological effects, especially when they involve reproduction or behaviour. The fundamental question for the agricultural and conservation biologist is: what effect does the pesticide have on the population? While the investigator is generally concerned with one particular species, or a small range of species in the ecosystem, no species lives in isolation, and so the species studied must be considered in relation to others within the ecosystem.

Predators always occur at lower numbers than their prey, and therefore, together with other species whose populations are small, they tend to be at greater risk from pesticides than abundant species. For the same reasons, they tend to recover more slowly from the effects of pesticide applications. In addition, predators are at a further disadvantage when they feed upon resistant prey species which have accumulated residues of pesticides in their fat. Thus the general effect of pesticides, like other pollutants, is to reduce species diversity, and through the differentially severe effects on predators they also allow an increase in the numbers of resistant phytophagous and saprophagous species. Hence pesticides tend to increase productivity at the lower trophic levels and to lessen it at high ones. Their effects on the productivity of whole ecosystems have been little studied but are likely to vary.

Pesticides affect a given species in the field both directly and indirectly. Direct effects are those due to the lethal or sublethal toxicity of the pesticide to the organism itself. Indirect effects are due to the toxicity of the pesticide to other species in the ecosystem which affect the organism, notably competitors, predators and prey. Ultimately, population numbers of a species reflect both direct and indirect effects and their interaction. Unless all the individuals in the population are killed by the pesticide its effects are likely to be complicated and difficult to unravel.

Two special methodological difficulties arise in studies on pesticide effects in the field. First, one often needs to know the toxicological effects of a pesticide on a species which cannot easily be kept or bred under laboratory conditions, and therefore one can only make inferences from laboratory species. Unfortunately, even closely related species can vary considerably in their response to the same chemical, and so extrapolations from toxicological studies on laboratory species to wild species can rarely be exact and may be misleading. For example, grey geese are much more sensitive than are pigeons, game birds and even Canada geese to the organophosphorus insecticide carbophenothion. Thus, while this substance has proved to be an excellent substitute for dieldrin cereal seed dressing (see p. 166) in most of England where there are few grey geese, its use in Scotland, where wintering geese are very numerous, has led to the deaths of many greylag and pinkfoot geese there. Accordingly, an arrangement with the manufacturers of the chemical to restrict the sale of carbophenothion to England had to be made; this has successfully prevented further deaths of grey geese in Scotland (Stanley and Bunyan, 1979). Incidentally, this case clearly demonstrates that no practicable registration scheme can be entirely foolproof: sale of pesticides must depend to some extent on experience in the field after registration.

The second methodological difficulty concerns experimental studies of populations. It is quite easy to carry out a controlled experiment on the effects of a pesticide on species with very dense populations, for example soil animals, but it is practically impossible to do such experiments on animals with widely dispersed populations, for example birds of prey. Conclusions on pesticide effects on these species have to be based on circumstantial evidence derived from rigorous observations on different responses to different pesticide situations in the field, and must be related to basic knowledge of the biology of the species.

SOME ECOLOGICAL EFFECTS OF PESTICIDES ON
ORGANISMS IN CROPS

No study has been made of all the effects of a pesticide application to a crop, but studies of pesticide effects on parts of the crop ecosystem give valuable insights into the processes involved. Two examples are given below.

In a series of papers, Dempster (1967; 1968a, b, c) described the interrelationships between the cabbage white butterfly, *Pieris rapae*, and other species in a crop of Brussels sprouts, and the effects of DDT upon them. He found that *Pieris* larvae were preyed upon by beetles, harvestmen and birds, and were parasitized by Hymenoptera and Diptera. They also suffered from a fungus and a granulosis virus. He constructed life tables for three generations of *Pieris* and found that there was a mortality of about 90 per cent between egg and pupal stages, that over half of this mortality occurred during the first two larval instars, and that it was due mainly to the ground beetle *Harpalus rufipes* and the harvestman *Phalangium opilio*. Birds took about 20 per cent of the larvae and their predation was only important during the last two larval instars. The effects of parasites and disease were insignificant. Bad weather had little effect on mortality, but it did affect the fecundity of *Pieris*.

DDT is a broad-spectrum insecticide of unusual persistence. Its concentration on the crop itself decreases as the sprayed leaves expand and new unsprayed ones develop. On the other hand, it remains concentrated in the soil, especially if it is ploughed in. Therefore, direct effects on animals living on the crop and on weeds are likely to be less than those on animals living partly or wholly in the soil. Dempster found that the numbers of the ground beetle *Harpalus rufipes*, millipedes, centipedes and arachnids decreased. DDT gave a good initial control of *Pieris*, but eventually caused an increase in the populations of this and other species, notably cabbage aphids, Collembola and the ground beetles *Trechus quadristriatus* and *Nebria brevicollis*.

The initial decline of *Pieris* was clearly due to the action of the pesticide (Figure 10.1). The subsequent increase of the species was due to immigrants laying eggs after the crop had been sprayed and to reduced predation on the larvae emerging from these eggs by *Harpalus rufipes* and *Phalangium opilio*, because these ground-living species had been particularly severely affected by the spray. DDT not only caused death, but also reduced the rate of feeding of those beetles which had survived treatment. Similarly, the increase of Collembola appeared to be due to deleterious effects on their predators. The increase of *Trechus* and *Nebria*, two species which appeared to be comparatively resistant to DDT, was probably due to the increase of the Collembola on which they fed. Many of the species feeding on the crop were able to recolonise it from surrounding areas, but the persistence of DDT in the soil prevented successful recolonization by ground-living species. As a result, the effects of DDT on the predators of *Pieris* continued to operate long after the application of the insecticide and

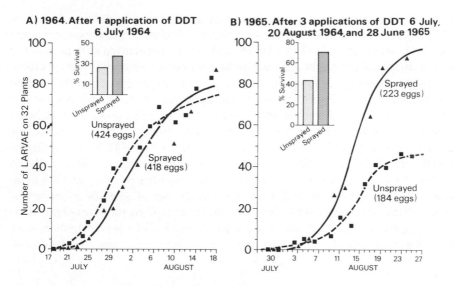

FIGURE 10.1　An illustration of the differential effects of DDT on crop and soil fauna. *Pieris* lives on crops; spraying with DDT to control it is effective for only a very short period in the first year (A). Because the soil-living predators of *Pieris* are affected by residuals in the soil, *Pieris* in fact increases markedly after repeated spraying (B). After Dempster (1968a).

enabled *Pieris* to increase even more in the second season (Figure 10.1). This admirable study illustrates the interplay of direct and indirect effects of an insecticide and the complexity of effects due to differential toxicity and differential persistence. It also showed that persistent organochlorine insecticides are inefficient tools for controlling brassica pests. Initially the synthetic pyrethroid insecticides seemed to provide useful substitutes for DDT, but unfortunately it has been discovered that some of them also have a differentially harmful effect on predatory anthropods, and so can encourage new pests. They are also extremely hazardous to aquatic organisms, and show cross-resistance with organophosphorus insecticides. For all these reasons they are proving less valuable as substitutes for DDT than was originally hoped.

Herbicides are used very extensively, but their ecological effects have been much less studied than those of insecticides. The farmer uses herbicides in order to reduce competition between crop and weed and to ease the handling of the harvested crop. Plants vary in their sensitivity to different herbicides: the use of one herbicide eliminates some species of weed, but enables other more resistant species to flourish. For example, growth regulating herbicides like MCPA have greatly reduced the population of the poppy in eastern England, but indirectly have caused a great increase in the population of the wild oat. Herbicides have caused conspicuous changes in the weed flora of countries like Britain, and these in turn must have caused great changes in the populations of phyto-phagous invertebrates and hence in predators and parasites dependent upon them. The work of Potts (1970; 1971; 1980) on the population dynamics of the grey partridge throws light on some of the changes which occur in cereal crops.

The grey partridge is a game bird of considerable economic importance, and much is known about the fluctuations in its numbers in Britain from game bag reports made during the last century. These show that its numbers have always fluctuated considerably, but that in recent years there has been a continuous general decline of a kind not previously described (Figure 10.2), the causes of which have been the subject of Potts' work. The partridge is essentially an herbivorous species which is preyed upon by a number of birds and mammal species as well as by man, and it is subject to a range of diseases. Mortality of young chicks was found to be the key factor in controlling the short-term varia-tions of partridge populations. The chicks require high-level protein diets at an early stage of their development, and they obtain them largely from insects.

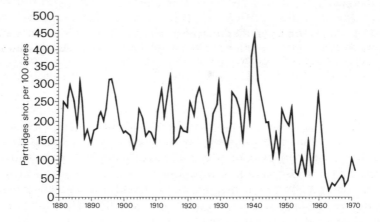

FIGURE 10.2.   Grey partridge numbers per 100 acres on 20 'best' estates; figures from game bag records. The recent general decline appears to be due to a number of changes in cereal growing practices such as the introduction of herbicide sprays, and the abandon-ment of undersown leys. After Potts, 1971.

Exceptional mortality was found to be due to a combination of two factors—shortage of insect food (mostly caused by inclement weather in April and May) and poor weather in June and July. Cold wet weather increased the need for food, this lengthened the search for it, and, if insects were not easily available, increased the danger of heat loss while the chicks were searching.

The decline of the partridge in recent years was found to be due to the combi-nation of below-average seasons, especially cold springs, and changes in cereal growing. One of these changes was the use of herbicides, which was shown to cause a 70 per cent reduction of the biomass of the insects living on weeds, and the other was the reduction in undersown leys, which support the overwintering state of the leaf-eating sawflies whose larvae are a particularly important food for young partridges. Other factors were not insignificant in some cases: if predators, especially carrion crows, stoats and weasels were controlled, and if grass strips were left to provide alternative food, shortage of food in the crop and bad weather had less harmful effects. In this case the use of herbicides of low toxicity has had important indirect effects not only on the invertebrate fauna, but also on a vertebrate species which is dependent on the invertebrates for a short time in its life history. This very promising study is still in progress.

During the first 12 years of the study, which is done on farms where pesticides are used extensively as a common practice, most of the arthropod species studied have declined although none of these is a pest species. The declines appear to be due to the indirect effects of herbicides and the direct effects of insecticides and insectidal fungicides. The work shows that an arthropod fauna with a high diversity has a larger proportion of entomophagous individuals than one with a low diversity. Indeed, the outbreaks of cereal aphids are correlated inversely with species diversity (Potts, 1970; 1971; 1980). As so often happens, a thorough autecological study on a wild species throws light on other aspects of the ecosystem, and has profound economic implications. For many years it has become obvious that the total effects of pesticides on agroecosystems should be studied. This entails making long-term studies on quite large areas of land with multidisciplinary research teams. That so little has been done so far seems mainly due to constraints imposed by the organization of science funding; no one organization is able to mount the operation on its own, and effective inter-departmental mechanisms for tackling the problem have not yet been devised.

Hitherto, the agricultural use of pesticides has been the most important in the United Kingdom. However, as forestry crops of conifers mature, the likelihood of serious outbreaks of forest pests increases. Already the larva of the pine beauty moth (*Panolis flammea*) has devastated hundreds of hectares of lodgepole pine (*Pinus contorta*) plantation in Scotland. This led to emergency treatment with the organophosphorus insecticide fenitrothion from the air. The potential risks to wildlife, including important salmon and trout fisheries, have led to joint research between the agricultural departments, the Forestry Commission, the Nature Conservancy Council and the Royal Society for the Protection of Birds in order to minimize such risks (Holden and Bevan, 1979). Two points of general interest arise from this work.

First, forestry in the United Kingdom provides great opportunities for comparing areas affected by pesticide treatments with untreated areas. Such a comparison is no longer possible in agriculture since nearly all agricultural land has been affected by pesticides—there are no longer 'natural' controls.

Second, work on the physics of pesticide application in connection with aerial spraying in Scotland has shown that by controlling droplet size and the way the pesticide is released from the aircraft, the application rate of active ingredient can be greatly reduced (Holden and Bevan, 1979). This suggests that the way pesticides are applied can be as important as their toxicity or persistence. It is well known that seed-dressed corn can be particularly hazardous to birds and mammals if exposed on the surface of the soil because it acts as a poison bait. Granules appear to be less hazardous. Drift and volatilization can cause damage to neighbouring crops, but so far very little research has been done on their environmental effects. Therefore, much more work should be done on application techniques both for agricultural and environmental reasons.

SOME ECOLOGICAL EFFECTS OF PESTICIDES ON ORGANISMS LIVING OUTSIDE THE SPRAYED AREAS

A small number of pesticides are broken down so slowly that they have time to spread outside the areas of application; if they have deleterious effects they can be classified as environmental contaminants. Some persistent pesticides, notably the dipyridyl herbicides paraquat and diquat and the triazine and substituted urea herbicides, remain bound up in the soil and do not appear to be released under normal circumstances; they seem to have little effect on the environment. Those which cause most concern as environmental pollutants are fat-soluble organochlorine insecticides, notably DDT and the cyclodienes aldrin and dieldrin. These compounds or their stable metabolites break down so slowly that they become transported throughout the world in air, water and in organisms. Their high fat solubility allows them to become concentrated in animal bodies and hence they get passed from prey to predator. Chemical analyses of persistent organochlorine insecticides in British birds and their eggs in the early 1960s showed that, on average, predators which fed on other vertebrates contained more pesticides than did herbivorous and insectivorous species (Moore and Walker, 1964). Thus if persistent organochlorine insecticides were having an effect on any bird populations, birds of prey were most likely to show these effects, and accordingly they have been the subject of much research in Britain, the United States and elsewhere. The studies of Ratcliffe, Prestt, Jefferies, Lockie, Newton and others have been particularly illuminating because so much background information is available about the species which they studied. The initial studies showed that two fish-feeding species, the great crested grebe and the heron, two bird-feeding species, the peregrine and the sparrowhawk, and two species which fed on both mammals and birds, the kestrel and the golden eagle, contained residues of DDE (the principal metabolite of DDT) and dieldrin which were above the average found in British birds as a whole. The responses of the six species differed greatly, a fact which underlines the danger of making sweeping generalizations about pesticide effects from particular instances.

The great crested grebe increased despite the introduction of pesticides: the lethal and sublethal effects on individuals which occurred locally were more than compensated in this case by some feature in the environment which favoured the general increase of the species. This feature was almost certainly the great increase in gravel extraction in postwar England, which produced numerous waterfilled pits, which provide a very suitable habitat for the species (Prestt and Jefferies, 1969).

Herons and their eggs have been found to contain higher levels of dieldrin and DDE on average than any other wild bird investigated. Eggshell thinning and abnormal breeding behaviour which caused males to throw eggs and young out of the nest were observed in heronries in districts subjected to particularly high persistent organochlorine insecticide contamination. Nevertheless, because this species will readily lay replacement clutches, the breeding success of the colonies was not reduced beyond the point at which replacements were available to compensate 'normal' adult mortality. Despite considerable pesticide contamination in recent years the total population has remained remarkably steady since 1928, when the first censuses were made of this species under the auspices of the British Trust for Ornithology. All declines recorded in the heron population of England and Wales have followed abnormally severe winters (Figure 10.3), but numbers have recovered to about 4500 pairs within a few years after each decline. So, in this species, despite severe subacute effects and some acute ones, the breeding population has remained unaffected by pesticides (Milstein, Prestt, and Bell, 1970; Prestt, 1970, Stafford, 1971).

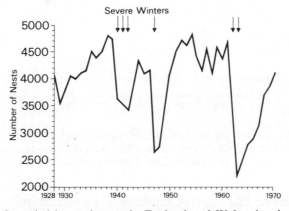

FIGURE 10.3. Occupied herons' nests in England and Wales showing the effects of severe winters and the absence of a distinct long-term trend. After Prestt, 1970.

The golden eagle population has not been affected by acute poisioning by persistent organochlorine insecticides. However, eggshell thinning and a decline in the number of pairs rearing young from 72 per cent to 29 per cent occurred in those areas where the species fed extensively on carrion sheep which had been dipped in dieldrin. In the eastern Highlands of Scotland, where eagles are known to feed on grouse and hares, the species bred normally and eggshells were of normal thickness. The golden eagle is a long-lived species, but eventually the inability to replace natural losses in the western Highlands would have probably caused the species to have died out had the use of dieldrin for sheep-dipping continued. Restrictions on the use of dieldrin as a sheep dip were made in 1966, and since that date the dieldrin content of eagle eggs has declined, and the reproductive success of the species in western Scotland has returned to normal. During the same period restrictions were put on uses of DDT. Subsequently the DDE content of eggs declined and their shells became thicker. In the eastern Highlands, where the eagles did not acquire significant amounts of dieldrin, reproduction remained normal throughout the period of observation (Lockie, Ratcliffe, and Balharry, 1969; Ratcliffe, 1970).

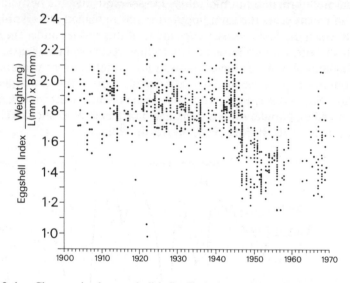

FIGURE 10.4. Changes in the eggshell index (relative weight) of eggs of the peregrine falcon in Britain. The start of the decline coincides with the introduction of DDT in the late 1940s. After Ratcliffe, 1970.

Ratcliffe showed that eggshell thinning in peregrines and sparrowhawks coincided with the introduction of DDT in the late 1940s (Figure 10.4). Nevertheless, the peregrine population was increasing at this time as a result of the protection being reimposed after wartime persecution, when it had been reduced in order to protect carrier pigeons used by the armed forces. The remarkable population decline in this normally very stable species, which still affects it to some degree, did not start until the late 1950s; it coincided with the introduction of aldrin and dieldrin as cereal seed dressings, and was almost certainly due to peregrines eating pigeons containing large but sublethal residues of these substances. The eating of only three such pigeons would be necessary to cause death. Since the restriction on the use of aldrin and dieldrin seed dressings in 1962 the peregrine has made a slow recovery except in areas where its prey is still contaminated. Thus in this species persistent organochlorine insecticides caused both acute and subacute effects, but only the acute effects had a serious effect on the population (Ratcliffe, 1972). The peregrine is one of the very few cosmopolitan bird species, and it is interesting to note that in the United States the decline of one of its races (*Falco peregrinus anatum*) appears to be largely due to sublethal effects of DDT (Hickey, 1969). The histories of the sparrowhawk and kestrel are similar to those of the peregrine; both showed sublethal and acute effects, the latter being the main cause of the declines. In both species recoveries in reproductive success and population numbers were recorded after restrictions had been made on the use of aldrin and dieldrin.

The research briefly summarized above is based on four types of information. First, chemical analyses of pesticide residues in birds and eggs; second, toxicological studies on several species which showed the range of acute toxicity of DDT, DDE and dieldrin to birds, and demonstrated that these substances could have subacute effects, including eggshell thinning, at levels encountered in the field—eggshell thinning had been shown to be mainly caused by DDE, the principal metabolite of DDT; third, field studies were made on the feeding habits and behaviour of the birds concerned; and fourth, total or sample population counts were conducted. Unfortunately, crucial controlled experiments on a bird of prey population are impracticable and have not been attempted in Britain. Nevertheless, if all the evidence from the four sources of information is carefully related, very little doubt remains that the unprecedented declines of some birds of prey in Britain were due to the use of persistent organochlorine insecticides, and their subsequent partial recoveries were due to partial restrictions on the use of these substances. In Britain, acute effects of the very toxic persistent organochlorine insecticides aldrin and dieldrin appear to have been much more important than sublethal effects of these substances or DDT, which is much less toxic. In the United States, on the other hand, where DDT was used much more extensively, the sublethal effects of DDT and its metabolites have been more important. An extreme case is that of the brown pelican. High levels of DDT in some colonies of this species have caused the birds to lay 'shell-less' eggs and so have caused total reproductive failure (Blus *et al.*, 1972; Keith, Woods, and Hunt, 1970).

THE EFFECTS OF POLYCHLORINATED BIPHENYLS AND OF PESTICIDE IMPURITIES AND MISUSE

Brief mention should be made of the polychlorinated biphenyls (PCBs), which have similar characteristics to insecticides like DDT. PCBs were invented before DDT and are used for a wide range of different purposes in many industries.

Like DDT, they have a very low solubility in water, a high solubility in fat, and are very persistent in the environment. Despite their long history of use, their presence in the environment was not recorded until 1966. Subsequent studies (Prestt, Jefferies, and Moore, 1970) showed that they were present in terrestrial, freshwater and marine environments. The presence of PCBs in raptorial birds in Great Britain is shown in Figure 10.5. Laboratory studies showed that they had low acute toxicities to mammals and birds, but recent work indicates that they can have measurable biological effects at low concentrations (Jefferies, personal communication).

There is no evidence to suggest that PCBs have had serious effects on the population of any organism in the United Kingdom. However, they can cause eggshell thinning (Newton, Marquiss, and Moss, 1979), they can have harmful effects on mammalian reproduction and there is evidence that seal populations have been affected by them (Helle, Olsson, and Jensen, 1976). PCBs were found in large amounts in the livers of guillemots which died in the Irish Sea bird disaster in 1969. Studies on the causes of this event are inconclusive, but they suggest that PCBs were only of secondary importance as a cause of death. The incident was important, since it drew attention to the general problem of industrial pollution of the sea. When it had been shown that PCBs were widely distributed in the environment, Monsanto, the principal manufacturer of PCBs in the United States and the United Kingdom, withdrew those uses of PCBs whose entry into the environment could not be controlled. Nevertheless, PCBs are very persistent substances and so are likely to remain in the environment for some time, but their levels should decline as the result of this action.

In recent years many organisms have come into contact with several new environmental contaminants at the same time, notably organochlorine insecticides, organomercury fungicides (Saha, 1972) and PCBs. Very little is known about the effects of combinations of these compounds. Research on their effects has caused governments and industry to keep a closer watch on other substances which have similar physical and chemical characteristics in case they are, or might become, significant environmental contaminants.

A rather different environmental problem is provided by pesticide impurities. Many pesticides contain small amounts of these; usually they are less toxic than the pesticide, or the amounts are too small to have biological effects. An exception is the dioxin impurity of the herbicide 2,4,5-T. This impurity is much more toxic than the herbicide. Laboratory and field evidence shows that it can cause teratological effects. An unusually impure form of 2,4,5-T was used in defoliation campaigns in Vietnam, and this drew attention to the hazards of pesticide impurities. In recent years improved analytical techniques have allowed more accurate assessments of dioxin content of 2,4,5-T to be made. The subject of possible hazards arising from ordinary use has been reviewed exhaustively in the United Kingdom, EEC and the United States. In the United Kingdom the dioxin content of all 2,4,5-T used is checked by the Laboratory of the Government Chemist. Both toxicological work and long experience in the field shows

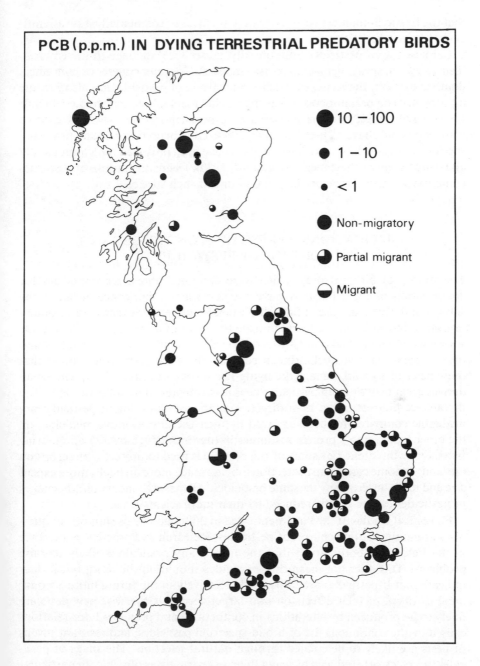

FIGURE 10.5. The map shows the widespread occurrence of polychlorinated biphenyls in areas well away from their industrial sources and the higher levels found in the more resident birds. After Prestt, Jefferies, and Moore, 1970.

that the risk to human beings is extremely small. Environmentally it is insignificant (Kilpatrick, 1979).

Careless use of pesticides undoubtedly causes local damage to the environment, as when spraying machinery is cleaned out in water courses or containers dumped in them. In recent years deliberate misuse of pesticides has also become a problem. The organophosphorus insecticide mevinphos, strychnine which is used to kill moles, and the stupefying agent α-chloralose, which is used to control pigeons, have all been used illegally to control predators. Rare protected species such as red kites and golden eagles have been poisoned both accidentally and deliberately by these means (Cadbury, 1980). Deliberate misuse of pesticides could have severe effects on the populations of such species.

IMPLICATIONS OF RESEARCH ON THE USE AND CONTROL OF PESTICIDES

The studies of birds of prey which were described above are important for conservation biology, but they have a much wider significance because they show that the ordinary use of fat-soluble persistent pesticides can affect populations of non-target species living outside the areas of application. When it was shown by Moore (1965) that fish-feeding seabirds contained appreciable quantities of persistent organochlorine insecticides there was justifiable concern that these pesticides could damage the living resources of the sea. Thus, while man derives great benefit from pesticides he is also threatened by some of them. The threats are indirect rather than direct: human casualties due to pesticides are negligible compared to the lives saved by their use in preventive medicine; on the other hand, man's protein resources in the form of freshwater and marine fish may be threatened by some of the pesticides used to protect plant crops on the land. In some cases crop protection is being made more difficult, more expensive and more uncertain by the same pesticides. All these secondary disadvantages of pesticides can be greatly reduced by their more scientific use.

Ecological studies of the kind mentioned in this chapter raise important questions about the use and control of pesticides. Research and practical experience in the field have both shown that the use of any pesticide is likely to raise problems. This does not mean that pesticides should not be used, but it does indicate that if pesticides are to be used effectively in the future more account must be taken of their effects on non-target species. Otherwise new pests are likely to be produced by alterations in competitive and prey/predator relations caused by the continuous use of broad-spectrum pesticides; and resistant strains of pests are likely to be created through natural selection. The snags of pesticides can be greatly reduced by using them as sparingly as possible, in particular by refraining from 'insurance spraying', by using persistent compounds as little as possible, and by developing more specific pesticides. Unfortunately, short-term economic pressures frequently prevent farmers from undertaking these commonsense measures, and chemical firms and agricultural advisers from supporting them.

The protection of consumers, contractors, domestic animals and wildlife from the deleterious effects of pesticides largely depends on assessments of risks which are based on acute and chronic toxicity tests. Insofar as direct effects on man are concerned, the procedure can be made to work efficiently simply by applying a large safety factor, but for animals in the field where many individuals are in contact with acute and subacute doses and indirect effects are important, overall effects cannot be predicted from toxicological studies alone; frequently the important effect is due to the combination of a pesticide and some other deleterious environmental factor. Therefore, insofar as effects in the field are concerned, the correct appraisal of a pesticide's value and its hazards must depend on field experience. Most of the predictions about pesticides which were based on toxicological tests and a limited number of field trials, have stood the test of time, although in the important, but exceptional cases of the persistent organochlorine insecticides and the organomercury fungicides, restrictions have had to be placed on their use as the result of experience in the field.

Most of the research on pesticides has been done to increase their value in agriculture, forestry and preventive medicine, to ensure their safety to man, and to reduce harmful side-effects to wildlife, including species such as fish which are important to man as food. Perhaps because of these practical considerations the extraordinary fundamental interest of pesticide effects is often overlooked. The speed with which resistant strains are selected and the delicacy and complexity of the competitive relationships which have been demonstrated by the use of selective pesticides are particularly notable. Pesticides are potentially valuable as tools in ecological research: for example the insecticide dieldrin has been used to dissect an ecosystem in order to test a hypothesis about the ecological role of a constituent species (Davis *et al.*, 1969). If more selective pesticides were invented, they would become much more useful for such purposes. Conservation biologists have hitherto been mainly concerned with discovering means of avoiding the harmful effects of pesticides, but some work has already been done which shows that herbicides can be useful as conservation management tools, principally in preventing scrub regrowth in plagioclimax grasslands (Moore, 1968). Similarly, introduced mammals which have become pests in vulnerable ecosystems on small islands can be reduced or eliminated by the judicious use of rodenticides.

The need for pesticides will increase so long as increases in the human population and raised standards of living cause an increased demand for higher outputs per hectare. If full use is to be made of pesticides, and if serious hazards to the environment are to be prevented, we need to understand their complicated effects much better than we do today. The necessary research must be interdisciplinary and is likely to be scientifically rewarding. Conservation of man and of wild organisms is, and should be seen to be, complementary.

REFERENCES

Blus, L.J., Gish, C.D., Belisle, A.A., and Prouty, R.M. (1972). Logarithmic relationship of DDE residues in Eggshell thinning. *Nature (London)*, **235**, 376–377.

Cadbury, C.J. (1980). *Silent Death*, Royal Society for the Protection of Birds, Sandy, pp. 1–25.

Davis, B.N.K., Moore, N.W., Walker, C.H., and Way, J.M. (1969). A study of millipedes in a grassland community using dieldrin as a tool for ecological research. In J.G. Sheals (Ed), *The Soil Ecosystem*, Systematics Association, London, Publ. no. 8, pp. 217–228.

Dempster, J.P. (1967). The control of *Pieris rapae* with DDT. I: The natural mortality of the young stages of *Pieris*. *Journal of Applied Ecology*, **4**, 485–500.

Dempster, J.P. (1968a). The control of *Pieris rapae* with DDT. II: Survival of the young stages of *Pieris* after spraying. *Journal of Applied Ecology*, **5**, 451–462.

Dempster, J.P. (1968b). The control of *Pieris rapae* with DDT. III: Some changes in the crop fauna. *Journal of Applied Ecology*, **5**, 463–475.

Dempster, J.P. (1968c). The sublethal effect of DDT on the rate of feeding by the ground-beetle *Harpalus rufipes*. *Entomologia Exp Appl*, **11**, 51–54.

Helle, E., Olsson, M., and Jensen, S. (1976). PCB levels correlated with pathological changes in seal uteri. *Ambio*, **5**, 261–263.

Hickey, J.J. (Ed) (1969). *Peregrine Falcon Populations: Their Biology and Decline*, University of Wisconsin Press, Madison, 596 pp.

Holden, A.V. and Bevan, D. (1969). *Control of Pine Beauty Moth by Fenitrothion in Scotland 1978*, Forestry Commission, Nuffield Press, Oxford.

Keith, J.O., Woods, L.A., and Hunt, E.G. (1970). Reproductive failure in brown pelicans on the Pacific coast. *Transactions of the 35th North American Wildlife Conference*, pp. 56–63.

Kilpatrick, R. (1979). *Review of the Safety for Use in the UK of the Herbicide 2,4,5-T*, Advisory Committee on Pesticides.

Lockie, J.D., Ratcliffe, D.A., and Balharry, R. (1969). Breeding success and organochlorine residues in golden eagles in West Scotland. *Journal of Applied Ecology*, **6**, 381–389.

Milstein, P. le S., Prestt, I., and Bell, A.A. (1970). The breeding cycle of the grey heron. *Ardea*, **58**, 171–257.

Moore, N.W. (1965). Environmental contamination by pesticides. In G.T. Goodman (Ed), *Ecology and the Industrial Society: A Symposium of the British Ecological Society*, Swansea, 1964, Blackwell, Oxford, 219–237.

Moore, N.W. (1968). The value of pesticides for conservation and ecology. In N.W. Moore and W.P. Evans (Eds), *Some Safety Aspects of Pesticides in the Countryside*, Joint ABMAC/Wild Life Education and Communications Committee, London, pp. 104–108.

Moore, N.W. and Walker, C.H. (1964). Organic chlorine insecticide residues in wild birds. *Nature (London)*, **201**, 1072–1073.

Newton, I., Marquiss, M. and Moss, D. (1979). Habitat, female age, organochlorine compounds and breeding of European sparrowhawks. *Journal of Applied Ecology*, **16**, 777–793.

Potts, G.R. (1970). Recent changes in the farmland fauna with special reference to the decline of the grey partridge. *Bird Study*, **17**, 145–166.

Potts, G.R. (1971). Agriculture and the survival of partridges. *Outlook on Agriculture*, **6**, 267–271.

Potts, G.R. (1980). The effects of modern agriculture, nest predation and game management on the population ecology of partridges, *Perdix perdix* and *Alectoris rufa*. *Advances in Ecology Research*, **11**, 1–79.

Prestt, I. (1970). Organochlorine pollution of rivers and the heron (*Ardea cinerea L.*). Papers from the *Proceedigns of the Technical Meeting of the International Union for Conservation and Natural Resources, 11th, New Delhi*, 1969, 1, pp. 95–102, International Union for Conservation and Natural Resources, Morges.

Prestt, I. and Jefferies, D.J. (1969). Winter numbers breeding success and organochlorine residues in the great crested grebe in Britain. *Bird Study*, **16**, 168–185.

Prestt, I., Jefferies, D.J., and Moore, N.W. (1970). Polychlorinated biphenyls in wild birds in Britain and their avian toxicity. *Environmental Pollution*, **1**, 3–26.

Ratcliffe, D.A. (1970). Changes attributable to pesticides in egg breakage frequency and eggshell thickness in some British birds. *Journal of Applied Ecology*, **7**, 67–115.

Ratcliffe, D.A. (1972). The peregrine population of Great Britain in 1971. *Bird Study*, **19**, 117–156.

Saha, J.G. (1972). Significance of mercury in the environment. *Residue Review*, **42**, 103–163.

Stafford, J. (1971). The heron population of England and Wales, 1928–1970. *Bird Study*, **18**, 218–221.

Stanley, P.I. and Bunyan, P.J. (1979). Hazards to wintering geese and other wildlife from the use of dieldrin, chlorfenvinphos and carbophenothion as wheat seed treatments. *Proceedings of the Royal Society of London B*, **205**, 31–45.

BIBLIOGRAPHY

Cooke, A.S. (1973). Shell thinning in avian eggs by environmental pollutants. *Environmental Pollution*, **4**, 85–152.

Fryer, J.D. and Evans, S.A. (Eds) (1970). *Weed Control Handbook, I: Principles*, Blackwell, Oxford, 494 pp.

Fryer, J.D. and Makepeace, R.J. (Eds) (1972). *Weed Control Handbook, II: Recommendations*, Blackwell, Oxford, 424 pp.

Martin, H. (Ed) (1964). *The Scientific Principles of Crop Protection, 5th Edition*, Edward Arnold, London, 376 pp.

Martin, H. (1969). *Insecticide and Fungicide Handbook for Crop Protection, 3rd Edition*. British Crop Protection Council, Blackwell Scientific Publications, Oxford and Edinburgh, 387 pp.

Mellanby, K. (1970). *Pesticides and Pollution, 2nd Edition*, Collins New Naturalist, London, 221 pp.

Moore, N.W. (Ed) (1966). Pesticides in the environment and their effects on wildlife. *Journal of Applied Ecology*, **3 (Suppl)**, 311 pp.

Moore, N.W. (1967). A synopsis of the pesticide problem. *Advances in Ecology Research*, **4**, 75–129.

Moriarty, F. (1969). The sublethal effects of synthetic insecticides on insects. *Biological Review*, **44**, 321–357.

Moriarty, F. (1975). *Pollutants and Animals. A Factual Perspective*. George Allen and Unwin, London, 140 pp.

Muirhead-Thomson, R.C. (1971). *Pesticides and Freshwater Fauna*, Academic Press, London, 248 pp.

Newton, I. (1979). *Population Ecology of Raptors*. T. and A.D. Poyser, Berkhamsted, 399 pp.

Ratcliffe, D.A. (1980). *The Peregrine Falcon*. T. and A.D. Poyser, Berkhamsted, 399 pp.

Rudd, R.L. (1964). *Pesticides and the Living Landscape*, Faber and Faber, London, 320 pp.

Worthing, C.R. (1979). *The Pesticide Manual. A World Compendium, 6th Edition*. British Crop Protection Council, Croydon, 655 pp.

CHAPTER 11

The Restoration of Mined Land

A.D. BRADSHAW

THE PROBLEM

Of all the activities of man that affect land, plants and animals, mining must be the most destructive. In the search for minerals of all sorts, from sand and gravel to base metals, not only are the vegetation, the animals and the soil eliminated but also often the whole landform. The original ecosystem is destroyed and in its place appears an empty pit, a sterile waste heap, or both.

Against this gloomy backcloth must be set various further points:

(1) It is impossible to do without mining. Ever since the Stone Age the land has provided us with a large proportion of the major materials which enable us to enjoy a comfortable existence. Although, some, such as copper and lead, may soon be exhausted, their substitutes such as aluminium and glass, also come from the ground. And for energy, although it is possible that instead of coal we shall eventually become reliant on other resources, both in Britain and in the world as a whole, coal resources will last for several hundred years longer.

(2) It is in the nature of all mining operations that they are only temporary uses of land. For geological reasons every mined resource is finite, so that it will eventually be mined out. Some mining operations, particularly surface mining for materials like coal or bauxite, where the deposits are no more than a few metres thick, the ground is mined and finished within a few months. The operation itself proceeds across further land, but then even this terminates (Bradshaw and Chadwick, 1980). Few deep mines have continued for longer than 50 years.

(3) For the same reasons it is in the nature of all mining operations that they eventually, as individuals, go bankrupt. When they are eventually mined out, they cease to earn income and can provide no money to pay for final restoration work unless this has been put on one side from earlier income.

(4) Although all present-day soils have been derived from original raw

mineral materials, the natural processes of soil development take a long time. The evidence provided by soil chronosequences suggests that 10 000 years is often necessary for the development of a mature soil profile from a pedological standpoint (see Warren, Chapter 2). A soil/plant system, however, develops more quickly, and soon there is an equilibrium amount of a well-weathered mineral soil that also contains organic matter and organically-bound nutrients. Nevertheless, even this takes about 100 years whether on natural soil materials (Crocker and Major, 1955; Olson, 1958) or on old mined land (Leisman, 1957; Smith, Tryon, and Tyner, 1971). Vegetation will take the same time to develop because we are dealing with primary and not secondary successions, in which the vegetation is controlled by the state of soil development.

(5) We depend on fertile land for nearly all our food. It would be difficult for the present human population to be maintained if the present amount of fertile land was substantially reduced. Yet land is a finite resource being diminished by the spread of industry and urbanization (Coleman, 1977), of which mining accounts for a loss of about 1600 ha year^{-1} in Britain and 50 000 ha year^{-1} in the United States.

(6) Mining activity has already destroyed large areas of land, much of which has not been reclaimed and has had little time to regenerate its own soil and vegetation. So we have a negative heritage of derelict land in Britain (*derelict land* is defined by the British government as land so damaged by industrial or other development that it is incapable of beneficial use without treatment) of approximately 55 000 ha in 1971. For the United States the figure is 1.5 million ha in 1974. At the same time, areas of dereliction lead to degradation of the land around them, as original land uses become impractical (Wallwork, 1974). It seems possible that, as a result, the final totals of damaged land may be twice these figures.

THE REQUIREMENTS

From these observations a number of conclusions follow, which should determine our attitudes to mining. For the United Kingdom many, although not all, of these conclusions have been emphasized in two reports by government committees on minerals and on sand and gravel (Department of the Environment, 1975a and b) reviewing the operation of the basic Town and Country Planning 1971 Act; these resulted in the Town and Country Planning (Minerals) 1981 Act. The conclusions have been echoed in a recent report on the coal industry (Commission on Energy and the Environment, 1981). In the United States the crucial step has been the Federal Surface Mining Control and Reclamation Act 1977.

Under present economic circumstances few countries are likely to refuse to allow the development of the valuable resources they possess. The impact of mining cannot normally be transferred, although sometimes wastes can be deposited away from the site of origin if economics permit, as when fuel ash from Trent Valley power stations using Yorkshire coal is transferred to Bedfordshire brick pits. Some resources such as sand and gravel or aggregates derived from hard rocks can be mined in a number of places. But if the mining is refused in one place as a result of public protest, it will only take place somewhere else.

Land must, therefore, be restored to its original state or reclaimed to a new use after mining. Here there are some surprises, because while it may seem obvious that full restoration of the original land use is what should normally be required, this may not always be practical, and also not necessarily always the most sensible use for the land. Mining is a total disturbance—the original land use is extinguished and often the original landowners have been bought out.

TABLE 11.1 The varieties of options which can be sensibly considered for different sorts of mined land.

		Waste heaps		Excavations	
		original	levelled	original	filled
Production	agriculture arable		+		+
	grassland	+	+	+	+
	horticulture	+	+		+
	forestry	+		+	
Amenity	parks	+		+	+
	camp sites	+		+	+
	golf courses	+	+		+
	educational areas	+		+	+
	nature reserves	+		+	+
Other use	water storage			+	
	refuse disposal			+	
	accommodation		+	+	
	housing		+	+	
	industry		+	+	

The topography may also have been radically changed. Although perhaps the land, as we shall see, could be restored to what it was, there could well be other land uses, which might not only be economically more sensible, but actually more useful to the general public. A gravel pit may best become an area for water sports or nature conservation, a quarry may be best as a campsite, a colliery spoil heap may be best as a wooded recreation area. There are a whole range of exciting and practical options (Table 11.1) which we would be foolish to forget.

Treatment of the mined land should be progressive. It is easy to fall into the trap of thinking that tidying up is something that is done when the job is finished. But if this is applied to mining, the result is that the effects of the mining, whether pits or waste heaps, will accumulate, so that they will provide an increasing long-term, environmental impact and loss of land. At the same time the total restoration costs occur at the end of the mining period, when the operation is in the worst position to be able to finance them. The financing of restoration should be carried by the industry itself when it is best able to pay for it. A separate consideration is that, unless there are special operational reasons, it is quite wrong that the mess of a mining operation should be allowed to inflict itself on people and the environment for a full 20 or 50 years, if that were the lifetime of the operation. If we believe that our children should enjoy the land as we did, it would be reasonable to argue that no adverse environmental impact should last longer than half a childhood, i.e. 5 years. In Germany, in particular, this approach is the norm (Bradshaw and Chadwick, 1980). But in Britain it is not so, although it is now being proposed for coal mining (Commission on Energy and the Environment, 1981) (Figure 11.1).

Wherever possible, the original soil should be retained. Since a biologically effective soil takes a long time to develop it is sensible not to throw away this valuable resource. This will include not only the topsoil, the A horizon, which contains most of the available nutrients, but also the subsoil, the B horizon, since this is weathered material which will usually be superior to underlying unweathered materials normally called overburden. However as we shall see, where the soil has been lost, an effective soil can be developed from raw wastes and overburden with assistance in not too many years, although whether or not it is as productive as the original soil depends on the nature of the starting material. So there may be a number of operations where it would not be unreasonable for the original soil to be lost. Much must depend on economics and the chosen after-use, although the economics must take into account the long-term rather than the short-term situation.

Whatever happens, the land must be restored in the sense that its biological potential is restored. For agriculture and forestry this means restoring its original productivity. But as the result of recent research and development, there are now many ways that this can be done (Bradshaw and Chadwick, 1980; Gemmell, 1977; Schaller and Sutton, 1978). What is crucial is that decisions on final land use and the means by which the land will be restored are made early in the life of the mining development, indeed at the planning stage. It is here that the value of the much debated environmental impact assessment procedure could be considerable (Roberts and Roberts, 1983). Despite the slight extra work involved, the procedure can ensure that all the options are considered and evaluated properly. In the end it may be possible to show that the mining operation, with appropriate restoration afterwards, leads to a more attractive, more useful, area of land than once existed. This will become apparent when we consider the restoration techniques which are available.

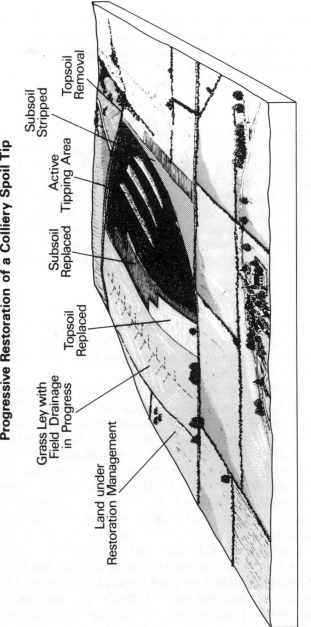

Progressive Restoration of a Colliery Spoil Tip

Subsoil
Stripped

Topsoil
Removal

Active
Tipping Area

Subsoil
Replaced

Topsoil
Replaced

Grass Ley with
Field Drainage
in Progress

Land under
Restoration Management

FIGURE 11.1. Redrawn from Commission on Energy and the Environment (1981).

COMPLETE REPLACEMENT

The first type of restoration we must consider is that in which all the soil which was present before mining is lifted, stored if necessary, and then put back after the mining operation has been completed exactly as it was beforehand, and the original vegetation reinstated.

Soil replacement

In a progressive strip mining operation where a thin deposit is being worked from the surface over a large area by an open cut, the surface coverings have to be stripped and subsequently redeposited. In the past, for economy, all the materials were excavated together and deposited afterwards in the simplest way possible, so that the end product was a series of parallel ridges of material. As machinery has become bigger (modern draglines can have a jib 90 m long) so the hills and dales have grown correspondingly—50 m crest to crest with materials at the natural angle of rest of about 40°. Such terrible dereliction still extends over many hundreds of hectares, in Kentucky, West Virginia and Pennsylvania in particular.

As legislation has improved, first the land was required to be partially, then completely, levelled. But with the soil lost, the fertility is also lost and the land will take several decades to recover its productivity. As this became clear, in America, Britain and other countries, legislation has been enacted to require the replacement of top soil, subsoil and overburden separately. This is not technically difficult, although it does cost money.

The topsoil is stripped first by box scraper, then the subsoil, either by box scraper or by bucket wheel excavator and conveyor belt (Peretti, 1977). These are then replaced in sequence, after the main overburden has itself been stripped and replaced and levelled. The soil stripping operation runs ahead of the main operation (Figure 11.2). It is not complex although considerable amounts of extra machinery are involved, and if a box scraper is used it has to circle round the operation from unmined to mined land, which means that if the operation is large it must travel considerable distances. At the start of the operation the soil and overburden from the first cut must be stacked on unmined land to allow the development of the open cut. It may or may not be replaced at the end; this will depend how far away is the final cut. In small operations, such as the customary open-cast coal mining in Britain, the material from the first cut must be replaced at the end, so therefore it has to be stockpiled for this (Brent-Jones, 1977).

Because all the soil is returned as it was, it is easy to presume that the final restored land surface will be as productive as it was originally, but this is not a safe assumption. The productivity of the soil depends not only on its nutrient content but also on its ability to retain water in dry periods and drain in wet periods, as well as on the freedom it provides for root growth. The enormous

Open Cast Mining Operations

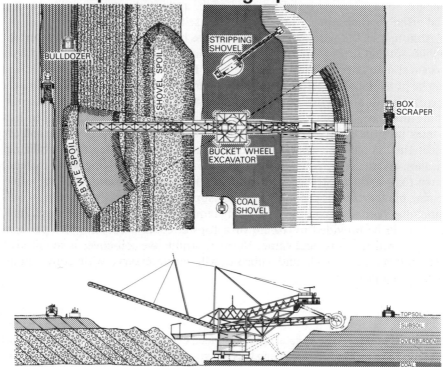

FIGURE 11.2. A modern open-cut coal-mining operation in which topsoil and subsoil are stripped and replaced separately.

mechanical disturbance can lead to loss of structure, even when each replaced layer is ripped and broken up by 80 cm deep tines drawn by a crawler tractor. It is not clear yet what occurs or how it can be obviated, but it is rarely possible to obtain better than 80 per cent of original yields (Jansen, 1981).

In open-cast mining in Britain, the complexities of small-scale operations, the need to use wheeled scrapers for much of the earth moving, the difficulty of achieving continuous replacement, and perhaps the humid conditions which mean that materials are readily consolidated, seem to conspire to cause at least 20 per cent and often even greater losses of productivity (Council for Environmental Conservation (Coenco), 1979). The detailed causes need investigation, particularly as since 1941 over 50 000 ha have been subject to open-cast mining.

Deep mining operations do not affect the land surface directly (except by subsidence), but can affect it considerably by waste disposal. In 1974 there were 13 000 ha of spoil heaps in Britain, placed over the original soil because it was easier to deposit them that way. In most open-pit surface mining operations, for metals, china clay etc., the waste materials, mainly tailings and waste rock, cannot be put back into the mining pit because they would prevent further mining, so they have to be left on the surface and usually occupy at least twice, and often six times, in the area of the pit itself.

Although in the past the soil under waste heaps has been lost, pressure of public opinion is now working to ensure that it is removed and replaced by a system similar to that shown in Figure 11.1. But such a system is expensive, and can be potentially dangerous because a covering placed on a sloping face where there are textural differences between the materials can set up a serious weakness at the interface, leading to landslips. A thin layer partially mixed in is safer than a thick layer in this respect. If therefore only the surface layers of the soil are used, they could well contain adequate nutrients for at least a stabilizing, medium-productivity, vegetation cover. Providing the underlying waste material is not toxic and can provide water and root hold, the most critical factor determining plant growth is probably nitrogen supply. As will be argued later, a self-sustaining vegetation cover will probably require about 1000 kg of nitrogen ha^{-1}. This could be provided by 10 cm of a topsoil cover containing 0.1 per cent nitrogen which is a typical value. But if a productive soil cover is required, a deeper layer of topsoil, and subsoil, will be necessary, with concomitant engineering problems.

Vegetation replacement

In agricultural land the vegetation is replaced every year or every few years and the seeds and techniques for doing this are standard agricultural practice. But in many parts of the world mining takes place in natural areas, where there is natural or semi-natural vegetation. Some sort of semi-natural vegetation can be resown, using commercial stocks of native pasture grasses and legumes, but in many places, for instance when pipelines are laid through moorland in Britain, or when ancient coastal dunes carrying a heath or forest vegetation are mined for zircon and ilmenite in Australia, it is the original vegetation that needs to be replaced; and seeds for it are not available commercially.

Ingenuity has provided very effective solutions. When the surface layers of existing soil are stripped they contain buried dormant seeds and fragments of plants, which can germinate and grow if the soil is replaced without delay. Some plants, particularly trees, but also some shrubs, do not have dormant seeds. But seeds can usually be collected at the appropriate season, by hand; in the bauxite mining regions of north Australia they are collected by aborigines. If these are sown in with the topsoil, and some sort of stabilizing cover-crop provided, the native vegetation will rapidly regenerate. A small amount of fertilizer is usually provided, although excess can encourage the growth of the wrong species. This technique is now applied, for instance, on over 600 ha year^{-1} of forest land being mined for bauxite in Australia and about 100 ha year^{-1} of sand dunes. The details vary from site to site, and some recalcitrant species may have to be raised as seedlings in nurseries and then planted out. But in essence the method is remarkably effective (Brooks, 1981; Hinz, 1979) and is now being developed in many parts of the world.

The same method is now being applied to moorland vegetation in Britain. The dormant seeds of heather species are concentrated in the top 5 cm of the soil profile. So this must be stripped separately from the lower layer of the surface soil, which is then also stripped and replaced (in a heathland podzol these would be the A_0 and A_1 horizons). If this double stripping is not carried out, the dormant seed is diluted and buried. Fairly substantial fertilizer treatment, mainly of nitrogen, can then be given, but on no account must grazing occur because this will damage the *Erica* and *Calluna* seedlings and encourage grass growth (Putwain, Gillham, and Holliday, 1982).

There is no doubt that these techniques can be applied to other communities, such as virgin woodlands, where the woodland herbaceous species have otherwise very poor powers of dispersal (Peterken, 1974). For grassland, turves can be transplanted, or broken up and scattered, or seed sown.

DIRECT TREATMENT OF MATERIALS

When the soil has not been removed and replaced, the exposed pit, the contorted overburden or the raw wastes will be left. They are hardly likely to have any characteristics of normal soils, although they may sometimes be physically reasonable in terms of available water capacity and penetrability to roots. They will inevitably be deficient in major plant nutrients, particularly nitrogen, and may be toxic because of heavy metal residues (or salinity, in arid climates). The exact characteristics, and consequent problems for plant growth depend on the nature of the material, which itself depends on the material being mined and on the country rock. The major characteristics of materials produced by different mining operations is given in Table 11.2. They are discussed in detail in Schaller and Sutton (1978) and Bradshaw and Chadwick (1980).

If we were to go over progressively to a routine of soil replacement in all mining operations, then these materials would cease to concern us. But this is most unlikely, because in many mining operations, soil covering will be impracticable, for instance in steep-sided pits or quarries, or where it is prohibitively expensive to retain the soil, or where it cannot be replaced safely. In some situations the overburden may with treatment become as fertile as the original soil, or in some cases even more fertile. This is true particularly in highly leached soils in humid climates such as in the china clay areas in Georgia (United States) and the bauxite areas of Jamaica. We must be aware of the treatments that are possible as well as their cost and effectiveness in the context of present-day mining.

But we must also be aware of their effectiveness for dealing with the backlog from past mining activity, the thousands of hectares we have already discussed. In some situations it is possible to import, from some other development,

TABLE 11.2 The physical and chemical characteristics of different types of derelict land materials.

Materials	Physical				Chemical				
	Texture and structure	Stability	Water supply	Surface temperature	Macronutrients	Micronutrients	pH	Toxic materials	Salinity
Colliery spoil	OOO	OOO/o	O/o	o/ ••	OOO	o	OOO/o	o	o/ ••
Strip mining	OOO/o	OOO/o	OO/o	o/ ••	OOO/o	o	OOO/o	•	o/ ••
Fly ash	OO/o	o	OO	o	OOO	o	•/•	•	o
Oil shale	OO	OOO/o	O/o	o/ •	OOO	o	OO/o •	o	o/ •
Iron ore mining	OOO/o	OO/o	o	o	OO	o	•	o	o
Bauxite mining	o	o		o	OO	o	o	o	o
Heavy metal wastes	OOO	OOO/o	OO/o	o	OOO	o	OOO/o	/ •••	o/ •••
Gold wastes	OOO	OOO	O	o	OOO	o	OOO	o	o
China clay wastes	OOO	OO	OO	o	OOO	o	o	o	o
Acid rocks	OOO	o	OO	o	OO	o	o	o	o
Calcareous rocks	OOO	o	OO	o	OOO	o	•	o	o
Sand and gravel	O/o	o	o	o	O/o	o	O/o	o	o
Coastal sands	OO/o	OOO/o	O/o	o	OOO	o	o	o	o/ ••
Land from sea	OO	o	o	o	OO	o	o/ •	o	o/ •••
Urban wastes	OOO/o	o	o	o	OO	o	o	o/ ••	o
Roadsides	OOO/o	OOO	OO/o	O/o	OO	o	O/o	o	o/ •

Source: from Bradshaw and Chadwick (1980) with kind permission of Blackwells

Notes

	deficiency			adequate	excess		
	severe	moderate	slight		slight	moderate	severe
	OOO	OO	O	o	•	••	•••

relative to the establishment of a soil/plant ecosystem appropriate to the material; variations in severity are due to variation in materials and situations.

unwanted topsoil and to use this as a covering. But usually this is impractical and we must cope with the excesses and deficiencies of the substrates shown in Table 11.2 as we find them. Because we are dealing with past dereliction one or two other substrates of interest have also been included, because they can be treated in the same way.

Physical problems

The first problem that must be dealt with is landform. If the material is in steep-sided heaps, its treatment cannot even be started without major earthmoving. This can be readily carried out by bulldozer and grader, and gradients set up commensurate with the land use proposed (Kohnke, 1950). But since all earth-moving is extremely expensive, it would have been better if the appropriate landforms had been produced at the time of deposition or formation.

The overriding problem of all materials is texture (the range of particle sizes) and structure (the arrangement of the particles into aggregates with intervening pores). Some materials, such as some mine tailings, have been ground to a material with a silty loam texture satisfactory for plant growth, but others have a high proportion of clay particles (such as colliery spoil) or coarse sand particles (such as china clay sand wastes) or rock (such as quarry wastes). None of them has a satisfactory structure because this only comes about by the accumulation of organic matter and the related activity of soil microorganisms (Russell, 1973). Any structure, particularly porosity, which might have been present when the materials were deposited, may have completely disappeared as a result of the combined effects of earthmoving equipment and rain—each has separate effects, but if they occur together the effects are compounded.

Nevertheless each problem can be treated. Judicious incorporation of other materials, other wastes derived from the operation, or locally produced materials such as sewage sludge, can transform texture. Structure can be temporarily improved by physical treatments such as tine or harrow cultivation. Major consolidation can be relieved by ripping with deep-set tines. However, such relief is only temporary and consolidation will begin again if plants are not quickly established to form pores by root growth and to contribute organic matter, which will attract soil organisms and further cause the bulk density of the soil to be lowered.

Improvements to soil structure bring about correlated improvements in stability. Water percolates into the soil instead of running over the surface and causing major surface erosion and gulleying. However the major contributor to soil stability is a continuous plant cover, which reduces both the capping effects of raindrops on the soil and the movement of water in bulk. It also eliminates erosion by wind. To provide stability, before the vegetation grows, mulches can be used, of chopped straw, wood fibre etc. to give temporary stability. These techniques, originally developed for eroding agricultural and forest land, are of considerable value for mine wastes (Kay, 1978; Schiechtl, 1980).

Little can be done about water supply except to ensure that plants are established in humid periods when there is adequate soil moisture, an obvious step but sometimes difficult to fit into the routine of mining operations unless requested with firmness. Extreme surface temperatures are a problem only in absence of vegetation cover. If the vegetation is properly established, with complete ground cover and deep root system, before the onset of extreme periods, the problems should be minimal.

Chemical problems

Nutrients

Inevitably mining wastes are deficient in major plant nutrients in forms which are available to plants. Because the materials are so new, there has been no opportunity for the mineral material to weather and release their nutrients so that they are transformed into more available forms. But at the same time mining wastes are usually of rather individualistic origins and do not necessarily contain a full range of all the mineral nutrients that plants need; some may be deficient in calcium, others in phosphorus or potassium. Nitrogen in particular is nearly always completely deficient, because it owes its origin in soils not to weathering of soil minerals but to biological processes by which gaseous nitrogen from the atmosphere is fixed by microorganisms and converted into ammonium, nitrate and organic compounds; there are no opportunities for these processes to occur in materials below ground.

Even if everything else were satisfactory plants would fail to grow on mining wastes and overburdens because of the lack of nutrients. The modern fertilizer industry is now able to provide these missing nutrients. Normally, in most mining wastes the addition of one or two heavy dressings of complete fertilizer, totalling about 500 of nitrogen, 200 of phosphorus and 200 of potassium (kg ha^{-1}), will stimulate excellent growth, although sometimes other nutrients such as magnesium and calcium may be necessary.

However, even if plant growth is excellent to begin with, it tails off and almost ceases a few months after the dressing. Experiments have shown that this is because nitrogen has become deficient; if further nitrogen is added growth recommences. Again growth will tail off and continued growth usually depends on continued application of nitrogen (Bloomfield, Handley, and Bradshaw, 1982) (Figure 11.3). Since nitrogen has been added at the start why should this be so?

Nitrogen occupies a special place among plant nutrients. It is not only required by plants in greater amounts than any other plant nutrient, but also, unlike other nutrients, it is stored in the soil only in organic matter. The supply of nitrogen to plants depends on there being a large capital store of nitrogen in the organic matter, a small fraction of which is released annually by its decomposition. It appears that the annual rate of release, or mineralization, in temperate

Effect of Nutrients on Reclaimed Colliery Spoil

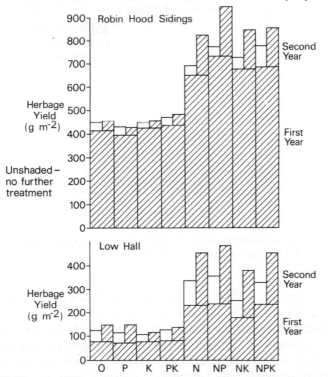

FIGURE 11.3. The effect of adding nitrogen, phosphorus and potassium on the growth of grass on reclaimed colliery spoil. Shaded = nutrient application; unshaded = no second treatment. From Bloomfield, Handley, and Bradshaw (1982).

climates is unlikely to be more than about 5 per cent and that plants need about 50 kg ha⁻¹ year⁻¹. This implies a minimum capital of about 1000 kg of nitrogen ha⁻¹. If the rate of mineralization were less, the capital required would be more (Bradshaw *et al.*, 1982). In fact, on naturally colonized derelict land materials 1000 kg of nitrogen ha⁻¹ does appear to be about the minimum amount of nitrogen that must be accumulated to get a soil/plant system which is self-sustaining (Roberts *et al.*, 1981). The organic matter which contains this nitrogen will also contain other nutrients, particularly phosphorus and sulphur, which will also be released by the decomposition processes and be made available to plants.

But how can this capital be accumulated? It would be possible to achieve it by continued addition of fertilizers, but this would be tedious and expensive. It is much more sensible to achieve it by the use of legumes which rapidly accumulate nitrogen (about 50–100 kg of nitrogen ha⁻¹ year⁻¹) even on mine wastes, by the nitrogen-fixing activities of their root-nodule bacteria (Figure 11.4). There is a

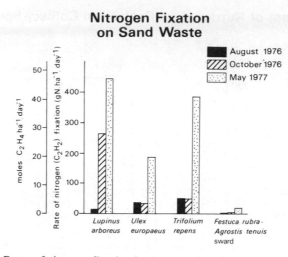

FIGURE 11.4.   Rates of nitrogen fixation in three species of legume and (for comparison) a grass sward, all growing on sand waste. From Skeffington and Bradshaw (1980).

wide range of legumes (Jefferies, Bradshaw, and Putwain, 1981) and other nitrogen-fixing species such as alder (*Alnus glutinosa*) and Russian olive (*Eleagnus sylvatica*) which grow well on mine wastes, although some such as the widely used white clover (*Trifolium repens*) do require the maintenance of adequate calcium and phosphorus. The value of all these species is not only that, after establishment, they accumulate nitrogen with no expenditure of labour or other costs except normal management, but also that the nitrogen they accumulate is in a form which is readily mineralized and made available to other plants (Table 11.3). If the nitrogen is not readily mineralizable the productivity of the vegetation will be restricted (Gildon and Arnold, 1981). It is

TABLE 11.3   The levels of mineralizable nitrogen in colliery spoil reclaimed to grassland, under naturally occurring white clover, *Trifolium repens*, patches of different ages (total mineral nitrogen (μg g^{-1}) accumulated after incubation for 10 days at 30°C.

	Age of clover patch			
	approx. 2 years	approx. 3 years	4−5 years	Mean
Clover absent	4.3	4.7	5.6	4.8
Clover present	85.4	141.5	23.4	83.5
Mean	44.9	73.1	14.5	
SE mean	ages	7.8		
	clover	9.3		

Source: Jefferies, Willson, and Bradshaw, 1981

significant that in naturally developing ecosystems, such as on moraines at Glacier Bay in Alaska, and on mine wastes in Cornwall, nitrogen-fixing species play a very important role in the development of the vegetation (Lawrence *et al.*, 1967; Palaniappan, Marrs, and Bradshaw, 1979).

Toxicity

Nutrients and other ions can sometimes be in excess rather than deficient, and this leads to toxicity. This is most obvious in situations where heavy metals, such as copper, lead and zinc, have been mined, or subsequently processed. Heavy metals are very toxic to plants, above about 10 ppm in solution or about 1000 ppm in soil; the exact level depends on the metal and the associated conditions. The metal contents of tailings vary depending on the extraction process; the amount of metal left may often be well above 1000 ppm and the tailings will therefore be toxic. In earlier processing as much as 1 or 2 per cent of lead or zinc, for instance, could be left behind. In open-pit operations, mining low-grade deposits, large amounts of rock may have to be removed whose metal content is less than that which is economic to extract; a common cutoff point is 0.2 per cent. As a result waste rock dumps may, when they weather and release this metal, become very toxic.

Metals do not break down and disappear, so metal-containing wastes remain toxic for long periods. This is obvious in the old lead/zinc mining areas of Wales and northern England where the waste heaps have remained uncolonized for many decades. There are three ways toxic metalliferous areas can be treated:

(1) In certain situations the wastes can be treated with organic matter, which complexes the metal and renders it unavailable (Goodman, Pitcairn, and Gemmell, 1973). But ultimately the organic matter decomposes and the toxicity returns.
(2) Naturally evolved metal tolerant plant populations already growing on some wastes can be used (Smith and Bradshaw, 1979). But only a limited range of tolerant plant material is available.
(3) The wastes can be covered by a layer of an inert non-toxic material (Johnson and Bradshaw, 1977). On this a normal vegetation can be established. This has the advantage of isolating the metals so that grazing can be permitted, but the earthmoving required is extremely expensive.

Another type of toxicity is that due to excess of hydrogen ions (acidity) or of hydroxyl ions (alkalinity). Acidity is extremely common on wastes produced by coal mining because these often contain high levels of pyrite (FeS) which, when it is exposed to air, oxidizes to produce sulphuric acid. This acidity can easily be treated by adding ground limestone to the waste. However, sufficient calcium must be added not only to treat the existing acidity but also that which will be produced by the pyrite not already weathered. The pyrite level in the waste must be measured, together with any carbonates present, so that the total limestone required can be determined (Costigan, Bradshaw, and Gemmell, 1981). As a result, on some wastes as much as 150 tonne ha⁻¹ of limestone are being used. Alkalinity is usually due to calcium hydroxide and can be treated by adding organic matter; this produces CO_2 which converts the hydroxide into carbonates.

Salinity occurs in some wastes because it was present in the rock being mined. But it can also be produced on weathering, when the acid produced by weathering of pyrite and other sulphides is neutralized by native carbonates, producing $MgSO_4$, etc. If there is excess pyrite the salinity can be combined with acidity. In humid climates the soluble salts causing the salinity are leached away in the passage of time by precipitation, and are only likely to be a problem in water courses. But in arid climates the soluble salts may accumulate at the surface, and be difficult to control except by covering with coarse materials which interrupt the soil capillaries and prevent the salts being conducted to the surface.

Biological problems

Mined land has been completely cleared of vegetation, and this has to be replaced. We have already seen that it is possible to replace the original vegetation cover by means of the different propagules carried in the surface soil. In these situations the soil provides an excellent environment for the development of seedlings and young plants. On recent mine wastes, however, there has been not only a lack of seed but also a difficult environment for establishment. There are biological problems to overcome.

The first problem is the choice of species. This must be related to the chosen final land use. But it must also be related to the environmental characteristics of the wastes that are likely to remain after the physical and chemical problems have been relieved. Within the range of grasses, legumes, and trees and shrubs that are commercially available there are certain species that are more likely to survive the initially poor conditions (Bradshaw and Chadwick, 1980).

These species are likely to be those that are tolerant of drought and nutrient deficiency such as *Festuca rubra*, *Poa pratensis*, *Pinus sylvestris* and *Betula pendula* among temperate grasses and trees. Unfortunately among temperate legumes the most tolerant are shrubs such as *Ulex europaeus* and *Lupinus arboreus* and these are difficult to use; useful species such as *Trifolium repens* and *Lotus corniculatus* require reasonable fertility. In tropical regions a range of species tolerant to drought and low nutrients such as the legumes *Stylosanthes scabra* and *Phaseolus atropurpureus* and the grass *Chloris gayana* have recently been developed and are readily available. In practice, because species are so idiosyncratic it is important to base a choice on local experience and field trials.

Wild species can be used if seed supplies can be organized. There is no reason why they should not be sown directly. This is commonplace in Australian bauxite and mineral sand mining. Alternatively seedlings can be raised and planted out. Some wild species, particularly gras‿es, can be planted as vegetative tillers. This is particularly useful for special material, for instance metal-tolerant populations (Hill, 1977).

In the early stages of establishment, seedlings are liable to erosion and desiccation. For this reason it has become commonplace to sow an ephemeral, rapidly growing species to act as a nurse. Cereals such as rye and sorghum and forage grasses such as hybrid sorghum and short rotation ryegrass are widely used. But it is important that these are sown at a very low rate leading to about 100 plants per m^2 or the good effects will be outweighed by excessive competition. Sometimes it may be better to use a fibrous mulch instead.

The sowing itself rarely presents problems on level surfaces. Standard agricultural techniques can be employed. But it is essential that the appropriate season is chosen and that the seedbed is adequate. On steep slopes agricultural machinery cannot be used. Here the technique of hydroseeding is valuable. The seed is sprayed onto the surface in a semiliquid mixture of water, mulch and, when conditions are appropriate, fertilizer. The exposure of the seed to desiccation is reduced by the mulch and by a rough soil surface. But in dry conditions the partly dissolved fertilizer can reach toxic concentrations in the immediate environment of the seed; it is therefore better to apply it when the seed has germinated; chemical stabilizers are often recommended but many of them inhibit seed germination. The whole technique has to be used with care to be effective (Roberts and Bradshaw, 1983).

The animal components of the ecosystem rarely have to be assisted if the vegetation and soil are developing properly. Soil animals colonize rapidly (Neumann 1973) and small mammals and birds appear as soon as the vegetation develops (Tables 11.4 and 11.5).

Aftercare

In the early stages of development of the new ecosystem, it will need more looking after than a fully developed ecosystem. In particular, attention must be paid to adequate nutrient supply, because this is the key to the whole system. Nutrient deficiencies are common, even where topsoil has been replaced (Brent-Jones, 1977) and they must be attended to immediately. If there are no legumes, repeated applications of nitrogen fertilizer will be needed; if legumes are included, their nutrient requirements must be looked after. It is crucial that the vegetation continues to grow rapidly, or its beneficial effects on soil stability and structure will not develop. Liming may also be necessary.

The growth of the vegetation itself must be managed. Pastures must be cut and grazed in ways which will maintain the sward in good condition and achieve a satisfactory legume balance. This may sometimes be difficult because of the ease with which the soil structure can be damaged. Trees may need weed control

TABLE 11.4 Breeding bird census of a restored lead−zinc−fluorspar tailings dam (1974−80) at Cavendish Mill, Derbyshire, England.

	Density* (breeding pairs/ha)		
	1975	1977	1979
Common partridge *(Perdix perdix)*	0	0	0.24 (2)
Curlew *(Numenius arquata)*	0	0	0.12 (1)
Meadow pipit *(Anthus pratensis)*	0.12 (1)	0.48 (4)	0.60 (5)
Pied wagtail *(Motacilla alba)*	0	0	0.24 (2)
Skylark *(Alauda arvensis)*	0.24 (2)	0.60 (5)	1.08 (9)
Wheatear *(Oenanthe oenanthe)*	0	0.12 (1)	0.24 (2)
Whinchat *(Saxicola rubetra)*	0	0	0.12 (1)
Yellow wagtail *(Motacilla flava)*	0	0.12 (1)	0.36 (3)

Source: Johnson and Putwain (1981).

*Values in parenthesis indicate the total number of breeding pairs on the tailings dam surface.

if there is too vigorous a growth of weeds or grass around them; and in a soil with initially poor available water capacity competition for water can be severe (Bradshaw and Chadwick, 1980).

In the maintenance of productivity in agricultural land, established swards are normally ploughed up after a few years, to incorporate the accumulated organic matter rapidly and to prevent deleterious change in the species composition of the sward; it also facilitates the incorporation of lime. This should be accepted practice in restored land. Good aftercare is crucial, and is one of the significant points of the Minerals (1981) Act.

NATURAL COLONIZATION

Any area of land, however poor, if left to itself will slowly become colonized by plants. Because land left by mining activity is extremely infertile, the process of colonization may be very slow and the ground may remain open for many decades. The process of primary colonization can be followed in detail (Usher,

TABLE 11.5 Estimated small mammal and predator densities on a revegetated* lead—zinc—fluorspar tailings dam at Cavendish Mill, Derbyshire, England.

| | Density (nos/ha) | | | |
| | tailings dam | | nearby pasture/ scrub | |
	1977	1979	1977	1979
Field vole *(Microtis agrestis)*	2.3	3.6	2.1	2.4
Woodmouse *(Apodemus sylvaticus)*	6.8	9.7	2.2	2.9
Common shrew *(Sorex araneus)*	3.2	11.2	3.4	5.4
Bank vole *(Clethrionomys glareolus)*	0.2	0.7	1.7	2.9
Weasel *(Mustela nivalis)*	0.3	0.8	0.1	0.2
Stoat *(Mustela erminea)*	0	0.3	< 0.1	< 0.1

Source: Johnson and Putwain (1981).

*Surveyed in September: figures are combined totals for grassland and scrub. The adjacent trapping area was selected to give a similar grassland: scrub area ratio to that on the tailings surface.

1977). At the same time species which are normally rare because they are intolerant of competition may find a refuge and prosper (Table 11.6) (Bradshaw, 1977). As a result many old mining sites are now of considerable biological interest, representing stages of biological succession which have otherwise disappeared from our more mature countryside. Out of 3000 sites of special scientific interest (SSSI), 75 are old mining areas (Ratcliffe, 1974). Some chemical waste disposal sites can also have very curious and interesting flora, such as the alkali wastes of the Lancashire region which maintain substantial populations of five different orchids (Greenwood and Gemmell, 1978) and add to our ideas on island biogeography (Gray, 1982). Metalliferous wastes have provided us with one of the best examples of natural selection and evolution in progress (Bradshaw and McNeilly, 1981).

Yet natural colonization can be slow, and insufficient to satisfy the requirements of tidiness and amenity. What may be attractive to biologists may be just a wasteland to other people. There is no reason why some steps cannot be taken to introduce species, either to add diversity or to provide landscape quality. The outstanding example is the Sevenoaks Gravel Pit Nature Reserve where the combined efforts of the gravel operator and the tenant have produced a remarkable haven for plants and wildfowl. At the end of the excavation islands and

TABLE 11.6   Plants of nature conservation interest recorded in five limestone quarries near Durham.

Uncommon and local British species	Species at or near the edge of their range in Britain	Species of local interest mainly on limestone
Astragalus danicus	Anacamptis pyramidalis	⨯ Anthyllis vulneraria
Botrychium lunaria	Acinos arvensis	Arabis hirsuta
Crepis mollis	Agrostis gigantea	Coeloglossum viride
Epipactis atrorubens	Cirsium eriophorum	Dactylorhyza purpurella
Hypericum x desetangsii	Crepis taraxacifolia	Gentianella amarella
Hypericum montanum	⊢Erigeron acer ⊢	Gymnadenia conopsea
Plantago maritima	Hieracium sublepistoides	Helianthemum chamaecistus
Salix nigricans	⨯ Leontodon taraxacoides⊢	Helictotrichon pratense
Salix phylicifolia	Rosa obtusifolia	⨯ Ononis repens
Sesleria albicans	Zerna erecta	Sieglingia decumbens

From: Davis, 1976.

beaches were formed, and a wide variety of plants introduced in mud and as plants. The birds came of their own accord, attracted by the environment (Harrison, 1974) (Figure 11.5). Such creative conservation could be developed in many other mining sites, many of which already have considerable nature conservation value (Davis, 1976). It is particularly effective for sites which began as bare wastes and on which productive agriculture is restricted by chemical problems (Johnson and Putwain, 1981).

CONCLUSIONS

Mined land is a fascinating challenge. Because the pre-existing ecosystems are extinguished, it is a challenge to the biologist and engineer to replace them as they were. Because the damage has so often been done in the past and nothing of the original soil and vegetation remains to be transferred, it is a challenge to the soil scientist, ecologist and agriculturalist to reconstruct an ecosystem from nothing, at minimal cost. Because the slate has been wiped clean and there are many other possibilities beside putting back what was there before, it is a challenge to the planner and ecologist to decide on imaginative solutions. And because in mined land the natural processes of soil and ecosystem development have been reversed, it is a challenge to conservationists to seize the opportunities provided by such unique sites and carry out creative nature conservation. There is great scope for imagination allied to action. The action provides an acid test of our understanding of biological systems. It is not enough to understand how ecosystems work; in mined land we actually have to construct them and make them work.

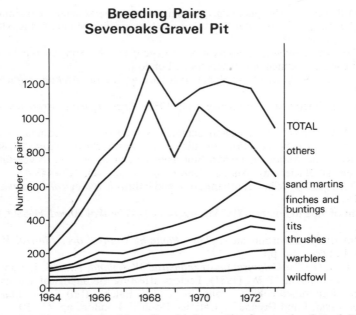

Breeding Pairs
Sevenoaks Gravel Pit

FIGURE 11.5. The number of breeding birds on the Sevenoaks Gravel Pit Nature Reserve after its establishment from a disused gravel working. From Harrison (1974).

In the end we are dealing with a problem of conservation. Land is a commodity which is finite; and yet much of the way we live causes its destruction. Now, by the proper application of ecological principles, land is something we cannot only use, but also recycle—a very practical form of conservation.

REFERENCES

Bloomfield, H.E., Handley, J.F., and Bradshaw, A.D. (1982). Nutrient deficiencies and the aftercare of reclaimed derelict land. *Journal of Applied Ecology*, **19**, 151–158.

Bradshaw, A.D. (1977). Conservation problems in the future. *Proceedings of the Royal Society of London, Series B*, **177**, 77–96.

Bradshaw, A.D. and Chadwick, M.J. (1980). *The Restoration of Land*, Blackwell, Oxford, 317 pp.

Bradshaw, A.D. and McNeilly, T. (1981). *Evolution and Pollution*, Edward Arnold, London, 76 pp.

Bradshaw, A.D., Marrs, R.H., Roberts, R.D., and Skeffington, R.A. (1982). The creation of nitrogen cycles in derelict land. *Philosophical Transactions of the Royal Society of London, Series B*, **296**, 557–561.

Brent-Jones, E. (1977). The agricultural restoration of open cast coal sites in Great Britain. In J. Essex and P. Higgins (Eds) *Papers of the Land Reclamation Conference*, Thurrock Borough Council, Grays, Essex, pp. 243–262.

Brooks, D.R. (1981). The planning and logistics of mineral sands rehabilitation. In *Proceedings of Environmental Workshop 1981*, Australian Mining Industry Council, Canberra, pp. 80–92.

Council for Environmental Conservation (Coenco) (1979). *Scar on the Landscape?* Council for Environmental Conservation, London.

Coleman, A.M. (1977). Land-use planning: success or failure? *Architects Journal*, **165**, 91–134.

Commission on Energy and the Environment (1981). *Coal and the Environment*, HMSO, London.

Costigan, P.A., Bradshaw, A.D., and Gemmell, R.P. (1981). The reclamation of acidic colliery spoil. I: Acid production potential. *Journal of Applied Ecology*, **18**, 865–878.

Crocker, R.L. and Major, J. (1955). Soil development in relation to vegetation and surface age at Glacier Bay, Alaska. *Journal of Ecology*, **43**, 427–448.

Davis, B.N.K. (1976). Wildlife, urbanisation and industry. *Biological Conservation*, **10**, 249–291.

Department of the Environment (1975a). *Aggregates—The Way Ahead*, HMSO, London, 118 pp.

Department of the Environment (1975b). *Planning Control over Mineral Workings*, HMSO, London, 448 pp.

Gemmell, R.P. (1977). *Colonisation of Industrial Wasteland*, Arnold, London, 75 pp.

Gildon, A. and Arnold, P.W. (1981). Factors affecting the biological activity of soils overlying deep-mined colliery spoils. In W.G.V. Balchin (Ed) *The Productivity of Restored Land*, Land Decade Educational Council, London, pp. 45–50.

Goodman, G.T., Pitcairn, C.E.R., and Gemmell, R.P. (1973). Ecological factors affecting growth on sites contaminated by heavy metals. In R. Hutnik and G. Davis (Eds) *Ecology and Reclamation of Devastated Land*, Gordon and Breach, New York, pp. 149–174.

Gray, H. (1982). Plant dispersal and colonisation. In B.N.K. Davis (Ed) *Ecology of Quarries*, Institute of Terrestrial Ecology, Cambridge, pp. 27–31.

Greenwood, E.F. and Gemmell, R.G. (1978). Derelict industrial land as a habitat for rare plants in S. Lancs (V.C.59) and W. Lancs (V.C.60). *Watsonia*, **12**, 33–40.

Harrison, J. (1974). *The Sevenoaks Gravel Pit Reserve*, WAGBI, Chester.

Hill, J.C. (1977). Establishment of vegetation on copper-, gold- and nickel-mining wastes in Rhodesia. *Transactions of the Institute of Mining and Metallurgy*, **86A**, 135–145.

Hinz, D. (1979). Rehabilitation of mined land and bauxite residues at Gove, N.T. In I. Hore-Lacy (Ed) *Mining Rehabilitation 1979*, Australian Mining Industry Council, Canberra, pp. 11–16.

Jansen, I.J., (1981). Reconstructing soil after surface mining of prime agricultural land. *Mining Engineering*, **March**, 312–314.

Jefferies, R.A., Bradshaw, A.D., and Putwain, P.D. (1981). Growth, nitrogen accumulation and nitrogen transfer by legume species established on mine spoils. *Journal of Applied Ecology*, **18**, 945–956.

Jefferies, R.A., Willson, K., and Bradshaw, A.D. (1981). The potential of legumes as a nitrogen source for the reclamation of derelict land. *Plants and Soil*, **59**, 173–177.

Johnson, M.S. and Bradshaw, A.D. (1977). Prevention of heavy metal pollution from mine wastes by vegetative stabilisation. *Transactions of the Institute of Mining and Metallurgy*, **86A**, 47–55.

Johnson, M.S. and Putwain, P.D. (1981). Restoration of native biotic communities and land disturbed by metalliferous mining. *Minerals and the Environment*, **3**, 67–86.

Kay, B.L. (1978). Mulch and chemical stabilizers for land reclamation in dry regions. In F.W. Schaller and P. Sutton (Eds) *Reclamation of Drastically Disturbed Lands*, American Society of Agronomy, Madison, pp. 467–483.

Kohnke, H. (1950). The reclamation of coal mine spoils. *Advances in Agronomy*, **2**, 317–349.

Lawrence, D.B., Schoenike, R.E., Quispel, A., and Bond, G. (1967). The role of *Dryas drummondii* in vegetation development following ice recession at Glacier Bay, Alaska, with special reference to its nitrogen fixation by root nodules. *Journal of Ecology*, **55**, 793–813.

Leisman, G.A. (1957). A vegetation and soil chronosequence on the Mesabi Iron Range spoil banks, Minnesota. *Ecological Monographs*, **27**, 221–245.

Neumann, V. (1973). Succession of soil fauna in afforested spoil banks of the brown-coal district of Cologne. In R.J. Hutnik and G. Davis (Eds) *Ecology and Reclamation of Devastated Land*, volume 1, Gordon and Breach, New York, pp. 335–348.

Olson, J.S. (1958). Rates of succession and soil changes on southern Lake Michigan sand dunes. *Botanical Gazette*, **119**, 125–170.

Palaniappan, V.M., Marrs, R.H., and Bradshaw, A.D. (1979). The effect of *Lupinus arboreus* on the nitrogen status of china clay wastes. *Journal of Applied Ecology*, **16**, 825–831.

Peretti, K. (1977). The compatibility of opencast mining and environmental protection. In J. Essex and P. Higgins (Eds) *Papers of the Land Reclamation Conference*, Thurrock Borough Council, Grays, Essex, pp. 231–242.

Peterken, G.F. (1974). A method of assessing woodland flora for conservation using indicator species. *Biological Conservation*, **6**, 239–245.

Putwain, P.D., Gillham, D.A., and Holliday, R.J. (1982). Restoration of heather moorland and lowland heathland with special reference to pipelines. *Environmental Conservation*, **9**, 225–235.

Ratcliffe, D. (1974). Ecological effects of mineral exploitation in the United Kingdom and their significance to nature conservation. *Proceedings of the Royal Society of London, Series A*, **339**, 355–372.

Roberts, R.D. and Bradshaw, A.D. (1983). Hydraulic seeding techniques—critique and guidelines. *Landscape Design* (in press).

Roberts, R.D., Marrs, R.H., Skeffington, R.A., and Bradshaw, A.D. (1981). Ecosystem development on naturally colonized china clay wastes. I: Vegetation changes and overall accumulation of organic matter and nutrients. *Journal of Ecology*, **69**, 153–161.

Roberts, T.M. and Roberts, R.D. (Eds) (1983). *Environmental Planning and Ecology*, Chapman and Hall, London.

Russell, E.W. (1973). *Soil Conditions and Plant Growth* (10th edition), Longmans, London, 849 pp.

Schaller, F.W. and Sutton, P. (1978). *Reclamation of Drastically Disturbed Lands*, American Society of Agronomy, Madison, Wisconsin, 742 pp.

Schiechtl, H. (1980). *Bioengineering for Land Reclamation and Conservation*, University of Alberta Press, Edmonton, 404 pp.

Skeffington, R.A. and Bradshaw, A.D. (1980). Nitrogen fixation by plants grown on reclaimed china clay wastes. *Journal of Applied Ecology*, **17**, 469–477.

Smith, R.A.H. and Bradshaw, A.D. (1979). The use of metal tolerant plant populations for the reclamation of metalliferous wastes. *Journal of Applied Ecology*, **16**, 595–612.

Smith, R.M., Tryon, E.H., and Tyner, E.H. (1971). *Soil Development on Mine Spoil*. Agricultural Experimental Station Bulletin 604. West Virginia University, Blacksburg, Va.

Usher, M.B. (1977). Natural communities of plants and animals in disused quarries. In J. Essex and P. Higgins (Eds) *Papers of the Land Reclamation Conference*, Thurrock Borough Council, Grays, Essex, pp. 401–420.

Wallwork, K.L. (1974). *Derelict Land: Origins and Prospects of a Land Use Problem*, David and Charles, Newton Abbott, 333 pp.

CHAPTER 12

Ecological Effects of Visitors and the Restoration of Damaged Areas

F.B. GOLDSMITH

INTRODUCTION

The ecological changes that result from high levels of recreational activity have been causing concern to those interested in and responsible for the management of seminatural areas, particularly in western Europe, North America and Australia (Goldsmith, 1974). Recreation in this context is defined as the pursuit of informal leisure in the countryside; its effects are to damage the vegetation by trampling, to accelerate the loss of soil and to disturb animals. The managers involved are from bodies such as local authorities, the National Park Authorities and the National Trust. In Britain the problem has been most acute at popular beauty spots such as Box Hill in Surrey and Ivinghoe Beacon in Buckinghamshire, Snowdon in Wales, the Cairngorms in Scotland, Kynance Cove and Lands End in Cornwall, Dovedale in the Pennines and Cheddar Gorge in the Mendips.

The most obvious damage is the loss of vegetation. Coastal and montane areas are particularly vulnerable because the vegetation has inherently slow rates of growth and recovery. The concern of the managers and the environmental lobby is increased when an unusual habitat type or a rare species is involved as on chalk grassland, coastal dunes or coastal heathland.

Some people (such as Brotherton, 1981) have recently suggested that the problem has been exaggerated. The problem is certainly very local, as even the metropolitan green belt of London has only about four sites that show serious damage, for there is a wide choice of areas available to visitors and the soils are mostly relatively fertile and the vegetation robust. The New Forest, on the other hand, with a lower but not inconsiderable visitor pressure, had such extensive areas of damage in the 1960s that it was thought to need a special management plan, in which the use of vehicles was restricted to roads by ditching, banking, posts etc., and camping restricted to a limited number of sites. The really important development in the last decade is that methods are now available to investigate the problem and techniques are available to correct it. Fortunately, on many sites the appropriate restoration management is a low-technology and low-cost one of manipulating the visitors as well as carrying out more expensive (but still cheap) works to the site itself.

A decade ago there was considerable debate about the determination of the carrying capacity for a particular site. Much of this debate was unproductive, for capacity is an illusory concept except in a bathtub. There is no single threshold, for the value depends on the activity of the visitors (footballers or old folk) and distance they travel, the condition and type of vegetation (active growth or dying-back, woodland or grassland), the weather conditions (wet or dry), and so on. We now know that it is possible to select one of several choices as a recreation management strategy:

(1) Concentrate visitors at one place (often referred to as the honeypot approach, as is done in some North American national parks).
(2) Disperse them around several foci (as in the New Forest) or all over the area.
(3) Direct them elsewhere (for example, from Old Winchester Hill National Nature Reserve to Butser Hill which is part of Queen Elizabeth Country Park).

All three approaches have been used to good effect and the capacities of the areas under pressure can be increased by judicious fertilizing, seeding or watering, etc. although it must be emphasized that these treatments can be just as damaging ecologically as the original trampling.

STUDIES ON THE EFFECT OF TRAMPLING

Trampling is merely another factor in the environment of species that respond to many other environmental, historical, biotic and management factors. To study these one has to select the key variables to measure and find controlled conditions where all variables are held constant, except the two under study. The problem has been encountered in a wide range of ecosystems. Ecologists and resource managers from the Arctic to the Tropics and the Orient to the New World have tackled it in a variety of ways. The two essential sets of variables are the number of people trampling, refined as frequency, intensity, pressure or shear-stress, and the response of the species, community or soil characteristic under study. Whilst trampling is the activity most often investigated, other forms of locomotion are increasingly frequently becoming the focus of attention. Horseriding was first studied by Perring (1967) on chalk grassland and the problems are known to all who walk footpaths and bridleways in areas of clay soils in the winter. More recently McQuaid-Cook (1978) and Weaver and Dale (1978) have examined the effects of both horseriding and hikers on mountain trails and found that horses loosen and move soil downhill on steep slopes whereas trampling by people compacts it on flatter areas. Keddy, Spavold, and Keddy (1979) looked at the effects of snowmobiles on old-field and marsh vegetation in Nova Scotia, Canada, and Smith (1981) discusses the nature conservation problems encountered with off-road recreation vehicles (ORRV).

THE RANGE OF INVESTIGATIONS

There were some early studies on footpaths in the 1930s, but from then on few detailed studies until the Conservation Course conducted its group project on the Isles of Scilly in 1969 (Goldsmith, Munton, and Warren, 1970). This was the first study to use questionnaire maps to show the detailed distribution of visitors pursuing different activities, the locations in which they stopped for different lengths of time, in relation to the distribution of areas of high ecological interest. Since then we have seen the publication of several reviews on the subject, including one commissioned by the British Ecological Society (Speight, 1973), another by the Council of Europe which covers much of the continental European literature (Satchell and Marren, 1976) and one with a Canadian perspective (Wall and Wright, 1977). Many researchers have reviewed the literature at the outset of their studies (Liddle, 1975a) and some have published multiauthor books (Lavery, 1974) or books summarizing primarily their own research, for example the Tourism and Recreation Research Unit at Edinburgh (Coppock and Duffield, 1975).

A wide range of research studies have been carried out on habitats from chalk grassland (Streeter, 1971), dunes (Liddle and Greig-Smith, 1975a and b), woodland (Kellomäki, 1973; and Kellomäki and Saastomoinen, 1975), mountains (Bayfield, Urquhart, and Cooper, 1981; Takahashi and Maenaka, 1977), tundra, uplands, peat-bogs (Hammitt, 1980; Slater and Agnew, 1977), and heathland (Burton, 1971) to lakes, reservoirs, loch shores (Rees and Tivy, 1978) and coral reefs (Woodland and Hooper, 1977). It is difficult to make generalizations about these but it is important to try to learn from this wealth and breadth of experience. Unfortunately many of the studies are superficial and some are merely anecdotal in their approach. Most studies define the problem but do not provide a basis for monitoring the deterioration or improvement of the situation in the future. Sadly, few projects have produced a ranking of species in order of their resistance to damage or in order of recovery after trampling (but see below). This kind of information is obviously of value to managers of sites and could be fairly easily prepared, although some caution has to be exercised as the ranking would change with environmental factors, for example, soil moisture or shade and in different geographical areas. Hardly ever does one read a definition of the problem, the research carried out, the recommendations for improvement and the subsequent restoration of the site as a logical sequence in one study. In this respect it is important to note how far-sighted the Countryside Commission was in initiating this sequential line of approach. One reason for the absence of this vertical approach to the problem and its cure is the specialization encouraged in many aspects of society today, including universities and government research agencies.

ANIMALS AND DISTURBANCE

There have been few studies of the effects of trampling on invertebrates (Buchanan, 1975; Duffey, 1975) or of insect collecting (Morris, 1967) or disturbance on large animals. Picozzi (1971) studied the breeding performance and shooting bags of red grouse in relation to public access in the Peak District and Watson (1979) considered the effects of ski developments on birds and mammals in the Scottish hills. Batten (1977) reported his observations on the effects of sailing on populations of mallard, tufted duck, pochard and smew on Brent reservoir in North London. Survival of the birds there depended on the presence of a large shallow or marshy area which was not accessible to boats. Batten also noted that larger flocks were more sensitive to disturbance than smaller ones, so that larger numbers might be accommodated by subdivision of the area, perhaps by screening. The Wildfowl Trust, with grant-support from the Sports Council and Nature Conservancy Council, are currently investigating the effects of waterborne recreation on bird populations (Tuite, 1980). Studies on large mammals in relation to recreation and tourism pressures have been few, but Jewell (1974) reviewed the subject in relation to East African national parks. In some areas, vehicles driving over open country are becoming a serious problem, and the continuous intensive viewing of some species, such as crocodiles, are expected to accelerate the collapse of the population. Lions on the other hand are highly tolerant of being surrounded by visitors in cars.

CHARACTERISTICS OF VEGETATION AND ITS ABILITY TO WITHSTAND TRAMPLING

Most farmers would be able to tell us that plants with their apical meristems or perennating buds near the ground (such as grasses and rosette herbs) are able to withstand grazing better than other species without that characteristic and a similar generalization holds good for trampling. The relationship between trampling resistance and growth rate or net primary productivity is less clear. Liddle (1975b) compared figures for community productivity and number of passes required to reduce the vegetation to 50 per cent of its original cover and found a significant log/log relationship between them:

$$\log_{10} \text{number of passes} = 1.36 \log_{10} \text{productivity} - 0.896$$

The advantage of increased productivity might be the increased rate of recovery of the vegetation whilst the initial resistance to trampling might be more a function of morphology of the species involved.

Studies recently completed at University College London (Jewell, personal communication) suggest that the growth rates of individual species affect their competitive potential and ability to withstand trampling. Competitive species such as *Lolium perenne* and *Plantago major* appear to be extremely tough and resilient, whilst *Festuca rubra, Holcus lanatus, Arrhenatherum elatior* and *Deschampsia flexuosa* are relatively vulnerable. Trampling increases tillering in most species and most grasses grow slightly more slowly on compacted soils. There are important differences between different soil types such as clays and sands in terms of the way they respond to compaction which has led to confusion in the literature. This is a technical subject which requires further research and will not be pursued further here. We should note, however, that the maintenance of a photosynthetic system (that is, an undamaged plant shoot) may be more important than the reduction in growth brought about by the degree of compaction of the soil.

RECOVERY AFTER DAMAGE

Increased sensitivity and reduced recovery can be anticipated for montane habitats and areas with shallow, nutrient deficient or excessively wet soils (Goldsmith, 1974). Studies of the impact of recreation on montane heath communities in the Cairngorms of Scotland (Bayfield, 1971a and b; 1980) have shown extremely slow rates of recovery. Bayfield's plea for greater attention to this problem encouraged Harrison (1981) to investigate this topic further in her study of publicly owned sites used for informal recreation in London's green belt. Five communities were subjected to controlled trampling in either winter or summer and then left for 6 or 7 weeks to recover. Figure 12.1 shows relative live cover after summer trampling as the horizontal axis and recovery as the vertical one. The sites fall into three groups:

(1) Acid grassland and *Calluna* heathland were seriously damaged at first and showed little recovery.
(2) Neutral/basic grasslands were less damaged at first but showed moderate levels of recovery.
(3) Acid grassland on gleyed soil suffered intermediate damage at first but recovered substantially.

The unmown acid grassland on gleyed soils (3) showed the best recovery whereas the *Calluna* heathland on podzolised pebble gravels (1) showed the poorest. Three other sites, on neutral or acid grassland, showed intermediate degrees of recovery. Resilience to trampling was believed to be enhanced by high soil fertility, the presence of certain species such as *Dactylis glomerata*, and the drainage characteristics of the site. Harrison also suggested that for acid and neutral grasslands in lowland England, quite short periods of rest could make good the impacts of trampling in both summer and winter.

Damage and Recovery from Trampling

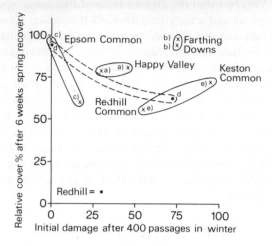

FIGURE 12.1.   Relative live cover of vegetation at the end of the winter treatment and relative live cover after 6 weeks' recovery. Initial damage and recovery are measured in terms of per cent live cover from a pair of replicated treatment strips. From Harrison (1981).

VISITOR DISTRIBUTION AND ATTITUDES

Site managers often wish to manipulate the distribution of visitors and so information about their origins and attitudes is useful. Many research workers claim to measure the levels of demand but actually measure levels of use which is not, of course, the same thing. In fact, for most studies it is the level of use which is important whilst the level of demand indicates the possible increase in use in the future. Direct observation and counting numbers of people per unit length of transect per unit time are still popular measures of visitor pressure and distribution and provide useful results. Hand-held 35 mm cameras, aerial photography and time-lapse photography also have a role to play and electronic people counters are part of the recreation ecologist's armoury. Questionnaire maps are less precise but can give a very useful overview of visitor distribution (Goldsmith, Munton, and Warren, 1970).

Visitor's attitudes are more difficult to determine but can be approached by means of semantic differential questionnaires. These were used in the Countryside Commission's Kynance Cove study (O'Connor, Goldsmith, and Macrae, 1980; see below) to monitor the visitors' reactions to onsite works. Five subtly different meanings for various characteristics of the car park, clifftops, steps to cove, footpaths and the beach were used in three different years. The number of visitors who considered the steps too steep dropped from 52 per cent to 28 per cent over the 2 years and those who considered them too narrow dropped from

47 per cent to 28 per cent, whilst those who found them unobtrusive or attractive went up from 67 per cent to 88 per cent. Visitor's attitudes are a complicated subject and their responses to questionnaires are affected by overall impressions of the area, their perception of the restoration effort and the interpretation programme offered. I suspect that the visitors at Kynance were 'tranquillized' by the interpretation provided and the fact that volunteers were working in their interest and their responses were therefore affected as much by this as by the characteristics of the steps *per se*.

MANAGEMENT OF VISITORS—THE IVINGHOE PROJECT

At Ivinghoe Beacon, in another study commissioned by the Countryside Commission on National Trust property, an area of chalk grassland was heavily scarred by white, chalk paths (Goldsmith and O'Connor, 1975). By means of questionnaire maps and twelve transects it was possible to establish the relationship between numbers of visitors per metre per hour and the width of bare ground and width of short turf. The shape of this line is difficult to ascertain but if we assume it to be linear we can draw an empirical wigwam-shaped diagram for site managers dealing with chalk grassland (Figure 12.2). It shows how bare ground first appears at about 10 000 visitors per transect per year and gets progressively wider as the number of visitors increases. The width of short

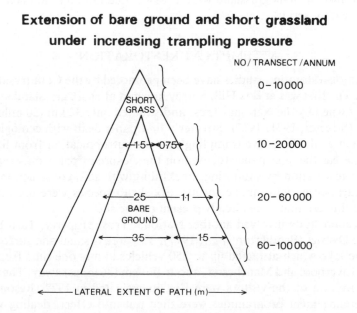

**Extension of bare ground and short grassland
under increasing trampling pressure**

FIGURE 12.2.   Data from chalk grassland at Ivinghoe Beacon. From Goldsmith and O'Connor (1975). Reproduced by permission of Applied Science Publishers.

turf on each side of the bare central zone however appears to remain relatively constant starting at 75 cm and increasing to about 150 cm with numbers of about 60 000–100 000 visitors per annum.

The Ivinghoe project involved the experimental rerouting of visitors. The normal route was considered the control for the experiment and alternative paths were used for the rerouting which was carried out on selected Sundays using temporary signs. The experimental rerouting consisted of a one-way system, a longer dogleg upward route and an attempt to concentrate everybody on a single direct ascent/descent route. The results were not particularly encouraging as only a small proportion of visitors responded to the signs on each of the rerouting days. This was probably because a high proportion of visitors were familiar with the site and knew which way to go and because even strangers could see the top of the Beacon Hill and therefore could see the direct route to their destination. It was concluded that it was impractical to incorporate a major rerouting in the restoration plan. Bare chalk can carry large numbers of people but whilst white chalk is obtrusive in the landscape the ecological damage is restricted to a relatively small area. In fact the principal ecological problem at Ivinghoe Beacon is not paths at all but the high rate of invasion by hawthorn scrub which is rapidly reducing the area of the chalk grassland. The most visible feature on a site (i.e. bare chalk) is not necessarily the most serious problem, as judged from the ecological viewpoint. The real problem is the rapid disappearance of chalk grassland with its associated flora and fauna as a result of the spread of scrub.

ATTEMPTS AT RESTORATION

Three major restoration studies have been promoted by the Countryside Commission. The first was at Box Hill, Surrey, an area of chalk grassland and other habitats owned by the National Trust and situated only 32 km (20 miles) from London (Streeter, 1971, 1977). Streeter's first paper dealt with ecological problems such as the excessive trampling and the eutrophication from litter and dogs, whereas the later paper focused on the closure of paths on steep slopes and their restoration by shuttering, backfilling with soil, construction of new steps and planting with native shrubs. Six shrub species were used, selected from the thirteen that were already present on the site.

The second study involved another National Trust property, Tarn Hows in the Lake District (Brotherton *et al.*, 1977). The area is small and surrounds an attractive lake which attracted up to 250 vehicles at any one time from as far away as Liverpool and Manchester, about 140 km (86 miles) away. The site and the distribution of the visitors were first investigated in 1972. Resource and visitor management programmes were then put into effect, dealing with the vegetation, the improvement of paths and the relocation of carparks. Because it was believed that the use of some areas should be reduced by about 60 per cent, fence lines were set up. At the same time trial plots were used to study the feasibility of using topsoil, seed, turves, fertilizer and landmesh, and areas of existing vegetation were spiked or treated with 12:18:18 compound fertilizer in either autumn or spring. The general conclusion was that if the pressure of trampling could be reduced and an adequate depth of uncompacted soil remained, then rapid revegetation would proceed spontaneously. Accordingly the visitors were rerouted, information about the site was provided, and after the nearest carpark was closed the number of visitors who went down the steepest slopes was reduced from 250 to around 100 per day. The report also contained an interesting discussion about the means of reducing the number of visitors which included the withdrawal of publicity, charging for entry and reducing the car-park capacity; 66 per cent of visitors indicated that they would be willing to pay, and 40p per car (1977) was considered to be an amount that would cover the cost of collection and would deter a significant number of potential visitors. Clearly sites such as Tarn Hows will need to be continuously managed in the future, and owners such as the National Trust will have to decide how this is going to be funded.

The third study involved Kynance Cove, a National Trust property referred to above, on the Lizard Peninsula in Cornwall (Goldsmith, 1977; O'Connor, Goldsmith, and Macrae, 1980). An average of 1400 visitors per day were attracted to the tiny cove in August 1975 and whilst no charge was being made by the owners, there was, as at several other sites (Dovedale, for example) a fee payable to a private individual for the use of the access road and the carpark. Between the carpark and the cove was an area of maritime grassland and Cornish heath, dominated by *Erica vagans*, with several unusual associated species. The area was a reserve managed by the Cornwall Naturalists Trust, an SSSI and was designated Grade 1 status by the Nature Conservancy Council. The unusual flora and fauna probably depended on (a) the serpentine rocks which produce soils with a high pH and little calcium, but significant amounts of magnesium, (b) the very mild southerly climate and (c) the high fallout of salt from sea spray.

The first phase of the project consisted of a year's survey work. The large number of visitors were found to have caused extensive damage to the turf, including the total destruction of 500 m^2 and the erosion of 215 m^3 of soil. The capacity of the site had clearly been exceeded, but the numbers of visitors could not be reduced because it was the policy of the National Trust, so far as is possible, not to restrict the number of visitors to its countryside properties. What was needed to quantify the capacity of the different types of site, was a detailed analysis of the movements of the visitors and their attitudes, and of the soils vegetation, drainage, slope and erosion. Transects were positioned along which the vegetation was recorded; questionnaire maps and semantic differential questionnaires were distributed and the returns analysed, and photoflux and pressure-switch counters were set up and the results interpreted.

Nine associations of plant species had apparently been derived from two original vegetation types as a result of different levels of trampling. Each successive group was evidently produced by approximately doubling the numbers of people per metre per hour on the group of species one stage nearer to the original vegetation. It appeared that at least a halving of visitor pressure was required across the whole area to change the vegetation of each zone one stage nearer to the original.

The second phase of the project involved the experimental rerouting of the visitors that included a new return path constructed by Conservation Corps volunteers and the improvement of the surface of the main existing path. Experimental plots were established to consider the feasibility of revegetating bare areas with local seed, commercial seed, clifftop turf or commercial turf. The plots were established in areas of both high and low salinity. The conclusions of this phase were that the visitors had to be attracted onto artificially hardened surfaces but that these had to be of local serpentine material so as not to change the unusual characteristics of the local soils. In some places, better drainage of the path surfaces was needed, but elsewhere the topsoil was moved to expose, and then to use, the underlying rock as path surface. The surplus soil was then used for revegetation elsewhere. It was recommended that in the final phase in which contractors would be used to restore the site only light machinery should be used and that it be restricted to the paths. A local contractor was very cooperative in this respect and his sympathetic attitude was vital to the overall success of the project.

For the restoration phase in the third year a project officer was appointed to monitor the flow of the visitors, to sample vegetation on transects, to work alongside the contractors, to maintain the interpretative display in a specially provided hut, and generally to supervise. For the revegetation local turf was used alongside the paths and elsewhere reseeding used local seed or material bought from weed-seed suppliers. The use of material from offsite was kept to an absolute minimum because of the high nature conservation status of the area. No fencing was used, even around the experimental plots, as it was considered that it might damage the cooperative attitudes that were being nurtured amongst both the visitors and the locals. Large stones removed during the construction of the path were however judiciously used along the edges of the main paths.

Evaluating the restoration

The restoration at Kynance was inexpensive, when the length of path, the steepness of the site and the number of steps that were required is considered, but it was largely successful. The serpentine chippings on the new paths were a little conspicuous but vital for ecological reasons, and whilst drainage had been carefully planned it was not overemphasized as, again, much of the ecological character of the site was due to impeded drainage (Goldsmith, 1982). The principal factors contributing to the success of the operation were:

(1) An initial understanding of the ecology of the site.
(2) The presence of a project officer during the restoration phase.
(3) A sympathetic contractor who used small and light machinery.
(4) Very valuable work by a dedicated group of Conservation Corps volunteers.

SOME GENERAL GUIDELINES FOR THE FUTURE

An understanding of the functioning of ecosystems, the requirements of each of their component species and the attitudes and probable movements of visitors are technical subjects which make recreation ecology a demanding interdisciplinary research subject. However, restoration of damaged sites requires common sense and a sympathetic attitude to the environment as well as the detailed study of both the resource and the visitors.

Whilst there are 350 papers defining the problems (Cole and Schreiner, 1981), the number dealing with planning for recreation and the restoration of damaged sites are relatively few. There are indeed the three projects mentioned above, some classic American ones dealing with campsites (LaPage, 1967), a book on forest recreation (Douglass, 1975), some specific projects in national parks such as the park'n'ride experiment in the Peak District (Miles, 1971), but one important subject area has been seriously neglected—the feasibility or practicability of trying to recreate a site from bare ground (Wells, 1981). In *Conservation in Practice* (Warren and Goldsmith, 1974) we called this creative conservation. Many ecologists instinctively shy away from it, because I think that if it turns out to be feasible the protection for areas threatened by planning applications would dwindle overnight, because the original species-rich vegetation could be recreated.

The subject of recreation ecology and restoration covers such a wide range of problems from trampling, boats disturbing birds and lakeside cottages polluting water to snowmobiles reducing the productivity of pasture, that it is difficult to make generalizations. Nevertheless, the following general rules might prove useful for restoration activities:

(1) Use local indigenous material (such as soil, seed, rock) whenever possible and avoid introductions, especially if the site is an ecologically important one.
(2) Work with rather than against the wishes of users (such as visitors).
(3) Minimize the use of signs which are already excessive in some areas and avoid 'DO NOT' type of instructions.
(4) Use volunteers whenever possible because they are highly motivated and inexpensive. They also produce a sympathetic, positive response amongst locals and visitors.

(5) Use small-scale machinery. The contractor at Kynance Cove was teased by competitors for the contract as the man who used 'Dinky' toys, but he won the contract and the job was well done. Small is beautiful when it comes to restoration work.

(6) Ensure that manmade features look natural by using curved or interrupted edges—avoid straight lines and right angles. Keep the appearance of facilities rustic, watch out for features which affect the skyline, and avoid evenly spaced steps, etc.

(7) Avoid fencing or overt means of restraining visitors whenever possible. If the path surface is easier to walk on than the more sensitive vegetation people will follow it.

(8) Good information and interpretation for the public engenders their understanding and support and reduces the level of vandalism.

(9) Most of the recommendations above will prove to be cost-effective and will therefore be well received by the site manager.

REFERENCES

Batten, L.A. (1977). Sailing on reservoirs and its effects on water birds. *Biological Conservation*, **11**, 49–58.

Bayfield, N.G. (1971a). Some effects of walking and skiing on vegetation at Cairngorm. In E. Duffey and A.S. Watt (Eds), *The Scientific Management of Animal and Plant Communities for Conservation*, Blackwell, Oxford, 652 pp.

Bayfield, N. (1971b). A simple method for detecting variations in walker pressure laterally across paths. *Journal of Applied Ecology*, **8**, 533–535.

Bayfield, N.G. (1980). Replacement of vegetation on disturbed ground near ski lifts in the Cairngorm Mountains, Scotland. *Journal of Biogeography*, **7**, 249–260.

Bayfield, N.G., Urquhart, U.H. and Cooper, S.M. (1981). Susceptibility of four species of *Cladonia* to disturbance by trampling in the Cairngorm Mountains, Scotland. *Journal of Applied Ecology*, **18**, 303–310.

Brotherton, D.I. (1981). Habitat restoration: why has not more been done. In E. Duffey (Ed), *Habitat Recovery and Restoration*, RERG Report No. 7, pp. 1–8.

Brotherton, D.I., Maurice, O., Barrow, G. and Fishwick, A. (1977). *Tarn Hows: An Approach to the Management of a Popular Beauty Spot*, Countryside Commission, Cheltenham, CP106, 34 pp.

Buchanan, K. (1975). Some effects of trampling on the flora and invertebrate fauna of sand dunes. *Discussion Papers in Conservation 13*, University College London.

Burton, R. (1971). Recent vegetational changes in the heathland of Cannock Chase, *North Staffs Journal of Field Studies*, **II**, 15–35.

Cole, D.N. and Schreiner, E.G.S. (1981). *Impacts of Backcountry Recreation: Site Management and Rehabilitation—An Annotated Bibliography*. USDA Forest Service Intermountain Forest Range and Station Report INT-121, Ogden, Utah, 58 pp.

Coppock, J.T. and Duffield, B.S. (1975). *Recreation in the Countryside: A Spatial Analysis*, Macmillan, London, 262 pp.

Douglass, R.W. (1975). *Forest Recreation, 2nd Edition*, Pergamon, 336 pp.

Duffey, E. (1975). The effects of human trampling on the fauna of grassland litter. *Biological Conservation*, **7**, 255–274.

Goldsmith, F.B. (1974). Ecological effects of visitors in the countryside. In A. Warren and F.B. Goldsmith (Eds), *Conservation in Practice*, Wiley, London, pp. 217–231.

Goldsmith, F.B. (1977). Rocky cliffs. In R.S.K. Barnes (Ed), *The Coastline*, Wiley, Chichester, pp. 237–251.

Goldsmith, F.B. (1982). Lessons from the Kynance Cove Project. In E. Duffey (Ed), *Habitat Restoration and Reconstruction*, RERG Report No. 7, pp. 50–56.

Goldsmith, F.B. and O'Connor, F.B. (1975). *Ivinghoe Beacon: Experimental Restoration Project*, University College London, 136 pp.

Goldsmith, F.B., Munton, R.J.C. and Warren, A. (1970). The impact of recreation on the ecology and amenity of semi-natural areas: methods of investigation used in the Isles of Scilly. *Biological Journal of the Linnean Society*, 2, 287–306.

Harrison, C. (1981). Recovery of lowland grassland and heathland in southern England from disturbance by seasonal trampling. *Biological Conservation*, 19, 119–130.

Hammitt, F.E. (1980). Managing bog environments for recreational experiences, *Environmental Management*, 4(5), 425–431.

Jewell, P.A. (1974). Problems of wildlife conservation and tourist development in East Africa. *Journal of the South African Wildlife Management Association*, 4(1), 59–62.

Keddy, P.A., Spavold, A.J. and Keddy, C.J. (1979). Snowmobile impact on old-field and marsh vegetation in Nova Scotia, Canada: an experimental study. *Journal of Environmental Management*, 3, 409–415.

Kellomäki, S. (1973). Ground cover response to trampling in a spruce stand of *Myrtillus* type (in Finnish). *Silva Fennica*, 7(2), 96–113.

Kellomäki, S. and Saastamoinen, V.L. (1975). Trampling tolerance of forest vegetation. *Acta Forestalia Fennica*, 147, 5–19.

LaPage, W.F. (1967). Some observations on campground trampling and ground cover response. *US Forest Service Research Paper NE-68*.

Lavery, P. (Ed) (1974). *Recreational Geography*, David and Charles, Newton Abbot, 335 pp.

Liddle, M.J. (1975a). A selective review of the ecological effects of human trampling on natural ecosystems. *Biological Conservation*, 7, 17–34.

Liddle, M.J. (1975b). A theoretical relationship between the primary productivity of vegetation and its ability to tolerate trampling. *Biological Conservation*, 8, 251–255.

Liddle, M.J. and Greig-Smith, P. (1975a). A survey of tracks and paths in a sand dune ecosystem. I. Soils. *Journal of Applied Ecology*, 12, 893–908.

Liddle, M.J. and Greig-Smith, P. (1975b). A survey of tracks and paths in a sand dune ecosystem. II. Vegetation. *Journal of Applied Ecology*, 12, 909–930.

McQuaid-Cook, J. (1978). Effects of hikers and horses on mountain trails. *Journal of Environmental Management*, 6(3), 209–212.

Miles, J.C. (1971). Rural park and ride: Goyt Valley traffic experiment. *Surveyor*, 4 June.

Morris, M.G. (1967). Insect collecting with special reference to nature reserves. In E. Duffey (Ed), *The Biotic Effects of Public Pressures on the Environment*, Monks Wood Symposium, 3, Nature Conservancy.

O'Connor, F.B., Goldsmith, F.B. and Macrae, M. (1980). *Kynance Cove: a restoration project*. Countryside Commission Publication, 128, 70 pp.

Perring, F.H. (1967). Changes in chalk grassland caused by galloping. In E. Duffey (Ed), *The Biotic Effects of Public Pressures on the Environment*, Monks Wood Symposium, 3, pp. 134–42.

Picozzi, N. (1971). Breeding performance and shooting bags of red grouse in relation to public access in the Peak District National Park, England. *Biological Conservation*, 5, 211–215.

Rees, J. and Tivy, J. (1978). Recreational impact on Scottish lochshore wetlands. *Journal of Biogeography*, **5**, 93–108.

Satchell, J.E. and Marren, P.R. (1976). *The Effects of Recreation on the Ecology of Natural Landscapes*, Council of Europe Nature and Environment Series No. *11*, Strasbourg, 117 pp.

Slater, F.M. and Agnew, A.D.Q. (1977). Observations on a peat bog's ability to withstand increasing public pressure. *Biological Conservation*, **11**, 21–27.

Smith, P.H. (1981). Cross-country vehicles and nature conservation. *Ecos*, **3(2)**, 22–27.

Speight, M.C.D. (1973). Ecological change and outdoor recreation, *Discussion Papers in Conservation*, **4**, University College London.

Streeter, D.T. (1971). The effects of public pressure on the vegetation of chalk downland at Box Hill, Surrey. In E. Duffey and A.S. Watt (Eds) (1971). *The Scientific Management of Animal and Plant Communities for Conservation*, Blackwell, Oxford, pp. 459–468.

Streeter, D. (1977). Gully restoration on Box Hill. *Countryside Recreation Review*, **2**, 38–40.

Takahashi, R. and Maenaka, H. (1977). Relationship between recreational densities and vegetational types of grassland of Wakakusayama Hill, Nova Park. *Japanese Institute of Landscape Architects*, **40(3)**, 24–37.

Tuite, C. (1980). Waterfowl, conservation and the effects of disturbance by water-based recreation: a review of the problem. In A. Fishwick (Ed), *Wildlife Conservation and Recreation*, RERG report 4, pp. 53–62.

Wall, G. and Wright, C. (1977). *The Environmental Impact of Outdoor Recreation*, Department of Geography Publication, 11, University of Waterloo, Ontario, 69 pp.

Warren, A. and Goldsmith, F.B. (1974). *Conservation in Practice*, John Wiley, London, 512 pp.

Watson, A. (1979). Bird and mammal numbers in relation to human impact at ski lifts on Scottish hills. *Journal of Applied Ecology*, **16**, 753–764.

Weaver, T. and Dale, D. (1978). Trampling effects of hikers, motorcycles and horses in meadows and forests. *Journal of Applied Ecology*, **15**, 451–457.

Wells, T. (1981). Creating attractive grasslands from seed in amenity areas. In E. Duffey (Ed), *Habitat Restoration and Reconstruction*, RERG report no. 7, pp. 9–16.

Woodland, D.J. and Hooper, J.N.A. (1977). The effect of human trampling on coral reefs. *Biological Conservation*, **11**, 1–4.

CHAPTER 13

The Creation of Species-Rich Grasslands

T.C.E. WELLS

INTRODUCTION

The British Isles contain a great variety of ecosystems of which almost all are, to a greater or lesser extent, influenced by man. The extent to which the natural vegetation has been modified or changed is constantly being reappraised (see Sheail, Chapter 18), but there can be little doubt that man and his grazing animals have played an important part in the formation of all of our native grasslands. When and how these were formed is still very much a subject for debate. In his classic work *The British Isles and their Vegetation*, Tansley (1953) described the chalk downlands 'as having been used as sheep-walks for many centuries and probably from Neolithic times' and from this arose the view, still widely held, that most if not all of our floristically rich grasslands are of great age, and can only be created after a long period of sheep grazing. This rather simple view, which may be valid in some cases, is now being challenged. There is increasing evidence of a much more complex pattern of land use in the past, whereby tracts of downland were ploughed up and sown with grain crops for a few years and then allowed to revert to grass. This seems to have been the case on the Porton Ranges, on the Hampshire/Wiltshire border, where a variety of grassland types differing in floristic composition were shown to be related to the period since the area was last ploughed (Wells *et al.*, 1976). Many present-day grasslands have been under cultivation within the last 200 years, and even some of the slopes, considered too steep for cultivation today, contain evidence of ploughing in the past, in the form of strip lynchets (Whittington, 1962) or ridge and furrow (Eyre, 1955).

The origins and age of meadows and pastures on clay and alluvial soils is equally obscure. Lane (1980) argued that prior to the sixteenth century six kinds of land produced feed for livestock. These included wastes which bordered the edges of villages and parishes; the ground cover on fallows, which may have had a catch crop of vetches; the stubble left on arable lands after the harvest;

woodlands which were browsed and or grazed for fruits and herbaceous plants; meadows, which are lands bordering running water and liable to flood; and pastures which are drier, and may or may not form part of the commons. Two distinct types of pasture and meadows can be distinguished on biological evidence: those which were solely grazed, and those which were mowed and grazed. It seems likely that wet, species-rich meadows lying in the flood plains of major rivers are ancient ecosystems that have been managed in the same way for centuries. In the case of Lammas meadows, as for example at Cricklade in Wiltshire and at Portholme in Cambridgeshire, we know that the time of year at which the hay was cut and the aftermath grazed was closely regulated by local tradition and enforced by law (Wells, 1969). From the scientific and conservation point of view the few remaining Lammas meadows are priceless, in the sense that they contain communities which have arisen over a very long period of time in response to particular management regimes—they provide reference points against which important theoretical concepts, such as the relationship between community diversity and stability, can be tested. They also provide a useful yardstick against which newly created grasslands (see page 217) can be measured.

Considerable changes in pasture and meadow management took place from the middle of the sixteenth century onwards, partly as a result of technical innovations introduced from the continent and partly in response to the demand, created by expanding towns, for livestock. In early times, meadows had been formed from arable land either by allowing seed from surrounding land to colonize the ploughland ('falldown') or by using the sweepings from the hay store. From about 1650 onwards, various legumes and grasses, such as lucerne (*Medicago sativa*), sanfoin (*Onobrychis viciifolia*), white clover (*Trifolium repens*) and rye-grass (*Lolium perenne*), were specially imported from the Low Countries and sown. For example, in 1685, Edward Fuller, a seed merchant at the May Pole in the Strand, London offered 'Clover-grass, Hop Cover, Sainefoin, La Lucern, Rye-Grass, French Furz, Dantzick-Flax' under the heading 'Seeds to improve Land' (quoted in Lane, 1980). Later writers, such as Ellis (1747), described in some detail ways of 'converting plowed land into a delicate thick planted profitable meadow', advocating the use of selected grasses and legumes and warning of the danger of 'laying down their plowed Ground with a promiscuous Mixture of common Grass-Seeds'. He also recommended improving 'old decayed Meadows' by broadcasting hay-seeds or clover into the meadows in February, followed by rolling 'to squeeze the seeds into the Earth'. For a more detailed account of grasslands and their history, the reader should refer to Chapter 1 in Duffey *et al.* (1974). The main point that needs emphasis here is that all lowland grasslands, whether floristically diverse or containing only a few species, have been created, either directly or indirectly, by the activities of man. Natural grasslands, in the sense of grasslands untouched by man, do not exist in western Europe.

THE NEED TO CREATE FLORISTICALLY ATTRACTIVE GRASSLANDS

Although grasslands occupy an estimated 49 163 km^2 (4.92 × 10^6 ha) in the United Kingdom (Natural Environment Research Council, 1977) only a small proportion are of conservation interest. Short or long-term leys ('rotation grass'), permanent grass that has been fertilized, and intensively used 'amenity' grassland, such as football pitches, lawns and school playing fields, usually consist of only one or a few species and are of little interest to the conservationist. In view of the importance placed on floristically rich grassland by the conservation movement, it is surprising that the area covered by them has never been calculated. For some grassland types, information is available. For example, in 1966, a special survey was undertaken of grasslands on chalk soils and a summary of the results is given in Table 13.1. Only 3.3 per cent of the chalkland was covered by permanent pasture, of which nearly three-quarters occurred in Wiltshire, which had 14 per cent of the total area of chalkland in England (Blackwood and Tubbs, 1970). Even these data have to be treated with caution as they only provided information on the quantity of that type of grassland and say nothing about the quality (or species-composition). In a botanical survey of grasslands in Huntingdonshire, made in 1973, only about 356 ha (3–4 per cent) out of 13 544 ha of permanent grassland and rough grazing were assessed as being of conservation interest (Wells, unpublished). The grasslands of botanical interest were distributed between 42 sites, of which 40 were less than 16 ha in size.

The main threat to floristically rich grasslands comes from agriculture. So-called marginal lands, previously thought too steep for machinery, have now been ploughed. Applications of inorganic fertilizers by light aircraft to grasslands on steep slopes and the increased use of selective herbicides have led to an increase in the productivity of grasses and a decline or total loss of many attractive broad-leaved plants. Water-tables have been lowered by water authorities, to control seasonal flooding, and this has enabled farmers to plough up ancient species-rich alluvial meadows. Much permanent pasture has disappeared under buildings as new towns, such as Greater Peterborough and Milton Keynes, have been developed, or on a smaller scale, as villages expanded. Other factors which have led to the decline in the area of species-rich grasslands include demands for gravel, sand and cement, the use of land for inland reservoirs (for example, Grafham Water and Rutland Water), for motorways and roads, and changes in the management of roadside verges by local authorities. Proposals to expand forestry in upland areas presage yet another threat.

METHODS FOR CREATING GRASSLAND USING NATIVE PLANT SPECIES

Much of the stimulus for research on the creation of plant communities has come from the need to restore or revegetate areas of derelict land left by mining

TABLE 13.1 Chalk grassland survey 1966: summary of data obtained.

County or topographical unit	Acreage* of chalk grassland	Acreage* of land on chalk	Chalk grassland percentage	Number of fragments	Number of fragments by size categories (acres*)					
					5–50	51–100	101–200	201–300	301–400	400+
Yorkshire Wolds	3939	262440	1.5	61	35	15	5	4	2	—
Lincolnshire Wolds	225	143280	0.2	12	12	—	—	—	—	—
Norfolk	44	332640	0.01	3	3	—	—	—	—	—
Cambridgeshire	170	156600	0.1	3	2	1	—	—	—	—
Chiltern Hills (Bedfordshire, Berkshire, Buckinghamshire, Hertfordshire and Oxfordshire)	2028	550080	0.4	59	47	7	4	1	—	—
Berkshire	1573	185400	0.8	59	50	7	2	—	—	—
Wiltshire	73085	464400	15.7	529	407	51	34	16	5	16
Dorset	8371	236880	3.5	145	101	28	10	2	1	3
Hampshire	5224	239760	2.2	119	95	13	7	1	—	3
Isle of Wight	2128	14040	15.2	24	13	6	2	—	2	1
North Downs (Kent and Surrey)	2226	437760	0.5	94	83	10	1	—	—	—
South Downs (Sussex)	8592	223200	3.8	117	83	13	11	2	1	7
Total	107605	3246480	3.3	1225	931	151	76	26	11	30

Source: Blackwood and Tubbs (1970), reproduced by permission of Applied Science Publishers.

* 1 acre = 0.405 ha.

and quarrying industries in the developed countries, notably in Britain and America. Recent reviews by Humphries (1979), Johnson and Bradshaw (1979) and Bradshaw and Chadwick (1980) provide a full, and up-to-date account of progress in this particular area of research, where special problems have been encountered such as extremes of temperature, acidity and alkalinity, high salinity and metal toxicities, shortage of water and nutrients. Much less attention has been given to establishing attractive grasslands on areas of land, where extremes of soil and water conditions do not prevail. The methods discussed below are intended for use on areas such as roadside verges, land for amenity purposes in towns, country parks, nature reserves and gardens where the objective is to provide a floristically diverse grassland which will both enhance the aesthetic quality of the environment and encourage the greatest diversity of wildlife.

ESTABLISHMENT OF GRASS/DICOTYLEDON MIXTURES ON CULTIVATED SOILS

In our present state of knowledge, it is not possible by sowing mixtures to recreate grasslands with the structure and floristic composition of long-established seminatural grasslands. On the other hand, it is possible to create grasslands which contain many of the species found in those grasslands, and with further research there is every reason to believe that the range of species can be increased.

In choosing species for seed mixtures the following criteria should be observed:

(1) The species should be regular members of the grassland community.
(2) They should not be rare.
(3) They should be relatively abundant in a variety of grasslands and preferably have a wide distribution.
(4) They should be perennial, preferably long-lived, and with an effective means of spread.
(5) A high proportion of the species used should have colourful flowers, and these should preferably also be attractive to insects.
(6) Highly competitive species, especially those known to form single species stands in the wild, should be avoided.
(7) The seed should have a high percentage germination over a range of temperatures and should not have dormancy problems or special requirements.

Not all of these requirements may be met, but it may still be desirable to include a particularly attractive species. For example, cowslip (*Primula veris*) will not germinate until it has received a winter cold treatment—nevertheless it is still worth including in a mixture because it provides colour in the early part of the year, and it is known from experimental sowings that it is able to establish even though it may germinate several months after other species. Economic factors may exert a strong influence on the choice of species. Studies by Wells

(1979a) have shown that hand-collected wildflower seed is expensive and likely to remain so even when semicommercial methods of seed production, such as those described in Wells, Bell and Frost (1981), are employed. Grass seed, on the other hand, is available commercially and is comparatively cheap, and for this reason alone grasses must clearly form the major component of any grass–wildflower mixture.

There are a large number of grass species and cultivars available for sowing in grass–wildflower mixtures. Valuable information oñ the characteristics of new cultivars is given annually by the Sports Turf Institute, Bingley (Anon, 1981) and by other organizations such as Rijksinstituut voor het Rassenonderzoek van Cultuurgewassen, Wageningen, Netherlands. These publications should be consulted for an up-to-date assessment of the best cultivars to use. In general, tall, highly productive species, such as tall fescue (*Festuca arundinacea*), cocksfoot (*Dactylis glomerata*), perennial rye-grass (*Lolium perenne*) and timothy (*Phleum pratense*) are unsuitable.

The commercially available grasses which appear compatible with most dicotyledons include many of the cultivars of red fescue (both *Festuca rubra* spp. *rubra* and *F. rubra* spp. *commutata*), common and fine-leaved sheep's fescue *F. ovina* and *F. tenuifolia*, common bent *Agrostis tenuis* and crested dog's tail *Cynosurus cristatus*. Native grasses not presently available in commercial quantities and which form a fine, yet not too competitive matrix, include meadow barley (*Hordeum secalinum*), yellow oat-grass (*Trisetum flavescens*), quaking-grass (*Briza media*) and crested hair-grass (*Koeleria gracilis*). Moffat (1981) provided a useful guide to the range of amenity grass mixtures available in Britain for sowing on specific soil types, many of which would be compatible with wild flower mixtures.

Although a large number of dicotyledons are found in grassland, only a small selection have been tested in artificial mixtures. Those that have established and prospered in a number of trials on chalk, clay and alluvial soils in Britain over the past 8 years are listed in Table 13.2, together with the recommended sowing rates. This list is meant to be used as a guide to the species most likely to succeed on a particular soil type, but it is not exhaustive.

Early results obtained from sowing mixtures of herbaceous species and grasses from a range of soil types are encouraging. Wathern and Gilbert (1978) sowed a mixture containing 26 dictyledons with three commercial grass seed mixtures (a) a *Lolium*-based mixture (*L. perenne* 75 per cent, *Poa trivialis* 15 per cent, *Dactylis glomerata* 5 per cent and *Trifolium repens* 5 per cent); (b) *Festuca* mixture (*F. ovina* 40 per cent, *F. rubra* 40 per cent, *P. pratensis* 10 per cent and *Agrostis tenuis* 10 per cent); (c) and *Agrostis/Festuca* mixture (*A. tenuis* 50 per cent, *F. ovina* 20 per cent, *F. rubra* 30 per cent) into a limestone subsoil. A fourth area (control) was sown only with the wild flower mix. After 2 years, 22 of the dictyledons had established in the *Festuca* sward, while slightly fewer were found in the other swards, the control area, with seventeen, containing

TABLE 13.2 List of 40 dicotyledons (herbs) which have performed well in mixtures in trials in England 1973−81, with recommended sowing rates.

	Clay soils	kg ha⁻¹ Chalk or limestone soils	Alluvial soils
Achillea millefolium	0.1		
Anthyllis vulneraria		2.0	
Betonica officinalis	0.1		
Campanula glomerata		0.01	
C. rotundifolia		0.01	
Centaurea nigra	1.0	1.0	
C. scabiosa	0.5	0.5	
Chrysanthemum leucanthemum	0.2	0.2	0.2
Clinopodium vulgare		0.1	
Conopodium majus			0.1
Daucus carota	0.5	0.5	0.5
Filipendula vulgaris		0.5	
Galium verum	0.3	0.3	0.3
Geranium pratense	1.0		
Hieracium pilosella		0.02	
Hypochoeris radicata	0.3		0.3
Leontodon hispidus	0.5	0.5	
Lotus corniculatus	1.0	1.0	1.0
Lotus uliginosus			1.0
Lychnis flos-cuculi			0.005
Medicago lupulina	2.0	2.0	2.0
Onobrychis viciifolia		2.0	
Ononis repens	1.0	1.0	
Ononis spinosa	1.0	1.0	
Pimpinella saxifraga		0.5	
Plantago lanceolata	2.0	2.0	2.0
P. media	0.2	0.2	0.2
Poterium sanguisorba	0.3	0.3	0.3
Primula veris	0.5	0.5	0.5
Prunella vulgaris	0.5	0.5	0.5
Ranunculus acris	1.0		
Rhinanthus minor	1.0	1.0	1.0
Rumex acetosa			0.2
Sanguisorba officinalis			0.3
Scabiosa columbaria		0.5	
Serratula tinctoria	0.5	0.5	
Silaum silaus	1.0		1.0
Silene alba	0.5		0.5
Thymus drucei		0.1	
Veronica chamaedrys		0.1	

TABLE 13.3   The frequency of sown herbs in different grass swards on limestone soils recorded 5 years after sowing. Swards at density of 15 g m^{-2} herbs at 10 seeds m^{-2}.

	Control	*Lolium*	*Agrostis/Festuca*	*Festuca*
Sown spp.				
Centaurea nigra	1.50	0.60	1.50	11.50
Primula veris	0.50	0.80	2.25	5.25
Lathyrus pratensis	4.25	0.80	1.50	31.00
Chrysanthemum leucanthemum	1.00	1.20	4.25	8.50
Galium verum	0.75	1.20	0.50	2.00
Anthyllis vulneraria	2.00	0.60	1.75	2.25
Lotus corniculatus	13.50	18.80	25.00	50.25
Poterium sanguisorba	0.50	0.40	2.00	5.25
Plantago lanceolata		0.40	1.00	6.00
Anthriscus sylvestris				0.25
Rumex acetosa	0.25	0.40		2.25
Ranunculus acris	0.50			
Prunella vulgaris	0.75	0.80		4.25
Medicago lupulina	0.75			10.50
Vicia angustifolia				1.50
Knautia arvensis	0.25			0.50
Achillea millefolium		0.60	1.50	3.50
Hypericum perforatum				0.25
Total frequency of sown herbs	26.50	26.60	41.25	142.00
Local frequency of bare ground	79.50	81.20	48.00	16.75

Source: Wathern and Gilbert (1978). Artificial diversification of grassland with native herbs. *Journal of Environmental Management*, 7, 29–42; reproduced by permission of Academic Press Inc. (London) Ltd.

least. Only four species failed to establish in any plot. After 5 years, the number of sown species still present declined to eighteen (Table 13.3). However, the total frequency of sown herbs increased markedly in each treatment during this period. The control and *Lolium* treatment had similar total frequencies* of sown herbs, 26.5 and 26.6 respectively, whereas the *Agrostis–Festuca* (41.2) and *Festuca* (142.0) swards had significantly higher total frequencies. The superior establishment of sown and invading herbaceous species in the *Festuca* grassland appeared to be related to the intermediate sward density which characterized this young grassland during the first 5 years. Unshaded germination sites remained available and considerable shelter was afforded by the grass plants.

In an experiment on a heavy clay soil at Monks Wood Experimental Station, Cambridgeshire (Wells, 1979b), eight grass–herb mixtures were sown in June

---

* Total frequency is the sum of the percentage frequency values of individual species determined by counting the number of 10 × 10 cm subdivisions of a 1 m^{-2} quadrat in which each species was rooted.

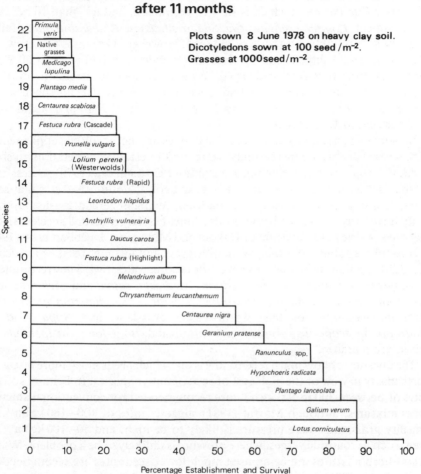

Percentage establishment and survival of 22 species after 11 months

Plots sown 8 June 1978 on heavy clay soil.
Dicotyledons sown at 100 seed /m⁻².
Grasses at 1000 seed/m⁻².

22 Primula veris
21 Native grasses
20 Medicago lupulina
19 Plantago media
18 Centaurea scabiosa
17 Festuca rubra (Cascade)
16 Prunella vulgaris
15 Lolium perene (Westerwolds)
14 Festuca rubra (Rapid)
13 Leontodon hispidus
12 Anthyllis vulneraria
11 Daucus carota
10 Festuca rubra (Highlight)
9 Melandrium album
8 Chrysanthemum leucanthemum
7 Centaurea nigra
6 Geranium pratense
5 Ranunculus spp.
4 Hypochoeris radicata
3 Plantago lanceolata
2 Galium verum
1 Lotus corniculatus

Species

Percentage Establishment and Survival

0 10 20 30 40 50 60 70 80 90 100

FIGURE 13.1.

1978; 23 herbs were sown at the rate of 100 live seed m⁻², and seven grasses at 1000 live seed m⁻². The establishment of sown and unsown species was estimated 2 and 11 months after sowing by counting the number of individuals rooted in randomly placed 20 cm² quadrats. The establishment and survival after 11 months of seventeen dictyledons and five grasses is shown in Figure 13.1. Eight dicotyledons had a percentage establishment of more than 50 per cent, the legume *Lotus corniculatus* and *Galium verum* reaching almost 90 per cent.

Establishment of grasses varied from 11–38 per cent, small seeded cultivars of *Festuca rubra* doing as well as or better than the much larger seeded Westerwolds rye-grass. *Primula veris*, which is slow to germinate, had less than 10 per cent establishment. *Poterium sanguisorba*, *Silaum silaus* and *Rhinanthus minor* failed to establish, probably due to use of old seed. However, 34 months after sowing, percentage establishment and survival of *Primula veris* had risen to 20.8 per cent, indicating that further delayed germination had occurred. In contrast, survival of *Hypochoeris radicata* had fallen from 78 per cent to 17.3 per cent, of *Centaurea nigra* from 55 per cent to 5.5 per cent and of *Plantago media* from 18.0 per cent to 1.6 per cent.

In North America, mixtures of native grasses and herbs, often supplemented with dwarf shrubs, are increasingly being used to establish vegetation on spoil resulting from open-cast mining. De Puit and Coenenberg (1979) successfully established 21 native species over a 2-year period, noting that at equal seed rates, drill seeding was superior to broadcast seeding in perennial grass productivity, primarily due to the better establishment of larger seeded species. In the glaciated prairie pothole region of Dakota and Minnesota, Duebbert *et al.* (1981) successfully established a range of native grass and legume species specifically as wildlife habitat, principally as cover for nesting waterfowl. They recommend *Agropyron elongatum*, *A. intermedium*, *Medicago sativa* and *Melilotus* spp. for 'cool-season grassland', *Andropogon gerardi*, *Sorghastrum nutans* and *Panicum virgatum* for tall, warm-season grassland and *Stipa viridula*, *Andropogon scoparius*, *Agropyron smithii* and *Bouteloua curtipendula* as mixed-grass prairie.

The question of how much seed to sow is one which needs much more research, particularly in view of the high cost of seed of native species. In Britain, sowing rates of between 300 to 500 kg ha^{-1} are recommended for conventional amenity grass mixtures, although Moffat (1981) suggests rates of 100–180 kg ha^{-1} for amenity grassland where pressure is likely to be high, and 50–100 kg ha^{-1} for areas such as roadside verges, where pressure is unlikely to be a problem. Where grass–herb mixtures are being used much lower seed rates are recommended. Wells *et al.* (1981) have established a variety of mixtures using seed rates of 30–45 kg ha^{-1}, in which grasses and herbs are sown in proportions of about 3:1, and the densities of some of these swards indicate that even lower seed rates could be considered. Duebbert *et al.* (1981) estimate that 215–430 'pure live seeds' per m^2 are needed to establish adequate stands of perennial grasses on prairie soils.

Many wild grassland species grow slowly, and establishment may also be slow. In many amenity areas it is essential that a vegetation cover is established quickly, both to prevent soil erosion and for aesthetic reasons. To meet this requirement, it is often necessary to sow a quick-growing annual 'nurse crop' which will germinate quickly, provide a green cover and die-back, allowing those species which will form the basis of the permanent grassland to replace it.

Westerwolds rye-grass (a form of *Lolium multiflorum*) has these characteristics and has proved successful on a range of soil types in Britain, although its value on fertile soils is more doubtful as it tends to 'swamp' the other species. Other grasses, such as rye (*Secale cereale*), sorghum (*Sorghum vulgare*) and common oat (*Avena sativa*), have also been used. Some of the fast-germinating, annual or biennial legumes, such as *Anythyllis vulneraria*, *Medicago lupulina* or *Trifolium arvense*, are possible candidates as nurse crops and this is a fertile area for further research.

Diversifying established swards

Grasslands are never static—they are always in a state of change and much effort, time and money is spent in trying to maintain a desired species composition, particularly in amenity grasslands such as lawns and bowling greens. In agriculture, sown grasslands are soon invaded by unwanted species and 'sward deterioration', as the agriculturalist calls it, begins. In view of the universality of such changes, it is surprising how little is known concerning the processes involved. In a stimulating essay on the maintenance of species-richness in plant communities, Grubb (1977) discusses the theoretical aspects of change in plant communities from a functional point of view. The most important point in grassland communities is that when any one plant individual dies, a gap is created and a new individual ultimately takes its place. It may or may not be of the same species and that is the crucial point. The gap may be filled by vegetative spread of surrounding species or it may be used by a species arriving as a seed. On the surface, this may seem a simple concept, but as Grubb (1977) has shown, the processes involved in the successful invasion of a gap by a given plant species are highly complex (Table 13.4) and difficult to unravel.

TABLE 13.4   Processes involved in the successful invasion of a gap by a given plant species and characters of the gap that may be important.

Processes	Characters
Production of viable seed	Time of formation
flowering	Size and shape
pollination	Orientation
setting of seed	Nature of soil surface
Dispersal of seed	Litter present
through space	Other plants present
through time	Animals present
Germination	Fungi, bacteria and viruses present
Establishment	
Onward growth	

Source: Grubb (1977). The maintenance of species-richness in plant communities: the importance of the regeneration niche. *Biological Review*, **52**, 107–145; reproduced with permission.

From the practical point of view of trying to diversify a predominantly grass sward by introducing dicotyledonous species, the essential thing seems to be to make the sward receptive to incoming propagules, whether arriving naturally or introduced deliberately. Various techniques are available, but all are in an early stage of development.

Herbicides and slot-seeding

Work at the Weed Research Organization has demonstrated the many possibilities that now exist for regulating sward composition using herbicides. Grass-suppressing herbicides such as paraquat, carbetamide, and propyzamide can be used in mixed swards to encourage the growth of *Trifolium repens* (Haggar, 1974). Low doses of dalapon can be used to suppress *Agrostis* spp. (Haggar and Oswald, 1976), asulam will selectively control *Holcus lanatus* (Watt, 1978) and *Poa* spp. can be controlled by ethofumesate (Haggar and Passman, 1978). The most promising development is the combined use of a herbicide, paraquat, with a new slot-seeder machine (Squires, Haggar, and Elliott, 1978; Figure 13.2).

Slot Seeding Procedures

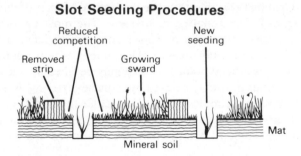

FIGURE 13.2. A shematic section through slot-seeded turf, taken across the line of the trenching. From Haggar (1980); reproduced by permission.

Turf cutting discs on the machine remove a narrow strip of turf, exposing mineral soil while the band spraying of herbicide reduces competition from surrounding vegetation. Seed of whatever species it is desired to introduce into the turf is dropped down a coulter, all in a one-process operation. Although intended primarily for introducing clovers into old pasture, this technique clearly has potential in the amenity field for diversifying floristically dull grasslands, and initial trials in Dorset appear promising (Haggar, 1980).

Cutting, rotovating and harrowing already established swards

The aim of these management practices is twofold: first, to provide areas of bare ground on which sown seed can germinate; and second, to reduce competition from established grasses. In an experiment to compare the establishment and survival of sown herbs in plots which had been cut, or rotovated, or sprayed with paraquat, Wathern and Gilbert (1978) found that in the control (uncut) and cut plots, seedling-survival was low because of severe competition from *Lolium perenne* and *Festuca rubra*. In contrast, in the shorter, more open paraquated and rotovated areas, percentage establishment and survival was over 50 per cent, *Lathyrus pratensis*, *Achillea millefolium* and *Plantago lanceolata* becoming particularly prominent after 5 years.

It is generally thought that throwing herb seed onto old grassland is wasteful but there has been little research on this aspect of establishment. Recent work on 'gap size' (Staal, 1979) has highlighted the fact that even small differences in the size of the gap, in the time of year when it was created, and the depth of soil exposed may have important consequences for the successful establishment of seedlings. Diurnal fluctuations in soil temperatures, which may vary in gaps of different sizes, are important for the germination of some species (Thompson, Grime, and Mason, 1977; Figure 13.3). Other factors, such as the effect of

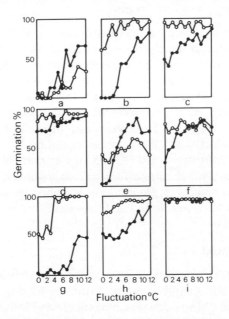

FIGURE 13.3. Germination responses to various amplitudes of diurnal fluctuations of temperature in light (o) and in continuous darkness (●) in nine herbaceous plants of common occurrence in pastures and arable land. Fluctuating temperatures were applied in the form of a depression below a base temperature of 22 °C extending for 6 in every 24 hours. Diurnal fluctuations were maintained throughout each experiment. Light intensity during the light period was 1846 μW/cm^{-2}. Prior to testing, seeds were stored dry at 5°C. (a) *Agropyron repens*, (b) *Deschampsia cespitosa*, (c) *Holcus lanatus*, (d) *Poa annua*, (e) *Poa pratensis*, (f) *Poa trivialis*, (g) *Rumex obtusifolius*, (h) *Stellaria media*, (i) *Lolium perenne*. From Thompson, Grime, and Mason (1977); reproduced by permission from *Nature*, **267**, 147–149. Copyright © 1977 Macmillan Journals Ltd.

aspect on soil moisture and soil temperatures, may limit the range of species able to germinate and establish at a particular site (Grime and Curtis, 1976).

Pot-grown plants

Many species of wild plants can be grown successfully from seed in peat pots using conventional horticultural methods. Using a bulb planter, it is possible to remove cores of turf and replace them with pot-grown plants of desirable wild species. Not only will these plants flower during the first season, but they should also be able to compete with the established sward. Because of high cost, this method may only be applicable to small areas, but little is known about the ability of these plants to spread from their 'islands' into the surrounding sward. Clearly, the management of the surrounding areas, and their receptiveness to incoming propagules are crucial to the success of this approach.

Use of turf transplants and turf fragments

Turf stripped from a calcareous grassland prior to quarrying operations has been used to revegetate a quarry floor with some success (Humphries, 1979), and although this does not increase the area of species-rich turf, it does ensure that it is not wasted. Gilbert and Wathern (1980) stripped turf (5–10 cm thick) from an old mineral working, cut it into pieces 15, 30 and 60 cm across, spreading the pieces of turf over recently graded subsoil, sowing the areas between with a *Festuca* mixture, at 15 kg ha^{-1}. Eight years on, the site is a tussocky grassland, the tussocks formed by the old turves with a vegetation 30–40 cm high while the interstices support a shorter sward (12–25 cm) with an 80–90 per cent cover. Species such as *Campanula rotundifolia*, *Viola riviniana*, *Galium verum*, *G. sterneri*, *Veronica chamaedrys*, *Potentilla erecta* and *Poterium sanguisorba* are among those which have established and flowered.

Use of seed-rich topsoil

It has been known for a long time that the soils of arable fields and pastures contain very large numbers of viable seed of certain herbaceous species (Brenchley, 1918; Champness and Morris, 1948; Chippendale and Milton, 1934). Studies of the viable seed content of grasslands of known history (Brenchley, 1918; Chippendale and Milton, 1934) indicate that buried seeds of some arable weeds can retain viability in undisturbed soils for periods in excess of 50 years. Recent studies by Dr J.P. Grime and his co-workers, in the Unit of Comparative Plant Ecology, Sheffield (see Grime, 1979 for extensive reference list) into the functional significance of the various types of seed banks in grasslands, provides an explanation for the sometimes conflicting results obtained when topsoil, with its associated seed-bank, is used to recreate floristically interesting communities.

They recognize four main types of seed bank, the essential features of which are summarized in Figure 13.4.

Seed bank type I contains species with transient seed banks present during the summer. Typical species include many of the relatively large-seeded (that is, more than 0.5 mg) perennial grasses such as *Arrhenatherum elatius, Bromus erectus, Dactylis glomerata, Festuca rubra, Helictotrichon pratense* and *Lolium perenne*. Smaller tussock grasses, such as *Briza media, Festuca ovina* and *Koeleria cristata*, also fall into this group. These species germinate soon after seed-fall, over a wide range of temperatures, and do not contribute to a persistent seed bank.

Temporal Pattern of Seed Bank Survival

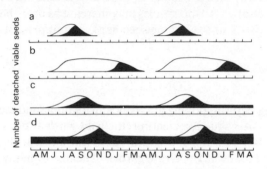

FIGURE 13.4.   Scheme describing four types of seed banks of common occurrence in temperate regions. □ Seeds capable of germinating immediately after removal to suitable laboratory conditions; ■ seeds viable but not capable of immediate germination. (a) Annual and perennial grasses of dry or disturbed habitats (such as *Hordeum murinum, Lolium perenne, Catapodium rigidum*); (b) annual and perennial herbs, shrubs, and trees colonizing vegetation gaps in early spring (such as *Impatiens glandulifera, Anthriscus sylvestris, Acer pseudoplatanus*); (c) winter annuals mainly germinating in the autumn but maintaining a small seed bank (such as *Arenaria serpyllifolia, Saxifraga tridactylites, Erophila verna*); (d) annual and perennial herbs and shrubs with large, persistent seed banks (such as *Stellaria media, Origanum vulgare, Calluna vulgaris*). From Grime (1979), reproduced with permission.

Seed bank type II contains species with transient seed banks that survive one winter. Species in this group, of which *Pimpinella saxifraga* is a good example, show a pattern of winter dormancy followed by germination in early spring in large numbers. These species often have large seeds which are shed late in the season when germination would be unfavourable. They are able to germinate at low temperatures in the spring and appear to represent a group with a specific adaption to delaying germination until the beginning of the growing season. Seed banks types III and IV contain species with persistent seed banks (that is, more than 1 year). Thompson and Grime (1979) recognize two main types of persistent seed banks: type (a) where many of the seeds germinate soon after

they are released (like type I), but where a proportion of the seeds fail to germinate and are incorporated into the persistent seed bank (examples are *Arabidopsis
thaliana, Arenaria serpyllifolia, Poa annua* and *Poa trivialis*; in type (b) few
seeds germinate in the period following dispersal, and the species maintains a
seed bank the size of which changes little with the season and is large in relation
to the annual production of seeds.

From the practical point of view, the use of the seed bank in the topsoil as a
means of producing floristically attractive grasslands is risky. Much depends on
where the soil comes from, what time of year it is dug, and the nature of the
vegetation growing on it. Storage of soil, as often happens when overburden is
removed, will lead to a loss of grasses and the probability of invasion by propagules of 'weed' species, such as *Rumex* spp. and *Urtica dioica*. The usefulness
of topsoil as a source of grassland species may increase as more becomes known
about the characteristics of seed in the seed bank, but at present its use cannot
be recommended.

CONCLUSION

The idea of creating natural-looking grasslands, with an abundance of colourful
flowers, is not new. A vogue for creating 'the wild garden' began in Britain in
the middle of the last century and continues to the present day, albeit most
exclusively confined to the large private garden. The loss of many areas of seminatural grasslands in Britain over the past 30 years or more has focused attention
on their value as a wildlife resource; it also stimulated the idea that it might be
possible to create species-rich grassland on arable land where the primary use
might not be for agricultural productivity. In America, attempts at restoring
prairie began in 1935 (Blewett, 1980), and some idea of what can be achieved in
recreating plant communities using a combination of restoration methods is
provided by Curtis and Greene prairies (Allsup, 1978). Similar encouraging
results have been obtained in shorter-term studies in Holland (Londo, 1977)
and in Britain. There seems to be no *a priori* reason why species-rich communities
cannot be created on a variety of soil types by introducing species, either as seed
into a bare soil, or as seed or vegetative propagules into established turf. One of
the crucial factors for success is the subsequent management of the grassland,
and this is an area where, despite the considerable amount of information
which already exists (Duffey *et al.*, 1974; Wells, 1980), more research is urgently
required.

REFERENCES

Allsup, M. (1978). Henry Greene's prairie, *Arboretum News, University of Wisconsin*,
    27(4), 1–4.
Anon, (1981). *Turfgrass 1982*, Sports Turf Research Institute, Bingley.

Blackwood, J. and Tubbs, C.R. (1970). A quantitative survey of chalk grassland in England. *Biological Conservation*, **3**, 1–5.

Blewett, T. (1980). Curtis prairie: a look backward. *Arboretum News, University of Wisconsin*, **29(4)**, 1–4.

Bradshaw, A.D. and Chadwick, M.J. (1980). *The Restoration of Land*, Blackwell Scientific, Oxford.

Brenchley, W.E. (1918). Buried weed seeds. *Journal of Agricultural Science, Cambridge*, **9**, 1–31.

Chippendale, H.G. and Milton, W.E.J. (1934). On the viable seeds present in the soil beneath pastures. *Journal of Ecology*, **22**, 508–531.

Champness, S.S. and Morris, K. (1948). Populations of buried viable seeds in relation to contrasting pasture and soil types. *Journal of Ecology*, **36**, 149–173.

De Puit, E.J. and Coenenberg, J.G. (1979). Methods for establishment of native plant communities on top soiled coal stripmine spoils in the northern Great Plains. *Reclamation Review*, **2**, 75–83.

Duebbert, H.F., Jacobson, E.T., Higgins, K.F., and Podoll, E.B. (1981). Establishment of seeded grasslands for wildlife habitat in the prairie pothole region. *Special Scientific Report, US Fisheries and Wildlife Service, Wildlife*, **234**.

Duffey, E., Morris, M.G., Sheail, J., Ward, L.K., Wells, D.A., and Wells, T.C.E. (1974). *Grassland Ecology and Wildlife Management*, Chapman and Hall, London.

Ellis, W. (1747). *The Farmer's Instructor; or, the Husbandman and Gardener's Useful and Necessary Companion*, J. Hodges, London.

Eyre, S.R. (1955). The curving plough-strip and its historical implications. *Agricultural History Review*, **3**, 80–94.

Gilbert, O.L. and Wathern, P. (1980). The creation of flower-rich swards on mineral workings. *Reclamation Review*, **3**, 217–221.

Grime, J.P. and Curtis, A.V. (1976). The interaction of drought and mineral nutrient stress in calcareous grassland. *Journal of Ecology*, **64**, 976–998.

Grime, J.P. (1979). *Plant strategies and vegetation processes*, John Wiley, Chichester.

Grubb, P.J. (1977). The maintenance of species-richness in plant communities: the importance of the regeneration niche. *Biological Review*, **52**, 107–145.

Haggar, R.J. (1974). Legumes and British grassland, new opportunities with herbicides. *Proceedings of the 12th British Weed Control Conference*, **1**, 771–777.

Haggar, R.J. (1980). Weed control and vegetation management by herbicides. In I.H. Rorison and R. Hunt (Eds), *Amenity Grassland: An Ecological Perspective*, John Wiley, Chichester, pp. 163–173.

Haggar, R.J. and Oswald, A.K. (1976). Improving ryegrass swards by a low dose of dalapon. *Technical Leaflet of the Weed Research Organization*, **1**, 1–2.

Haggar, R.J. and Passman, A. (1978). Some consequences of controlling *Poa annua* in newly sown rye grass leys. *Proceedings of the British Crop Protection Conference: Weeds*, **1**, 301–308.

Humphries, R.N. (1979). Some alternative approaches to the establishment of vegetation on mined land and on chemical waste materials. In M.K. Wali (Ed), *Ecology and Coal Resource Development*. Based on the International Congress for Energy and the Ecosystem held at the University of North Dakota, June 1978, Pergamon, New York, Oxford, pp. 461–475.

Johnson, M.S. and Bradshaw, A.D. (1979). Ecological principles for the restoration of disturbed and degraded land. *Applied Biology*, **4**, 141–200.

Lane, C. (1980). The development of pastures and meadows during the sixteenth and seventeenth centuries. *Agricultural History Review*, **28**, 18–30.

Londo, G. (1977). *Natuurtuinen en-parken*, B.V.W.J. Thieme, Zutphen.

Moffatt, J.D. (1981). Techniques no. 37: Grass seed mixes—their choice, growth and development. *Landscape Design*, **133**, 33–37.

Natural Environment Research Council (1977). Amenity grasslands—the needs for research. *Publication of the Natural Environment Research Council (GB)*, C, **19**.

Squires, N.R.W., Haggar, R.J., and Elliott, J.G. (1978). A one-pass technique for establishing grasses and legumes in existing swards. *Technical Leaflet of the Weed Research Organization*, **2**, 1–4.

Staal, L. (1979). *An investigation into the pattern of the regeneration within gaps and the type of seed bank on a north-facing slope and a south-facing slope at Millersdale near Litton*, thesis, University of Sheffield.

Tansley, A.G. (1953). *The British Islands and their Vegetation*, Cambridge University Press, Cambridge.

Thompson, K., Grime, J.P., and Mason, G. (1977). Seed germination in response to diurnal fluctuations of temperature. *Nature, London*, **267**, 147–149.

Thompson, K. and Grime, J.P. (1979). Seasonal variation in the seed banks of herbaceous species in ten contrasting habitats. *Journal of Ecology*, **67**, 893–921.

Wathern, P. and Gilbert, O.L. (1978). Artificial diversification of grassland with native herbs. *Journal of Environmental Management*, **7**, 29–42.

Watt, T.A. (1978). The biology of *Holcus lanatus* (Yorkshire fog) and its significance in grassland. *Herbage Abstracts*, **48**, 195–204.

Wells, D.A. (1969). The historical approach to the ecology of alluvial grassland. In J. Sheail and T.C.E. Wells (Eds), *Old Grassland. Symposium of the 5th Monks Wood Experimental Station, 5th,* pp. 62–67.

Wells, T.C.E., Sheail, J., Ball, D.F., and Ward, L.K. (1976). Ecological studies on the Porton Ranges. Relationships between vegetation, soils and land-use history. *Journal of Ecology*, **64**, 589–626.

Wells, T.C.E. (1979a). Habitat creation with reference to grassland. In S.E. Wright and G.P. Buckley (Eds), *Ecology and design in amenity land management*, Wye College and Recreation Ecology Research Group, Wye, pp. 128–145.

Wells, T.C.E. (1979b). *Interim report to the Nature Conservancy Council on establishment of herb rich swards*. Natural Environment Research Council Contract Report to the Nature Conservancy Council. CST Report no. 240.

Wells, T.C.E. (1980). Management options for lowland grassland. In I.H. Rorison and R. Hunt (Eds), *Amenity Grassland—An Ecological Perspective*, John Wiley, Chichester, pp. 175–195.

Wells, T.C.E., Bell, S.A., and Frost, A. (1981). *Creating Attractive Grasslands using Native Plant Species*, Nature Conservancy Council, Shrewsbury.

Whittington, G. (1962). The distribution of strip lynchets. *Institute of British Geographers Publication*, **31**, 115–130.

Conservation in Perspective
Edited by A. Warren and F.B. Goldsmith
© 1983 John Wiley & Sons Ltd.

CHAPTER 14

Evaluating Nature

F.B. GOLDSMITH

The criteria and processes for selecting and managing the more important sites in the countryside are central to a book on nature conservation. The omission of the topic from *Conservation in Practice* (Warren and Goldsmith, 1974) was not because it had not been 'invented' at that time for Ratcliffe had published his criteria for *Nature Conservation Review* in 1971 and we had conducted a group project with our postgraduate conservation course on the theme in 1972. The real reason was that nature conservationists were then arguing amongst themselves about the differences between ecological and conservation evaluation, whether criteria should be weighted in any way, whether they could be added together, and whether we wanted any kind of index at all. It would be pleasing to be able to report now that all the wrinkles have been ironed out and that we can now set before you a definitive procedure for making conservation evaluations. Regrettably this is not the case but there is now a large group of well-intentioned people who have also committed their thoughts to paper, or who have embarrassed themselves in planning enquiries by not being able to answer questions adequately. This chapter aims to be a guide through the quagmire and to leave the reader more cautious, and better-informed.

EVALUATION FOR WHOM?

After discussions by the Ecological Affairs Committee of the British Ecological Society, Goldsmith (1973) published a statement in the *BES Bulletin* about the assessment of ecological value. We wished to know who was already involved in such evaluations and how they were being made. Replies came from six different regions of the Nature Conservancy Council, planning authorities, a county trust for nature conservation, a polytechnic, a museum, the soil survey and from overseas. All recognized that there was a problem but none suggested an ideal solution, although the Nature Conservancy Council had by then published the criteria (Ratcliffe, 1971) that they were to use in their *Nature Conservation Review*. At the time this review only dealt with 1 per cent of the land surface, but has since been extrapolated to SSSIs representing another 5 per cent. We had also asked 'for what' and 'for whom' were the evaluations being made: to improve an amenity for the public, to maintain gene banks, for educational purposes, to protect rare species, to prevent development, for research, etc? In the following years I have received the impression that we have elaborated on the methodology without really answering the question of for whom the evaluation is being prepared (Goode, 1981).

Mabey (1980), amongst others, has argued that nature conservation has served to cater principally for an elite minority and that government agencies and trusts have selected reserves on pseudoscientific grounds for ecological specialists. On the other hand the interests of the general public are often served by the area immediately around them which carries no designation and yet we have all seen the general level of ecological interest of the countryside and of urban areas decrease steadily during this century suggesting that at least some existing statutory designations and legislation are ineffective.

Let us briefly consider what the general public really wants. The majority live in urban or suburban environments and their principal interest is in their immediate vicinity. They want to occasionally enjoy the 'feeling' of being in the countryside in their locality. They want to see pretty flowers, birds and butter-flies as frequently as possible. They may not have to be golden eagles or purple saxifrages but a bullfinch or a ragged robin is just as good so protection for, and access to, local areas is very important.

Most of our ecological evaluation work, however, is prepared for qualified planners, expert scientists or at least nature buffs. It focuses on rare species, pristine habitats or areas where the two occur together. This kind of assessment is important, but will not gain maximum public support. I suspect that large numbers of people join the RSPB because we all see birds from our homes, on the way to work and whenever we wander in the country. It is true that many are brightly coloured, and they move, but it is also important that they are regularly seen and appreciated by everybody.

In this context it is worth considering a recent American publication on wild-life values (Shaw and Zube, 1980). They discuss the ambiguities of trying to define value and suggest that it can be partially rationalized by distinguishing between (a) economic measures of value, (b) social and psychological inter-pretations and (c) ecological measures. I think that this is a useful distinction to make. In this country we have focused on (c) because the training of biologists makes the recording of plants and animals relatively easy. Our evaluations are acceptable if we realize that they only give us an answer to (c). If we have been measuring (c) and assuming that it gives us the answer to (b) we are deceiving ourselves.

The importance of the consumer, principally the general public, has come to our attention as he becomes more vocal, joins groups such as Friends of the Earth and fights local issues, for example in public enquiries. In two recent instances local groups have led the Nature Conservancy Council and won major conservation issues (Grove, 1981). For example, an area outside the SSSI of Moseley Bog in Birmingham was the subject of an enquiry and was shown to be at least as important as the SSSI itself, and Amberley Wildbrooks in Sussex was successfully defended against land drainage by the water authority whilst the Nature Conservancy Council, at least initially, was largely unaware of the interest of the site. The social and psychological interpretations of wildlife value have been largely overlooked whilst increasingly detailed assessments of ecological measures of value have been devised.

ECOLOGICAL CRITERIA

Ratcliffe's (1971, 1977) ten criteria devised for the *Nature Conservation Review* were size, diversity, naturalness, rarity, typicalness, fragility, recorded history, position in an ecological geographical unit, potential value and intrinsic appeal. They have since been critically evaluated by various workers, such as Adams and Rose (1978), Usher (1980), Spellerberg (1981), and by a British Ecological Society working party (unpublished).

The main criticism is that they contain a mixture of ecological criteria which can be more-or-less precisely measured, such as size, diversity or richness, and rarity, and conservation criteria which are themselves value judgements of ecological or social and aesthetic criteria. Examples in the latter category are potential value and intrinsic appeal. For example, the intrinsic appeal of moorland dominated by heather depends on whether you were brought up in Bromley or Sutherland.

Naturalness is stated to be a criterion although many of our national nature reserves (NNRs) and sites of special scientific interest (SSSIs) are selected because they are the product of a traditional form of land use such as the coppicing of woodlands or the sheep grazing of chalk grassland. As such they are of anthropogenic origin rather than being truly natural systems. Presumably typicalness is introduced to try to overcome this problem, such areas ceasing to be natural but becoming representative of a former land use. Whether we are talking about representativeness or naturalness is immaterial because both are virtually unmeasurable.

Diversity is an unfortunate term as it has different connotations for different people. It is preferable to distinguish between the richness or number of species and the diversity of habitats. However some very interesting areas are quite poor in species but nevertheless generally agreed to be very important, for example certain categories of heathland such as those of the south-west dominated by *Erica vagans*, and Wastwater in the Lake District is oligotrophic and species-poor. Richness is important when it suits us but otherwise we ignore it. We do not like agriculturally improved grassland although in the Malham SSSI it has a mean of 12.7 higher plant species per 0.25 m^2 quadrat when the more natural acidic grassland has only 6.9 species per quadrat (Usher, 1980). I am sure that the Nature Conservancy Council asks farmers not to lime or fertilize their grasslands, although in this situation it increases species-richness. Similarly, some manmade habitats, such as abandoned urban railway sidings can be very species-rich.

Size appears to be a simple, straightforward criterion, and it is known that bigger areas contain more species and the species are supposedly less prone to extinction. This is a basic tenet of island biogeographic theory but many people now believe that size is little more than a surrogate for habitat diversity; the larger the island, or wood, the greater the number of habitat types it is likely to contain and therefore the greater number of species. Should we artificially create rides, ponds, or coppice in order to diversify the habitats and increase the number of component species?

Rarity is what triggers the excitement of many naturalists and Adams and Rose (1978) suggest that a species becomes more interesting or important as it becomes rarer. Was the large blue butterfly most valuable in the last year that it occurred in Britain, and has the fulmar become less interesting as its population has increased in the last few decades?

Fragility is another criterion referred to in the *Nature Conservation Review*, where it is used in the sense of an area with high inherent sensitivity, such as a wetland, a coastal dune or some montane habitats which are potentially physically unstable. The distinction between fragility and vulnerability is confusing but I presume that an area is inherently fragile but is vulnerable to an extrinsic threat such as a change of land use, for example gravel extraction. This suggests that fragility is a desirable characteristic whereas vulnerability or threat is often the real reason for designation or acquisition. The relationship between fragility and a concept popular amongst theoretical ecologists, stability, is obscure. Stability has been the subject of mathematical simulations of species interactions and computerized predictions and, if it is an inherent characteristic of ecosystems, it is presumably more closely related to fragility than it is to vulnerability. Stability does not justify a place as a characteristic for defining ecological value, because there is no consensus that it is important.

Research and educational potential are also difficult to use as practical criteria as they are as much a function of the location of the site as of its inherent ecological interest. However they are nevertheless important. A site such as the William Curtis Ecological Park on Hays Wharf beside Tower Bridge is valuable because of its central location, but it is totally manmade and ranks at the bottom of the scale in terms of naturalness. Conversely a bog in the Outer Hebrides is likely to have a low educational potential because of its inaccessibility to most people, and presumably most of the locals are not interested in something so familiar to them. Research potential also tends to be a function of distance, although in this instance from a university or research station. Meathop Wood, the International Biological Programme research site near to the Institute of Terrestrial Ecology research station at Grange-over-Sands, has been analysed in great detail, whereas conservationists were not able to convince the Teesdale Reservoir public enquiry of the importance of that site in spite of its agreed high ecological value. One reason why the conservationists lost was because there was little published evidence of research carried out on the arctic–alpine species or the site which were stated to be so important. However, Moor House National Nature Reserve nearby, an area of relatively poor blanket bog, had been studied by numerous University of Durham MSc and PhD students because accommodation and laboratories were available onsite. Which is the more important site as judged by the criterion of research potential?

Recorded history is a criterion which is of dubious value, because first, it cannot in any way be construed as an inherent characteristic of the site and, second, it is very difficult to measure or quantify. Is a sixteenth-century record worth more than a nineteenth-century one, and if so are two eighteenth-century records worth more than one sixteenth-century record?

Potential value is another fascinating criterion. Presumably it implies that appropriate management can sometimes, perhaps often, increase the ecological value of a site. If this is the case, does it not conflict with naturalness? Presumably the more the area is managed, the less natural it becomes, so as the site increases in value using one criterion it may decrease in value as measured by another one.

These various criteria from the *Nature Conservation Review* have been discussed both within the Nature Conservancy Council and by other ecologists for nearly 10 years and have been slightly modified in the Nature Conservancy Council document on the Selection of Sites of Special Scientific Interest (Anon; 1980). The 3000 biological SSSIs are now selected or assessed on the basis of seven habitat types in geographical subdivisions approximately 50 km × 50 km, about the size of the English counties. The criteria are now divided into three groups (Table 14.1).

Candidate sites should be compared with others of the same habitat or formation in a particular geographical subunit using the criteria in order, and where difficulties arise the appropriate specialist in the chief scientist's team should be consulted. It is pointed out that typicalness is not to be equated with

TABLE 14.1

Principal criteria (in order of importance)	Ancillary criteria	Criteria to determine number of sites
1 Typicalness	Recorded history	Rarity
2 Naturalness	Position in an ecological unit	Fragility
3 Diversity	Potential value	
4 Size	Intrinsic appeal	

the average example of a formation but with the 'best'. This appears to be the replacement of one value judgement (ecological value) with another, equally vague, one (best). Naturalness is explained by distinguishing between acceptable management (however this is judged) and interference, which is unacceptable. Diversity is said to provide a convenient measure of naturalness and hence of typicality, which prompts me to hope that this removes the necessity to use the latter two. Diversity is used to indicate species-richness but must be based largely on the number of species of flowering plants (as opposed to animals, fungi, ferns, bryophytes and lichens?) until more is known about animal and other taxa. The size criterion is variable depending on the habitat under consideration; for example, habitats should exceed the following sizes to be worthy of designation:

Open water	0.5 ha
Peatland	1.0 ha
Woodland	5 ha
Lowland grassland	10 ha
Heathland	10 ha
Upland grass/heath	50 ha

Very rare habitats or formations are those that are confined to five or fewer sites in Great Britain. The number of more common formations, that is habitats with a total area of over 10 000 ha in Great Britain, requiring designation as SSSIs depends on whether they are under threat or not. Between one and five sites per county of each habitat type should be notified.

Any confusion with these criteria stems from the fact that we are trying to produce a single rigid approach to evaluation. This is difficult because we must, first, prepare a scientific description of what is there and then, second, assess its value for conservation. This involves both an impartial or scientific ecological description followed by a naturalist's or consumers evaluation of the site. Size, richness and rarity are criteria that fall into the former category and the others (such as potential, research, and educational value) into the latter. The first three tend to be more important in the more recently evolved criteria for SSSIs.

Margules (1980) took a group of nine expert assessors to visit eight sites in Yorkshire. They used a total of eighteen different criteria to determine the relative interest of different sites (Table 14.2).

The relationships between variables in pairs were analysed using regression and principal components analysis. These showed that there is no obvious overall ranking of criteria used by different assessors but that criteria can vary depending on size of site with fragility, threat and rarity being important for small sites and representativeness, area and naturalness being more important for larger sites (Figure 14.1). They also demonstrate that whilst most criteria are 'scientific' some such as threat of human interference are 'social' or 'political'. Of the scientific criteria some can be determined by visiting only the site in question,

TABLE 14.2

Most frequently used			Less frequently used
Diversity	Area	Potential value	Replaceability
Rarity	Naturalness	Wildlife reservoir potential	Amenity value
Representativeness	Threat of interference	Management factors	Recorded history
Uniqueness	Scientific value	Position in ecological geographical unit	Educational value
Ecological fragility			Availability

for example, diversity and area, whilst others require extensive survey work in the surrounding area, for example, rarity, uniqueness, representativeness and naturalness.

IS A STANDARD APPROACH PRACTICABLE OR DESIRABLE?

The need for a single, standardized approach is exemplified by the recent production of county structure plans at one scale, criteria for national nature reserves and sites of special scientific interest at another, and the International Biological Programme (IBP) survey of conservation sites at an international

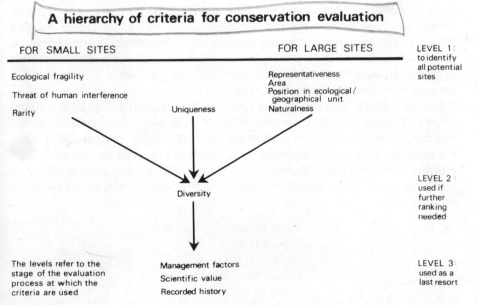

A hierarchy of criteria for conservation evaluation

FOR SMALL SITES   FOR LARGE SITES   LEVEL 1: to identify all potential sites

Ecological fragility

Threat of human interference

Rarity

Uniqueness

Representativeness
Area
Position in ecological / geographical unit
Naturalness

LEVEL 2: used if further ranking needed

Diversity

The levels refer to the stage of the evaluation process at which the criteria are used

Management factors
Scientific value
Recorded history

LEVEL 3: used as a last resort

FIGURE 14.1. The levels (on the right hand side) refer to the stage of the evaluation process at which the criteria are used. From Margules (1980).

scale (Clapham, 1980). It is obviously desirable that these should be prepared in a comparable, if not standardized, way. However, consider the problems of scale and the way in which criteria must change from region to region, for example land dominated by heather is more valuable in southern England than in Sutherland. It is also important that one habitat, such as chalk grassland, should have the same value on each side of an administrative boundary, such as the boundary line between Kent and Surrey. Whether it does have the same value depends on whether the purpose of the survey is national (when it will) or regional at the county level (when it might not). For these reasons it is impractical to describe a single generalized method.

WHAT TECHNIQUES ARE AVAILABLE AND FOR WHICH PURPOSES ARE THEY APPROPRIATE?

Having made several cautionary points it is now appropriate to review briefly a selection of methods that have been designed in relation to a particular objective or area of study.

Although Hills (1961) is often quoted as being the pioneer of ecological evaluation methods it is really Helliwell (1969, 1971, 1973) and Tubbs and Blackwood (1971) as well as Ratcliffe's (1971) work that started the ball rolling. Tubbs and Blackwood's approach was simple and rapid being devised principally for a county council planning department. Using aerial photographs and field-work they evaluated land in terms of the relative rarity and species diversity of the habitats present and produced 'ecological zones' of standard value. Grade I land, for example, having abundant unsown vegetation including non-plantation woodland. Even today Tubbs (personal communication) believes that this approach is more practicable than methods that have been published since because of its time and manpower efficiency.

Helliwell (1969, 1971) adopted a totally different approach as he attempted to give values to trees and woodlands based on largely subjective judgements which were later converted into arbitrary monetary values. The values for trees, for example, were based on crown area, useful life expectancy, importance of position in the landscape, presence of other trees, form, species, and special or historical value. His system for valuing wildlife habitats was based on seven factors:

> direct return, capital value
> genetic reserve
> ecological balance
> educational value
> research value
> natural history interest
> local character

Whilst cost-benefit analyses of recreational activities are frequently prepared and money can be justified as a common system of measurement, conservation evaluations involve a variety of objectives, various interested users and a complex and varied resource. It is difficult to see how evaluations for structure planning and planning control with their statutory and legal responsibilities could be converted into monetary terms. It could be argued that the importance of different taxonomic groups might be assessed from the membership of the organizations which study them (for example, Royal Society for Protection of Birds: approximately 250 000 members; Botanical Society of the British Isles: 2300 members) but membership is affected by the marketing efficiency of the organization as well as the size, movement and attractiveness of the organisms concerned.

Kent (1972) produced a method for assessing the potential of marginal land in agricultural areas although the objectives were recreation and landscape as well as wildlife. The characteristics he measured were area, shape, habitat type, plant species richness, quality of surrounding agricultural land, access, distance from nearest similar feature, present use of site, landscape quality and ownership. The data for the sites were then grouped together using cluster analysis with a weighting for habitat types which was considered to be the most important characteristic. The method has the failing that it has a variety of loosely related objectives, could employ any number of site characteristics and requires value judgements to make the final assessment.

Many evaluations have been carried out in relation to specific habitats such as woodland (Peterken, 1974), limestone pavements (Ward and Evans, 1976), linear habitats (Yapp, 1973), coastal habitats (Ranwell, 1969), urban areas (Marsh, 1978), and farmland (Helliwell, 1978; Kent and Smart, 1981). Others have focused on particular objectives such as alternative routes for roads or motorways (O'Brien, 1977; Yapp, 1973). Some have placed emphasis on different criteria such as habitat (Anon., 1980), indicator species (Peterken, 1974), rarer species (Ward and Evans, 1976) and the age of a plant community (Smith *et al.*, 1975). Hooper (personal communication) has suggested that county trusts should use as their 'ultimate criterion' the number of additional species contributed by a site to the total protected by the trust. In this way the members could choose between two possible new candidates for acquisition. Many produce numerical values for several criteria and then add them together (Yapp, 1973) or multiply them (Goldsmith, 1975; O'Brien, 1977) although most workers would now emphasize that the weighting of different criteria is not recommended and that the comparison of like-with-like is acceptable, but not like-with-unlike. For example, if one needs to compare an area of woodland with an area of moorland as potential sites for a power station the two cannot be directly compared. One can only compare the woodland with other woodlands to indicate its importance, and then the moorland can be judged in relation to other areas of moorland. The shape, distribution, margins and spacing of a habitat or sites is also very often important. For example Hooper (1971), Diamond (1975) and Game (1980) have suggested this for nature reserves, and Moore and Hooper (1975) and Helliwell (1976) for woodlands.

A CASE STUDY

Having explored the philosophical background to our subject and considered some techniques let us now take a case study for more detailed analysis. Walthamstow Marshes are 36 ha (90 acres) of relect Lea Valley wet meadowland only $6\frac{1}{2}$ km (4 miles) from St Pauls Cathedral in London. On one side is the River Lea and Hackney with high-density housing and blocks of flats alongside, on two sides are railway lines and on the fourth, Walthamstow Reservoirs, a designated site of special scientific interest (Figure 14.2). The Marshes are the subject of a planning application to extract aggregate to a depth of 9 m (30 ft) and afterwards turn the site into a sailing lake. Although the Lea Valley has plenty of open water, there is a national and regional need for open water, as there is of course for aggregates.

Most local people do not want another sailing lake but enjoy this semiwild area where children can ride bikes, get the feel of the countryside and burn off some of their surplus energy. They started a 'Save the Marshes' campaign and a local botanist, Brian Wurzell, started documenting the species found on the site. Currently he has listed 350 species of plants of which many are ruderals or introductions but there is a core of about 90 species characteristic of wetland habitats. The area is also important for birds with a list of about 100 visitors, and the sight of snipe in winter or bearded tits in the reed-bed so close to the centre of London is not to be denigrated. When the Nature Conservancy Council were approached with a request to designate the area an SSSI, they refused because of the degree of disturbance to the area. The flora was studied in more detail and several hybrid sallows, sedges and docks were found, an uncommon horsetail (*Equisetum littorale*) and adder's tongue (*Ophioglossum vulgatum*). The two local authorities, Waltham Forest and Hackney, as well as the Greater London Council, objected to the planning application but the applicant, the Lea Valley Regional Park Authority, stood firm.

What is the ecological status of this area of one-time Lammas grazing land? Does it justify protection? Should our criteria for evaluation be different in the urban context, and if so, what is the appropriate statutory designation? The Marshes make an interesting comparison with the Reservoirs; the latter are manmade, of no botanical significance, and important for a rather small number of bird species and are designated a site of special scientific interest. The Marshes on the other hand would come out with a high score for richness and diversity, especially if we bear in mind the SSSI criteria which place the emphasis on plants. In terms of naturalness, the Marshes do better than the Reservoirs, but because of the human disturbance they also do well in terms of

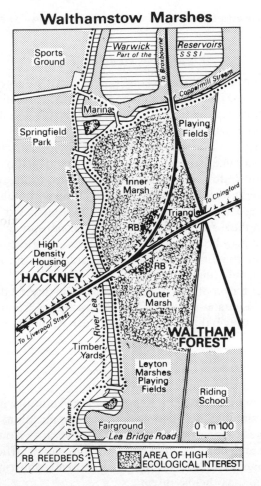

FIGURE 14.2. Sketch map of Walthamstow Marshes (based on 'Save the Marshes' Campaign Report). The planning application concerned the Inner and Outer Marsh which are owned by the Lea Valley Regional Park Authority. RB = reed bed.

ecological fragility and threat of human interference. They have rarities which occur in very few other sites in the London area. Their biological documentation is better than most existing SSSIs and their recorded history is reasonably good. Of course their best feature is their educational potential with schools and colleges all around. Any other capital city in the world would be delighted to have a site like this on its doorstep, especially as it has the potential for enjoyment which some consider to be the ultimate criterion.

If we accept the *Nature Conservation Review* criteria, give them equal weighting, and do not combine them in any way because there is no agreed basis for any other procedure, then we can only compare two sites in terms of the following:

$$V = f(c, c_2, c_3, c_4, c_5 \ldots c_n)$$

where V = index of conservation value, $c, \ldots, c_n$ are the criteria. On this basis the Marshes are more important than the Reservoirs. If we disagree we have to accept that statutory evaluations are totally subjective and it is a question of what the appropriate officer likes and dislikes.

The reason that the Marshes are not a site of special scientific interest is because they are not in pristine condition and are used by too many people, two characteristics which do not appear in any published evaluation procedure. If however our criteria were modified for use in urban and suburban areas where there is a real need for sites like Walthamstow Marshes or Moseley Bog in Birmingham then these areas could be given some appropriate statutory designation. Therefore it may be that 'local nature reserve' status could be the appropriate designation, redefined with legal safeguards and some means of funding for the management required.

The ultimate criterion of Site of Special Scientific Interest designation is that it is an area so designated by the Nature Conservancy Council, and the final irony in this case study is that it is a Regional Park Authority which is the villain in the case.

Postscript

At the time of going to press the Regional Park Authority has withdrawn its planning application but local people fear a new application, this time dealing with only half the Marshes; and the Nature Conservancy Council have recommended that the area be designated an SSSI. A Regional Officer of the Nature Conservancy Council once told me that there are really only *two* criteria—one is 'opportunity' and the other is 'threat'.

REFERENCES

Adams, W.M. and Rose C.J. (Eds) (1978). The selection of reserves for nature conservation. *Discussion Papers in Conservation 20*, University College London, 34 pp.
Anon. (1980). *The selection of sites of special scientific interest: an explanatory paper by the Nature Conservancy Council*, unpublished MS, 9 pp + appendices.
Clapham, A.R. (Ed) (1980). *The IBP Survey of Conservation Sites: An Experimental Study*, University Press, Cambridge, 344 pp.
Diamond, J.M. (1975). The island dilemma: lessons of modern biogeographic studies for the design of nature reserves. *Biological Conservation*, 7, 129–146.
Game, M. (1980). Best shape for nature reserves. *Nature*, **287**, 630–632.
Goldsmith, F.B. (1973). Comment. *Bulletin of the British Ecological Society*, IV, 1, 2.
Goldsmith, F.B. (1975). The evaluation of ecological resources in the countryside for conservation purposes. *Biological Conservation*, **8**, 89–96.

Goode, D.A. (1981). Values in nature conservation. In C. Rose (Ed), *Values and Evaluation*, Discussion Paper in Conservation No. 36, University College London, pp. 28–37.

Grove, R.H. (1981). The use of disguise in nature conservation: the evidence from three case studies, *University College London Discussion Paper No. 32*, London, 38 pp.

Helliwell, D.R. (1969). Valuation of wildlife resources, *Regional Studies*, 3, 41–47.

Helliwell, D.R. (1971). A methodology for the assessment of priorities and values in nature conservation. *Merlewood Research and Development Paper 28*, Institute of Terrestrial Ecology, Grange-over-Sands, 39 pp.

Helliwell, D.R. (1973). An examination of the effects of size and isolation on the wildlife conservation value of wooded sites. I. Birds, *Merlewood Research and Development Paper*, 49, 9 pp. (1974) II. Plants, *Merlewood Research and Development Paper*, 59, 8 pp.

Helliwell, D.R. (1976). The effects of size and isolation on the conservation value of wooded sites in Britain. *Journal of Biogeography*, 3, 407–416.

Helliwell, D.R. (1978). Survey and evaluation of wildlife on farmland in Britain: an 'indicator species' approach. *Biological Conservation*, 13, 63–73.

Hills, G.A. (1961). *The Ecological Basis for Land-use Planning*. Ontario Department of Lands and Forests Research Report No. 46.

Hooper, M.D. (1971). The size and surroundings of nature reserves. In E.A. Duffey and A.S. Watts (Eds), *The Scientific Management of Animal and Plant Communities for Conservation*, British Ecological Society, Symposium 11, Blackwell, Oxford, pp. 555–562.

Kent, M. (1972). A method for the survey and classification of marginal land in agricultural landscapes, *Discussion Papers in Conservation*, 1, University College London.

Kent, M. and Smart, N. (1981). A method for habitat assessment in agricultural landscapes. *Applied Geography*, 1, 9–30.

Mabey, R. (1980). *The Common Ground: A Place for Nature in Britain's Future?* Hutchinson, London, 280 pp.

Margules, C. (1980). *A study of the assessment of potential conservation sites*, duplicated report, Department of Biology, University of York, 20 pp.

Marsh, P. (1978). Formula for the needs of man and nature. *New Scientist*, 12 January, 84–85.

Moore, N.W. and Hooper, M.D. (1975). On the number of bird species in British woods. *Biological Conservation*, 8, 239–249.

O'Brien, M. (1977). Road development and its ecological evaluations. *Discussion Paper in Conservation*, 15, University College London, 42 pp.

Peterken, G.F. (1974). A method for assessing woodland flora for conservation using indicator species. *Biological Conservation*, 6, 239–245.

Ranwell, D. (1969). *A semi-quantitative index for comparative biological value of sites*, unpublished report, Institute of Terrestrial Ecology, Norwich, 3 pp.

Ratcliffe, D.A. (1971). Criteria for the selection of nature reserves. *Advancement of Science, London*, 27, 294–296.

Ratcliffe, D.A. (Ed) (1977). *A Nature Conservation Review*, Volumes 1 and 2, Cambridge University Press, Cambridge.

Shaw, W.S. and Zube, E.H. (1980). *Wildlife Values*. Centre for Assessment of Non-commodity Natural Resource Values, report No. 1, University of Arizona, Tucson, 117 pp.

Smith, D.W., Suffling, R., Stevens, D., and Dai, T.S. (1975). Plant community age as a measurement of sensitivity of ecosystems to disturbance. *Journal of Environmental Management*, 3, 271–285.

Spellerberg, I.F. (1981). *Ecological Evaluation for Conservation. Studies in Biology*, 133, Edward Arnold, London, 60 pp.

Tubbs, C.R. and Blackwood, J.W. (1971). Ecological evaluation of land for planning purposes. *Biological Conservation*, **3**, 169–172.

Usher, M.B. (1980). An assessment of conservation values within a large site of special scientific interest. *Field Studies*, **5(2)**, 323–348.

Ward, S.D. and Evans, D.F. (1976). Conservation assessment of British limestone pavements based on floristic criteria. *Biological Conservation*, **9**, 217–233.

Warren, A. and Goldsmith, F.B. (1974). *Conservation in Practice*, Wiley, London, 512 pp.

Yapp, W.B. (1973). Ecological evaluation of a linear landscape. *Biological Conservation*, **5**, 45–47.

Conservation in Perspective
Edited by A. Warren and F.B. Goldsmith
© 1983 John Wiley & Sons Ltd.

CHAPTER 15

Management Plans

J. BRIAN WOOD

INTRODUCTION

The management of reserves for the purpose of biological conservation is now a widespread practice, but the methodologies applied and the philosophies upon which they are based are frequently diverse. One major constraint upon the type of management that is practised is imposed by the relative scarcity of natural and seminatural areas that may be available for reserves. In particular when reserves are small, their managers are tempted to indulge in quite intensive manipulation of natural processes so as to maximize the conservation value; but the wisdom of a policy of extensive interference is rarely questioned when much of the land surface, outside reserve areas, is even more intensively managed for non-conservation purposes.

Very high human population densities have developed over much of western Europe, aided by the abundant supplies of cheap raw materials which emanated from the empires of European nations, which also provided markets for manufactured goods. The contemporaneous exploitation of Europe's own natural resources, the demand for land for factories, housing and ancillary services, and the subsequent attempts at self-sufficiency in temperate farm and forest products, has left very little of the land surface unexploited. Such areas of land as have remained undeveloped are particularly associated with sites which present formidable problems to developers. Thus, certain coastal areas have remained largely unspoilt, as the dynamic nature of their ecosystems can only be subdued by massive feats of engineering. This is a boon for industrial man who, in his masses may annually seek a respite from a mainly artificial environment, perhaps subconsciously returning to his evolutionary home (Hardy, 1960), and for the less ubiquitous but still numerous members of his society who seek a closer communion with nature. In places, even rather large tracts of coastlands have remained intact, particularly where they consist of largely unconsolidated water or windborne sediments.

What remains from this onward rush to develop is now jealously guarded by conservation agencies of various guises and often with diverse aims and ideologies, but is frequently managed so as to retain features believed to be natural or seminatural and perceived to be desirable. One such area occurs where the river Rhône flows into the Mediterranean sea, forming a very extensive delta region known as the Camargue, famed for its white horses, black bulls and massive wildfowl populations. Most of the Camargue is subjected to measures aimed at retaining the natural and traditional features for which it is renowned. Virtually the whole delta falls within the boundaries of the Parc Regional de Camargue, and is thus under special developmental planning control. Within this, several wildlife reserves exist to conserve specific natural features (Figure 15.1): the largest is the state-controlled Réserve Nationale (13 117 ha) and there are also smaller reserves owned and managed by local authorities or trusts. No one of these reserves itself contains a full range of natural features considered to be typical of the Camargue, but instead, each tends to be representative of particular component ecosystems. Whilst a considerable degree of cooperation may exist between the managers of the various reserves, intriguingly, the individual approaches to management are sometimes strikingly different.

Of the smaller reserves in the Camargue, perhaps the most well known is that which occupies the former estate of the Tour du Valat. From 1948 until 1980 this was privately owned and managed by its owner, Dr Luc Hoffman, but is

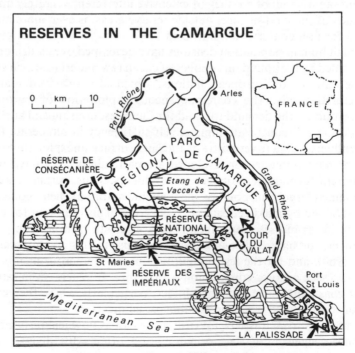

FIGURE 15.1.

now owned by a trust, the Fondation Sansouire, under the directorship of the former owner; it has thus retained a continuity of aims and management despite a change in ownership. Although the reserve at Tour du Valat is mostly typical of the higher, less saline parts of the Camargue and it contains no dunes of marine origin, salt-water pools or natural riverine woodland, its managers aim to provide habitats suitable for a wide range of birds occurring in the delta and, to this end, practise diversification of its ecosystems, mostly by controlling the distribution of water and by allowing varying amounts of grazing by horses and cattle. This management has been remarkably successful in enhancing the number of bird species regularly using the reserve, but many of these species find only a part of their day-to-day requirements from within its boundaries; if it were to become isolated its species complement would consequently fall dramatically. Furthermore, if a reduction in financial resources led to an enforced reduction in the management of its ecosystems, this would also lead to a loss of species, as the reserve returned to a more normal, unmanaged state (see Wood, Chapter 8). Certainly one rarely appreciated consequence of this type of management has been its tendency to replace the former relatively random changes in certain aspects of the environment with regular changes under management control, thereby increasing resource-predictability for the birds, and hence regularizing their occurrence and distribution, but at the same time reducing the naturalness of the system. Moreover, management to produce a diverse bird fauna has occasionally had undesirable consequences: the only extensive bed of reed (*Phragmites australis*) on the reserve was eliminated following deliberate inundation with salt water in order to encourage diving ducks; it is very slowly returning now that this management practice has been curtailed.

The boundary of the Réserve Nationale is partly coincident with that of the reserve at Tour du Valat, but Eric Coulet, the present director of the Réserve Nationale, has a very different approach to management. Although very extensive, the reserve under his charge is largely covered by water (65 per cent of the total area) comprising the Étang de Vaccarès and several smaller, interconnected *étangs* between it and the sea. These lagoons receive a freshwater input from the river Rhône (nowadays very attenuated and coming by way of agricultural irrigation of adjacent farmlands) and have outflows to the sea (now controlled by sluices). They are occasionally flooded by sea water in winter and also receive undetermined groundwater inputs of varying salinity. As a consequence, a salinity gradient exists from relatively fresh water in the north to more saline to the south, but substantial fluctuations occur as a result of evaporation during summer and flooding by rainfall and fresh or salt water in winter. Although most of the inflowing water courses and the coastal sluices lie outside the boundary of the reserve, and thus outside its direct management control, the unpredictable fluctuations that ensue are somewhat akin to those that must have occurred when the lagoon system had natural connections with adjacent waters. Eric Coulet is philosophical about this. He recognizes that a natural system of the type found in the Réserve Nationale is essentially dynamic in nature, and moreover, he knows that no planned scheme of water management would be likely to mimic successfully the random changes that would have affected the area in the past. His approach is thus to accept the changes in salinity and water level that now occur as an inherent feature of the reserve. By doing so he must also accept that numbers and distributions of plants and animals will also change in an unpredictable fashion. No attempts are made to stabilize fluctuations nor to encourage additional species to the reserve; to do so would be considered as increasing the unnaturalness of the area and contrary to the operation of natural selective forces, which produce evolution.

Thus, in this particular instance the managers of two adjacent reserves, both aiming to conserve the wildlife of their area, have come to adopt opposing attitudes to management. One is attempting to maximize the species-richness of his reserve and the predictability of occurrence of species, whilst the other accepts dramatic fluctuations as a natural occurrence together with the species turnover that must ensue. Fortunately, at present their efforts tend to complement one another, although considerably different resource inputs are required to support the different approaches. As we cannot accurately predict the future it would be foolish to try to assess which approach to management is the more realistic and likely to be the more successful in the long term. But it should be necessary, at least, for both managers carefully to examine their respective attitudes, to set down the aims that they hope to achieve and to explain the reasons for their actions. Neither possesses the wildlife he conserves, both are merely present guardians of a national and international heritage that, hopefully, will outlive them. And, if each should happen to be succeeded by a manager with the oppostite approach to management, how well would the wildlife heritage fare then?

THE NEED FOR PLANS AND PLANNING

If all wildlife reserves were very big, conservation managers could perhaps afford to adopt a policy of non-interference and still expect a high proportion of their reserve's biota to persist almost indefinitely; such an approach has been tried with the very large national parks in America (Blacksell, 1981). However, it would be unrealistic to expect even an extensive series of reserves to support both the large variety of species that have evolved and, at the same time, large populations of all species. It is evident that the greater the proportion of the world's resources that are channelled to human use, the smaller will be the proportion remaining for use by other life-forms, and hence the fewer their numbers and species. Since most conservationists recognize this fact, it becomes imperative to make a selection, and the usual selection is that which favours the retention of the maximum number of species at the expense of a reduction of numbers within species. Such a strategy is clearly realistic in terms of conservation of the greatest variety of genes, although it may lead to some overall reduction in heterogeneity of genotype within species (Franklin, 1980). To achieve such an adjustment in the balance of nature it is necessary to interfere with natural systems. Thus, the price to be paid for this type of strategy is at least three-fold: a reduction in the size of populations produces reduced heterogeneity within species, a stabilization of population sizes to prevent the extinction of species reduces natural selective evolution, and management resources must be continually invested to achieve this. Even in this simplest of situations it is thus necessary to plan ahead, to devise techniques for the manipulations to be made, to record results and constantly to remake choices as the remaining natural variability changes the circumstances under which management is operating. The alternative, which is less acceptable to most people, is to suffer a much greater loss of genetic material by the irreversible extinction of species; the magnitude of the latter would be in some way inversely proportional to the fraction of the earth that we could afford to set aside as wildlife reserves.

Although the type of choice discussed above is usually bewilderingly complex in all but the very simplest ecological situations, it is usually only one of many choices that a reserve manager is called upon to make. The pressure on land is such that it is rather rare for biological conservation to be the only land use that must be catered for on any reserve. Most usually a range of amenity and recreational uses must also be accommodated, not least of which is the cultural enjoyment by people. Often, potentially conflicting land uses may present severe constraints to management. At least around the boundaries of reserves they are likely to be formidable influencing factors, so that they need perhaps the closest consideration in small reserves, but few managers can afford to ignore factors such as the quality and quantity of water flowing through a reserve, airborne pollutants, or indeed the movement of organisms across the boundaries of the reserve (in both directions). Neighbours must be consulted, laws obeyed, policies formulated, techniques evolved and compromises achieved, before the manager can even afford to devote his attention to the manipulation of biota so as to best attain his conservation objectives. To be satisfied efficiently, all these peripheral considerations need to be encompassed within an orderly plan, which should indicate priorities and, when considered together with the biological management of the reserve, staffing levels and resource availability, should indicate how much the manager can hope to achieve in a given timespan. Moreover, the present attributes of the reserve, temporal changes and the results of management activities, all need to be recorded so as to enable updated reassessments to be made and management changes accomplished as they are necessary in fulfilment of the overall management aims. In the short term and on a limited scale, such a plan could be formulated and retained entirely within the memory of the reserve manager. However, because conservation management is more usually extremely long-term and often a cooperative venture, a more tangible plan is required in most cases. Documents of this sort must therefore provide for at least the following:

(1) To record all relevant features of the reserve and surrounding areas, updated as features change.
(2) To set out the reasons for management and the aims it hopes to achieve.
(3) To discuss the possible ways in which aims may be fulfilled and resolve the particular methods to be used and the reasons for the particular choice.
(4) To foresee conflicts and problems that may arise and suggest means of avoidance or amelioration.
(5) To plan in advance the use of manpower and resources.

In addition, plans have the advantage of both outliving their authors and also ensuring that managers think out the consequences of their actions before taking measures which may prove to be possibly detrimental in the long term.

ONE APPROACH TO THE FORMULATION OF MANAGEMENT PLANS

During the past 5 years staff and students of the Ecology and Conservation Unit at University College London have been actively involved in preparing management plans for wildlife reserves. Initially, a format was devised which would be applicable to national nature reserves in Great Britain and this has been used subsequently as the basis for plans for local naturalist trust reserves in England, the Tour du Valat reserve in France and national parks in Spain and Tunisia. During this time the format has been slightly modified and refined, but has retained all of the main original characteristics. The format is thus well tried and tested and could justifiably be claimed to have a fairly universal applicability. Examples of management plans using this format are provided by Stedman (1979), Wood (1981), and (Wood and Hollis 1982), the initial development of the format is discussed by Wood and Heaton (1976) and Wood and Warren (1978) provide a handbook to assist in the preparation of plans.

The University College London (UCL) format is designed to follow what is seen as a logical sequence that must be taken by reserve management staff, in order to arrive at an optimum use of resources in completion of the work necessary on a reserve (Figure 15.2). Because this type of decision-making is essentially a sequential process, changes in any of the factors taken into consideration will inevitably require changes to be made in the subsequent parts of the plan. Three major categories of material can be discerned:

(1) description;
(2) policy statements, evaluation and prediction;
(3) prescription.

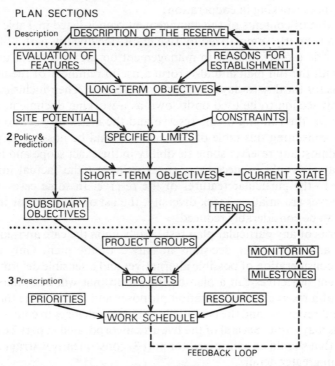

MANAGEMENT PLAN FLOW DIAGRAM

PLAN SECTIONS

1 Description

DESCRIPTION OF THE RESERVE

EVALUATION OF FEATURES

REASONS FOR ESTABLISHMENT

LONG-TERM OBJECTIVES

SITE POTENTIAL

CONSTRAINTS

2 Policy & Prediction

SPECIFIED LIMITS

SHORT-TERM OBJECTIVES

CURRENT STATE

SUBSIDIARY OBJECTIVES

TRENDS

PROJECT GROUPS

MONITORING

3 Prescription

PROJECTS

MILESTONES

PRIORITIES

RESOURCES

WORK SCHEDULE

FEEDBACK LOOP

FIGURE 15.2.

Accordingly, such a plan could consist of three parts, divided into these categories and bound as separate documents according to their relative permanence. The descriptive section will be relatively permanent, whilst prescriptions will require frequent updating and consequent revision: policy and prediction is likely to be of intermediate permanence. If several staff are involved in the management of a reserve, it is possible for different sections of the plan to be prepared by different people. The descriptive section (part 1) is relatively straightforward, consisting of a compilation and summarizing of all relevant factual material. However, the time taken in assembling the description may often be much greater than that involved in all other stages of plan preparation; it can be safely entrusted to a junior member of staff. The final prescriptive section of the plan must always be prepared after part 2, and normally requires the expertise of the day-to-day manager of the reserve. He may need to consult with the author of part 2 but, for the most part, the assembly of the working schedule is a relatively straightforward procedure. Most thought needs to be devoted to part 2 (policy and prediction), as this forms the heart of the plan and must contain cogent arguments for why particular policies should be favoured, the reasons for the setting of certain objectives, and the tolerance limits that are to be allowed. It may be best for this section of the plan to be the product of several authors working in cooperation.

A full table of contents of the management plan is listed in Table 15.1. The entire plan is briefly summarized by five maps, included in part 3 of the plan. These provide a quick insight to management on the reserve for anyone unfamiliar with the full plan and also form a useful reminder of the major considerations for the day-to-day manager of the reserve. Their inclusion in part 3 enables this section to be used on its own as a working document, providing guidelines for the running of the reserve and the order of work to be undertaken. By consulting this table of contents it would be possible to produce a plan for almost any reserve; some flexibility in the exact scope and hence content of part 1 would be necessary, so as to adjust the factual information according to the particular features of the reserve. In some cases, especially where reserves are small and lack diversity, the list of contents for this part of the plan can be considerably reduced.

The merits of any particular plan rest largely upon the care and skill devoted to policy and prescriptive decisions in part 2 of the plan. Only through a thorough consideration of possible alternatives and a sensible determination of management objectives can a plan be drawn up that will both maximize the potential of a reserve for conservation purposes and also obviate the necessity for frequent revisions, and the potentially damaging changes in course of actions which thus may ensue. Several of the features incorporated in part 2 of the plan, and some that should be considered for part 3, consequently warrant individual attention in greater detail.

THE EVALUATION OF THE FEATURES OF A RESERVE, ITS CONSERVATION INTEREST AND SITE POTENTIAL

A review of sites of conservation importance in Great Britain has recently been published (Ratcliffe, 1977) which attempts to be comprehensive and which provides an evaluation of each site on a comparative scale and through reference to a set of ten criteria. For British national nature reserves the evaluation thus made, provides a sufficient 'reason for establishment' in the plan and would be quoted as such; this thus provides the logic for commencing part 2 of the management plan with a statement of this reason. However, most parts of the world have not been the subject of a review of this type and, moreover, the comprehensiveness and evaluatory techniques employed in *A Nature Conservation Review* are themselves open to criticism (Adams and Rose, 1978; Margules and Usher, 1981; Usher, 1980). Conservation evaluation is an important topic in itself, warranting an entire chapter elsewhere in this book (see Goldsmith, Chapter 14), but certain aspects of evaluation need a consideration in the special context of management planning.

TABLE 15.1 Contents of the management plan.

In particular, it is necessary to determine the value of the main features of any reserve in terms of two potential principal uses. The subdivision of wildlife conservation into three main types as proposed in Adams and Rose (1978) would seem most logical, and their scheme is followed here. The aims of conservation are seen to be either to perpetuate the maximum number of species (for their own sake or for potential future use by man) or to conserve species, habitats or entire ecosystems so as to provide cultural value (for enjoyment, education or research). These two may in turn benefit the third principal conservation aim, to preserve the biogeochemical cycles upon which the existence of all life depends. Only the first two need concern us in the context of plans for individual reserves.

Whilst it is likely that many reserves will serve both species survival and cultural uses, few will be equally well equipped to fulfil each of these roles, and separate evaluatory criteria will be required to determine the relative worth of potential reserves for each end-use. For example, simple ecosystems containing rather few species could be of immense value for educational or research purposes, but would be less valuable for species conservation. Thus, where both functions are to be served by a reserve, it would seem best to perform separate evaluations for each. Hopefully, in many cases, separate parts of a reserve may be best suited to each function and can be zoned accordingly; when this is not the case, it may prove necessary to weigh one set of values against the other, and resolution of this necessarily subjective appraisal must rely heavily on the stated objectives of the reserve. However, even when considering one of the two functions suggested, it will often be difficult to avoid entirely subjective assessments: is a representative ecosystem of more or less worth than a unique one, and to what extent can naturalness be sacrificed for the sake of diversity? If the perpetuation of the maximum number of species were the sole aim of wildlife conservation then it might be possible to apply relatively rigorous objective evaluations. Adams and Rose (1978) have suggested some values that could be measured for this purpose.

In some ways the assessment of the conservation interest of a reserve is more readily undertaken than an evaluation based on apparently objective scientific criteria. At least it is almost universally accepted that what people find of interest depends quite largely upon personal judgements and is thus subjective in nature; there are consensus views as to what type of features represent the best value, but it is possible to argue that almost any physical or biological feature is valued highly by someone and special interest-groups may be willing to argue the case for their own preferences with considerable force. Nor is it unknown even for representatives of official conservation organizations to claim high value for features because they are unusually barren or depauperate in species complement (Grove, 1981). Notwithstanding this extreme example, most people seem to value highly species-rich environments. Fortunately, this conservation interest may complement the aim of perpetuation of many species, although quite often the most interesting diversity is generated by an unnatural reduction in the scale of pattern of distribution of organisms. However, the peculiar values ascribed to extremes of abundance and the preference for certain groups of organisms are much less tractable issues. Both very numerous and extremely scarce species may be particularly sought-after. To many people the sight of a wood full of bluebells or a lagoon packed with flamingoes may be as equally rewarding as the quest for a rare orchid or a glimpse of a vagrant bird migrant, and each extreme of abundance may even serve particular research purposes. But, if the wood were full of nettles or the vagrant were a bacterium, there would be much less enthusiasm. Thus, particular groups of organisms, or sometimes certain species, are highly rated, whilst others rate hardly at all. Large mammals, birds, butterflies and most flowering plants seem to be most favoured and hence ecosystems which may support an unusual preponderance of any of these tend to be preferred by conservationists. Thus, artificial systems such as chalk grassland are perpetuated and acquired as reserves and are often valued far more highly than more natural systems. In Figure 15.3 an attempt is made to synthesize the prevalent cultural assessment of the worth of wildlife features. Since individual taste is involved in these evaluations it has to be recognized that a management plan must here rely on widely held views, or an individual assumption of how these views would be weighted.

EVALUATION OF SPECIES

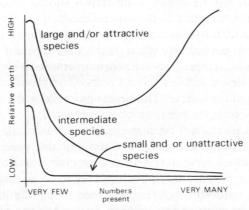

FIGURE 15.3. The currently prevalent practice for evaluating the conservation value of a species.

If consideration is given to the potential for a reserve to develop or to be modified so as to provide new opportunities for cultural use or so as to support additional species, further decisions need to be made. Here, the manager must attempt to weigh present, often tangible, features or values against future unknowns. The assessments will be more realistic if the techniques which may produce the desirable changes are well known, or the natural vegetation succession which may be allowed is well-documented. If faced with a choice at time of purchase, few conservation organizations would be likely to opt for an area with potentially high future-value in preference to one with high present-value. But there may be cases where existing reserves could be readily modified so as to enhance their value for biological conservation. In such cases potential future-value can figure highly in the process of management decision-making, although it is important to be aware of the problems of creating new situations in which continuous management inputs are required so as to maintain a reserve in a new and more highly desirable state. Local naturalist trusts are particularly vulnerable to decisions of this type, as their management resources are frequently very small and often overcommitted.

THE DETERMINATION OF OBJECTIVES

The objectives for any reserve are consequent upon policy decisions made by the owning or managing organization but they are rarely stated precisely. Unless the objectives are carefully thought out and clearly defined, there is a real risk that the management undertaken will change with time and with the inclinations of the incumbent manager, and serious compromising of the original purposes of the reserve can ensue (Goriup and Wescott, 1977). This is not to say that objectives should invariably remain those determined at the time of acquisition; but any flexibility in objectives should be controlled by conscious decisions, not tolerated as the consequences of lax management.

There is often a wish to maximize the range of conservation benefits that a reserve may afford, particularly when reserves are few and potential uses are numerous. However, few reserves are able to tolerate as wide a range of uses as the manager may desire without serious risk of compromising one or more of the principal aims of the reserve. The almost universal wish to preserve as many species as possible on a reserve is an objective which, if strictly enforced, imposes the severest disciplines on management. Conservation is rife with unwritten rules concerning acceptable and unforgivable practices and beset by ecological myths which serve to reinforce these practices (Rose, 1978). Few are willing to recognize that the manipulations they so commonly undertake often merely reduce the naturalness of reserves, even though naturalness is an attribute highly rated by most conservationists. Since the species we may wish to preserve are themselves the product of evolution, and as evolution itself is a dynamic process, the suppression of change by management can become counterproductive. Thus, in the strictest sense, reserves set aside for the conservation of species in natural communities should be subjected only to minimal management, and all of this management should be aimed at reducing unnatural (usually induced) impacts on the reserve. But, as the area of the earth that we can afford to set aside for conservation purposes is only of small extent, reserves managed by the above-stated objectives could only hope to preserve a fraction of our planet's biota.

If we are to perpetuate a much greater proportion of our flora and fauna it thus becomes imperative that manipulative techniques are used so as to ensure the survival of at least a few representatives of each species. The intensity of the techniques applied would need to relate inversely to the total area set aside for conservation purposes and to its distribution. With very small areas available, very intensive management of the type found in zoological or botanic gardens would be necessary. If more land could be devoted to reserves, manipulation could afford to be less intensive, although it would be necessary to coordinate efforts on an international scale. If survival of the maximum number of species is the goal, there is little point in one country spending resources on protecting a much-valued rare species if that same species is common and readily protected in another part of the world. At present there is little international cooperation of this sort. Individual nations take great pride in their own flora and fauna (as do individual local naturalists' trusts at the county level) and protect rarities often at great expense; this is only justifiable in cultural, not ecological terms. Similarly, many commonly accepted conservation practices (such as the elimination of alien species) are really only artefacts.

A clear understanding of the distinction between ecological and cultural goals is needed. Usually reserves are intended to provide benefits to the community that supports their creation, and so manipulations of biota for cultural reasons are acceptable. But, both the logic of the management plan and the credibility of the management authority will ultimately suffer if management attempts to hide such cultural decisions beneath a blanket of scientific objectivity.

The stated objectives for any reserve need to indicate if the principal aim is to conserve as many species as possible on the reserve, or to adjust the species complement in such a way as to increase cultural benefits. In the case of the former goal it is then necessary to decide between minimum management which will be mostly aimed at reducing human impact, or manipulative management, so as to favour species which the particular reserve is best able to conserve. If cultural benefits are to be of principal concern it is necessary to state the priority of uses, discuss possible zonation to cater for various types of use and analyse likely demands, so as to produce rational objectives. In every case it is imperative that alternatives are considered and the reasons for the preferred objectives are stated. Failure to do so is likely to lead ultimately to a more variable management of the reserve, with consequent detriment to its biota.

Multiple use of a reserve is more difficult to resolve. Many reserves may be able to serve a primary function of species conservation and, at the same time, support some cultural uses. In this instance the stated objectives must make it quite clear that cultural use is to be subordinate to species conservation, and that the latter must never be compromised in order to benefit the former. It is all too easy to introduce gradual changes in management policy which are compounded through time and eventually lead to a complete overturning of the original objectives. Very rarely will it be possible to establish reserves whose primary objective is to provide cultural benefits but which may also fulfil a useful role in conserving natural ecosystems. This may occasionally prove possible with very extensive reserve areas, but most commonly the solution to problems associated with this type of dual use is resolvable only in terms of a zonation of uses, leading to possible fragmentation and devaluation of the reserve.

SPECIFIED LIMITS FOR FEATURES EVALUATED AS IMPORTANT

In all reserves where management involves manipulation of the flora and fauna, or where physical features such as water levels are under direct human control, it is important to recognize that changes in status brought about by these manipulations will affect not only the target organisms but also other organisms with which these species may interact. Thus, for example, stabilization of the population size of a large herbivore species will have marked effects on the vegetation, producing adjustments in both productivity and species composition. If the plant species thus affected are themselves important features of the reserve, it becomes imperative that the manager recognize the secondary changes that may be produced and take account of these in determining his primary courses of action.

Because the ramifications of any action are likely to be numerous, and the links between different elements in ecosystems are imperfectly understood, the best that any manager can hope to include in a management plan is an indication of interactions between all the important features of the reserve. This information is most easily digested if presented in tabular format (Wood and Warren, 1978).

In practice, most managers find it extremely difficult to draw up a table of specified limits. However, the problems of decisions are not a valid reason for omitting this process, as this stage in management planning forms an essential link between the preceding evaluations and objectives and the succeeding schedule of work. It is oversimplistic to express a desire for all important features to increase in status, for this is unlikely to be attainable without reductions in features valued less highly, but which may themselves be parts of the ecosystems that provide support for the more highly valued items. Where decisions concerning the areas of alternative types of vegetation are concerned, it may be a relatively straightforward procedure to specify certain limits for each alternative; but, when attempting to manage an animal population which may show marked natural fluctuations in size, specified limits, particularly the lower tolerable limit, are much harder to determine. It is, nevertheless, important to set such limits, as variations in status beyond the specified limits are the signals for the need to revise management actions, so as to restore the preferred status.

When it is considered important to retain as many elements of naturalness as possible on a reserve there is, however, a real danger of attempting to over-restrict the variations of status of individual features. All ecosystems are subjected to both cyclical and random changes and attempts to smooth their effects can lead to unwanted consequences. Ultimately, most managers must come to realize the interdependence of features on their reserve with those outside its boundaries, with consequent unmanageable ramifications. If nothing else, a tabular specification of tolerable limits, should at least promote an acceptance that no reserve can hold indefinitely a complete range of features that owners or managers would ideally like to see there, and may thus curb the overenthusiastic attempts at manipulation with the aim of constantly adding desirable features. The ultimate attainment of such policies would be to create reserves that required constant high levels of management and contained artificially rich but equitable assemblies of species, yet bearing only superficial resemblance to the natural communities of the area, and with reduced links between the features within the reserve and those beyond its boundaries. More sensibly, the thinking reserve manager will recognize the limitations of such policies, set himself realistic limits for the important features on his reserve, and set about enhancing these by forging stronger links with its surroundings, be they physical, or, more often, social, by seeking to attain a greater understanding and more sympathy for his own objectives by the users of surrounding areas.

THE SUBDIVISION OF RESERVES FOR MANAGEMENT PURPOSES

All but the very smallest and structurally simplest of reserves contain a wide diversity of features that it is hoped will be preserved and have a potentially vast range of problems and conflicts which must be considered by management. Because it is extremely difficult to consider all possible interactions at the same time, and thus resolve all the many implications of a particular course of action, it is necessary to simplify matters to a level at which only a manageable number of interactions are considered. For the reserve as a whole this may be achieved by considering only the most direct interactions between organisms and between these and their environment, although this has the disadvantage of possible oversimplification to the point of absurdity and the risk that quite minor but nevertheless important issues will be overlooked. As a consequence there is a real risk of these minor issues negating the original purpose of the management action. The alternative, which many reserve managers adopt, is to simplify complex issues by spatial separation, assigning specific parts of a reserve to specific features or issues, such as compartmentalization. This often works quite well on a limited scale, but particularly when management is quite intensive, there is the profound risk that it will lead to fragmentation of the reserve and a consequent loss of valuable features. This process has the most serious potential consequences for the species with the biggest areal requirements; quite often these are the most spectacular elements of the biota and consequently of greatest cultural value. In the management plan format developed by the MSc Conservation Course at University College London (Wood and Warren, 1978) an alternative approach was suggested, which uses a grouping of management activities so as to help avoid fragmentization.

Management activities are most easily controlled if they are dealt with as individual jobs or projects. Our system enables records to be readily located from the initial planning stages through to completion of the work and recording of its results. A combination of records from all projects will also enable budgets to be formulated for finance, manpower and equipment. However, it will rarely be possible to undertake all projects that may be considered desirable on a reserve, as resources will usually be a constraint. Nevertheless, it is worthwhile to list all potential projects in order to indicate the comprehensiveness of the management plan. Ultimately, a limited number of these will be fulfilled, and this will be determined according to a scheme for assigning priorities to projects; how many are completed will result from the quantity of available resources.

Having listed all the potential projects that might be undertaken on a reserve, it will become evident that these tend to fall naturally into groups of projects with similar purposes. These groups may relate to essential management that must be undertaken in order to maintain the fabric of the reserve, to the resolution of a particular problem or to the management of a particularly important natural feature of the reserve. In some instances, only a circumscribed part of the reserve may be affected by such a collection of potential projects, in other cases the whole of the reserve may be involved. The recognition that natural groupings of projects occur, and the incorporation of this into the formal structure of the management plan, provides a useful alternative to the divisive process produced by compartmentalization of management activities.

Project groups

We have termed collections of projects for a reserve 'project groups' (Wood and Warren, 1978) and recognize that almost all reserve management plans will need to consider at least three such groups. The essential project groups will be:

(1) Obligatory management, containing all projects necessary in order to maintain the fabric of the reserve in fulfilment of legal requirements and obligations to neighbours.
(2) Public access and education, containing all items directly concerned with the management of visitors.
(3) Extensive survey, to ensure comprehensive monitoring and record natural and induced changes for the whole of the reserve, and possibly also for adjacent areas.

Each of the remaining project groups for a reserve is likely to be concerned with the management of a specific feature of importance on the reserve. The work thus entailed may therefore sometimes affect only a circumscribed part of the reserve (for example, when a certain type of vegetation is the important feature) or may involve the whole of the reserve (for example, if wintering waterfowl were to be considered an important feature of a mainly wetland reserve). Interactions between projects undertaken within different project groups are thus likely to occur. Potential conflicts of this type may be recognized by a tabulation of all projects within each project group and an individual consideration of how each project may interact with the features forming the focus of all other project groups; conflicts can be readily listed in such a tabular presentation. Within each project group individual projects should have more in common with each other than with projects assigned to other project groups. If this is not the case it is an indication that project groups have been unrealistically determined. However, since a system of project groups does not involve a rigid areal delineation, the scheme adopted should maintain a degree of flexibility, enabling the fairly straightforward division or combination of project groups so as to afford a more realistic allocation of management activity, if changing circumstances dictate this. Providing that this does not also involve a complete re-evaluation of features or a change in management policy, a reallocation of projects amongst project groups will not compromise the original aims of management.

Although it may be imagined that the grouping of projects is either an extremely complex or entirely arbitrary process, in practice this allocation is no more problematical than the original evaluation of the relative merit of individual features. Project groups are normally unlikely to be created solely on the basis of managing a single species, unless this is a species of extreme importance on the reserve. Because most management actions will affect a range of species, and specific projects will promote similar responses from species with similar ecologies, groups of important species are more likely to form the basis of a project group. Thus, manipulation of water levels in winter may promote reactions from a wide range of wintering waterfowl, not just a particularly important duck species. This would be recognized by creating a wintering waterfowl project group, although individual projects within this group might nevertheless be aimed at managing the most important single species. In this way, complex overdivision of management activities is avoided, because this would be counterproductive.

Thus project groups are essentially a system for collecting related activities into convenient categories for management purposes. They enable staff to classify management activities, manpower requirements and costs into a limited number of categories. Above all, project groups allow related problems to be viewed as an entity, enabling overlaps and conflicts between individual projects to be realistically resolved, and gaps in the plan of management to be more readily identified. In recent plans, and particularly when dealing with complex interactions, we have found it helpful to draw up schemes indicating the major links

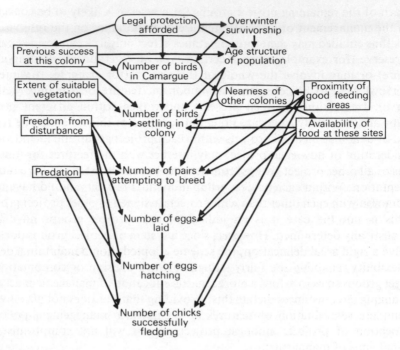

FIGURE 15.4. An example of the scheme for the management rationale of a project group which aims to manage breeding waterbirds. (Factors which may be controlled by management are placed in rectangular boxes, factors which can be controlled only indirectly are placed in oval boxes).

between the features upon which a project group is based and other elements in the ecosystems to which they belong. An example of such a scheme is illustrated in Figure 15.4. The use of this type of pictorial summary of interacting processes can greatly assist the identification of critical pathways through which management can most effectively operate, and avoid the failure to consider all the important interactions.

SOME FINAL CONSIDERATIONS

The production of a management plan involves a great deal of work, frequent consultation between interested parties, and the resolution of numerous problems. As such, the very writing of a plan may go a long way towards resolving the numerous issues concerned in the management of a reserve. However, it should never be seen as being an end in itself. Plans only come to fruition when their recommendations are translated into actions, and it will be the consequences of these actions that themselves ultimately determine the success or failure of the planned attempts at wildlife conservation. But this should not be interpreted as advocacy of active management in all reserves. Indeed there is a real risk in adopting the attitude that once a reserve is acquired it must be actively managed. What a management plan seeks to do is to assess the attributes of a reserve, explore the consequences of particular courses of action, and resolve what, if any, management is required in order to best fulfil the objectives of the reserve. If this process identifies the best course of action as being one of non-interference, then the reserve manager should have sufficient strength of will to adopt this strategy. Indeed, if it proves impossible adequately to identify the consequences of a particular course of action it would be prudent to adopt a cautious approach and take no action at all, or at least undertake only a limited manipulation at first (perhaps on a limited-area trial basis) in order better to resolve the consequences of more extensive management.

If we return briefly to a consideration of our original example of the two adjacent, but very differently managed, reserves in the Camargue, we can see that each manager is probably doing the best he can towards ensuring the conservation of the features of his reserve, within his own philosophy of conservation. But has each manager adequately considered alternative approaches and their consequences? Surely, one part of conservation is the theme of keeping open the maximum number of future options, and another is the development of a more sustainable exploitation of the environment, both fostered by the dissemination of a greater understanding of our environment and more sympathetic public attitudes. It may seem ironic that, of these two reserves in the Camargue, the one perhaps best suited to maintain a wide range of species is the one where unpredictable changes are accepted, and the reserve where management has produced a greater equitability and species richness, that could provide enhanced cultural value, is the reserve least accessible to a wider public.

REFERENCES

Adams, W.M. and Rose, C.I. (Eds) (1978). The selection of reserves for nature conservation. *Discussion Papers in Conservation*, **20**, University College London, 34 pp.

Blacksell, M. (1981). A truly natural national park. *Geographical Magazine*, **53**, 334–340.

Franklin, I.R. (1980). Evolutionary change in small populations. In M.E. Soulé and B.A. Wilcox (Eds), *Conservation Biology: An Evolutionary-ecological Perspective*, Sinauer Associates Inc., Massachusetts, pp. 135–149.

Goriup, P.D. and Wescott, G.C. (1977). Comments on the relative importance of objectives and management in nature conservation. *Environmental Conservation*, **4**, 231–232.

Grove, R.H. (1981). The use of disguise in nature conservation: the evidence from three case studies. *Discussion Papers in Conservation*, **32**, University College London, 38 pp.

Hardy, A.C. (1960). Was man more aquatic in the past? *New Scientist*, **7**, 642–645.

Margules, C. and Usher, M.B. (1981). Criteria used in assessing wildlife conservation potential: a review. *Biological Conservation*, **21**, 79–110.

Ratcliffe, D.A. (Ed) (1977). *A Nature Conservation Review*, Cambridge University Press, Cambridge.

Rose, C.I. (1978). A note on diversity and conservation. *Bulletin of the British Ecological Society*, **9(4)**, 5–6.

Stedman, N. (1979). An appraisal of the UCL management plan format for use by a county trust. *Discussion Papers in Conservation*, **21**, University College London, 32 pp.

Usher, M.B. (1980). An assessment of conservation values within a large site of special scientific interest in North Yorkshire. *Field Studies*, **5**, 323–348.

Wood, B. (1981). Management proposals for the Tour du Valat/Petit Badon reserve of Fondation Sansouire. *Conservation Reports*, **11**, University College London, 65 pp.

Wood, B. and Hollis, G.E. (Eds) (1982). A management plan for Sebkhet Kelbia, Tunisia. *Conservation Reports*, **12**, University College London, 204 pp.

Wood, J.B. and Heaton, A.M. (1976). Management planning for National Nature Reserves. *Conservation Reports*, **9**, University College London, 53 pp.

Wood, J.B. and Warren, A. (Eds) (1978). A handbook for the preparation of management plans: Conservation Course Format, revision 2. *Discussion Papers in Conservation*, **18**, University College London, 41 pp.

Conservation in Perspective
Edited by A. Warren and F.B. Goldsmith
© 1983 John Wiley & Sons Ltd.

CHAPTER 16

Urban Nature Conservation

LYNDIS COLE

INTRODUCTION

At first glance the urban areas of Britain, with all their manifestations of man's subjugation of nature, would appear an unlikely candidate for nature conservation. Fifteen years ago the term 'urban nature conservation'—if heard at all—was considered by many, if not by most of those active in the field, as a freak sideshow to the rural theatre of conservation activity and research. Why is it, therefore, that the promotion of nature in urban areas is now seeing a rapid rise in popularity?

The reasons are many. The interest stems in part from developments in landscape architecture and land reclamation, but more importantly from the inextricable link with changing attitudes within the nature conservation movement itself. Richard Mabey, in his book *The Common Ground* (1980), argued that the enjoyment of nature and therefore its conservation, should not just be concerned with the rare and spectacular but with the common; nature conservation while embracing scientific research should not lose sight of our individual and very personal associations with nature. With 80 per cent of the British population now living in urban areas and with many inner-city dwellers unable to reach the countryside, it follows that nature should be actively encouraged within the urban environment so that it is not the experience of the few but of the many. This thinking is closely allied with that in environmental education and the belief that a clear understanding of, and concern for, nature is best instilled through direct and frequent experience—quotidian nature.

Another reason for the surge of interest in urban nature conservation is that there is a growing realization (as fully debated during the passing of the 1981 Wildlife and Countryside Act) that rural habitats are fast dwindling. Few would suggest that the conservation of urban seminatural habitats, of which there are a surprising variety, could make up for this loss. Yet these urban sites do have one potential advantage—they are frequently managed specifically for public enjoyment rather than commercial production and are therefore not directly subject to changing policies in agriculture and forestry.

The three major objectives of urban nature conservation can thus be defined as follows: to conserve and press for the appropriate management of urban sites of intrinsic natural history value; to increase the habitat diversity of formalized areas of public open space; and to create new wildlife habitats, either on a temporary or permanent basis, on downgraded and derelict sites within the inner city. In each case, the concern is as much for people as it is for wildlife and heavy emphasis is placed on public enjoyment, community participation and school activities. Ultimately, urban conservation is concerned with the dispelling of any perceptual division between urban and rural areas.

URBAN FORM AND INFLUENCE

Fordham (1975) has estimated that urban areas cover some 1.8 million hectares in the United Kingdom. Within these built-up areas an enormous number of factors influence the range of, and potential for, wildlife. There are the inhibiting effects of water pollution (Harrison and Grant, 1976) and air pollution, where research has concentrated on lichen flora (Gilbert, 1965; Laundon, 1970; Rose and Hawksworth, 1981) and birds (Burton, 1976; Cramp and Gooders, 1967; Gooders, 1968). Then there is the benign influence of the urban 'heat island', which, for example, encourages urban blackbirds to breed 3 weeks earlier than their rural counterparts (Batten, 1973).

However, when considering the range of habitats found in urban areas, and the potential linkage between these habitats and rural areas, the major influence is inevitably the form of the urban area in question.

The growth of London conforms neatly to a concentric ring model: successive waves of expansion from a central core. As this expansion continued, remnants of the countryside became enmeshed in the urban fabric; either by reason of conscious planning, as in the case of the royal parks; or by reason of ownership, topographical constraint, or government legislation including the Metropolitan Commons Acts of 1866, 1878 and 1899, which made enclosure and loss of common land practically impossible (Stamp and Hoskins, 1963). Examples include remnants of the Great North Wood in the grounds of Dulwich School, fragments of low-lying marsh at Walthamstow (see Goldsmith, Chapter 14), and the commons of Wimbledon and Wandsworth.

Today London's parks, commons and public open spaces, including those Victorian parks such as Finsbury and Southwark, whose sites were acquired under the Metropolis Management Amendment Act (1856, section 10), cover some 16 890 ha (Greater London Council (GLC) 1979) or approximately 11 per cent of the Greater London Council area. And this figure excludes the large Victorian cemeteries which alone cover a further 905 ha within a 16 km radius of St Pauls Cathedral (Brain, 1981), all private open spaces and the numerous outer fresh-water and canal feeder reservoirs. The figure of 16 890 ha also excludes the extensive areas of dormant and vacant land, which in 1971, prior to the major dockland dereliction, the Greater London Council conservatively estimated at some 6540 ha (Cantell, 1977).

Despite the increasing erosion of the urban fabric through dereliction, London can still be described as consisting of discrete open spaces surrounded by development. On the other hand, a city such as Stoke-on-Trent, like so many other midland and northern conurbations, can be described as consisting of relatively discrete settlements, separated by belts of enmeshed urban fringe, river valleys and industrial dereliction. The general open fabric of Stoke, and its potential for wildlife habitats, stands in contrast to the tight packaging of London (Figure 16.1).

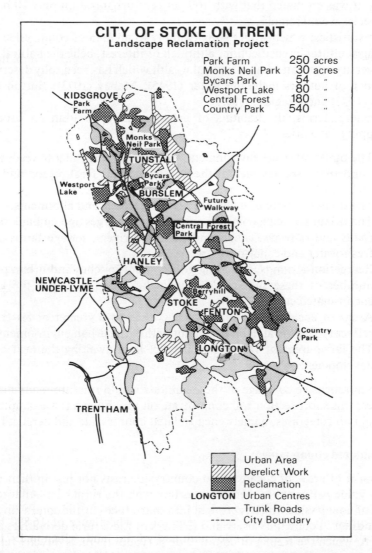

FIGURE 16.1. The urban form of Stoke-on-Trent in 1974. Reproduced by permission of Land Use Consultants.

THE URBAN WILDLIFE RESOURCE

There is a general theory that biological diversity is inversely proportional to the degree of urbanization. This theory is substantiated by a number of ornithological studies. Batten (1972), taking the historical records of the Brent Reservoir area in north London, has demonstrated a decline in breeding bird species with increasing development. In 1833, when the newly constructed Brent Reservoir lay beyond the urban edge, 72 bird species regularly nested in the area. By 1970, with 65 per cent urbanization, the number of breeding bird species had fallen to 47, and it was estimated that with 100 per cent urbanization only 20 breeding bird species would remain.

Such statistics, which would equally apply to an area of countryside undergoing agricultural intensification, although of interest, belie the natural wealth and diversity of urban areas, a natural wealth which has been fully described by a number of authors including Fitter (1945); Mabey (1973); Burton (1974); Simms (1975) and Teagle (1978).

In general terms, the habitats of urban areas fall within six broad and overlapping categories:

(1) The built and street environment, the home of the ruderal weeds and the opportunist species such as the house sparrow, feral pigeon and brown rat.
(2) The domestic system, including private gardens and allotments.
(3) The urban servicing complex, including the verges and embankments of roads and railways, canals and canalized rivers, sewage farms and the fresh-water and canal-feeder reservoirs.
(4) Recreation grounds, including golf courses, pitches and urban parks (a number of these sites could equally be classified under the heading encapsulated countryside).
(5) Areas of encapsulated countryside, which to a greater or lesser extent reflect typical rural habitats caught within the urban environment.
(6) The large areas of vacant and derelict land awaiting the next cycle of development.

A comprehensive coverage of these habitats would run into many hundreds of pages; this short section will concentrate on providing a few examples from the final two categories, namely encapsulated countryside and derelict land.

Encapsulated countryside

The charm of areas of encapsulated countryside rests not just in their natural history value *per se* but in their obvious link with the rural past—that tangible feeling of countryside which is brought into sharp focus by the contrasting urban surroundings. Today the variety and richness of these rural derivatives and the degree to which the feeling of countryside pervades there, is directly related to the management and use to which these areas have been subjected.

The urban park, although often a direct derivative of the countryside, reflects, like the building facade, the fashions, demands and moods of the period during which it was established. But despite horticultural design and maintenance, small remainders of the countryside survive, such as the spectacular growth of cow parsley (*Anthriscus sylvestris*) along the banks of the Grand Union Canal in Regent's Park and within the woodland area of Holland Park in west London. In terms of natural history, the trees, shrubberies and lakes of these parks can provide a valuable habitat for birds and insects. The heronry of Regent's Park is well known, as are the occasional rare migrant visitors such as osprey, avocet, hoopoe and little auk which have graced Hyde Park and Kensington Gardens. On the other hand the macrolepidoptera of Regent's Park, of which some 100 species have been recorded (De Worms, 1972) often pass unnoticed.

In striking contrast to these semiformal areas are the few sites which have remained remarkably unscathed by urban influence. Wimbledon Common in south-west London, despite tremendous recreational pressure survives as typical heathland, scrub and woodland with the few remaining bog and pond areas supporting a wide diversity of wild flowers including marsh pennywort (*Hydrocotyle vulgaris*), trifid bur-marigold (*Bidens tripartita*) and skullcap (*Scutellaria galericulata*). The West Field Bog of Hampstead Heath boasts six species of sphagnum moss and the rare wood horsetail (*Equisetum sylvestris*). In Birmingham, Sutton Park supports some 400 species of flowering plants of which 27 are found nowhere else in the old county of Warwickshire (Readett, 1971).

Of equal interest are the small patches of urban ancient woodland, of which two notable examples are Dulwich Woods in south London and Saltwells Wood in Dudley Metropolitan Borough. Both these woodlands are dominated by sessile oak (*Quercus petraea*) which, in the case of Saltwells Wood, regenerates freely and is one of the best examples of natural regeneration to be found in the whole of the West Midlands. In terms of ground flora, the Dulwich Woods support the richest ground flora within 10 km of St Paul's Cathedral, including such typical woodland species as bluebell (*Endymion nonscriptus*), primrose (*Primula vulgaris*), ramsons (*Allium ursinum*) and wood anemone (*Anemone nemorosa*), while Saltwells Wood offers a number of rarities to the West Midlands conurbation.

Urban cemeteries are another example of direct derivatives from the countryside. Initially the subject of grandiose landscape schemes, many of these sites in more recent years have been left to run wild, as their owners—usually cemetery companies—faced bankruptcy; once a cemetery is full, no further revenue is available. Typical examples are Nunhead and Tower Hamlets cemeteries in south-east London. Here a wide range of woodland trees, hedgerow and meadow herbaceous species thrive, including bittersweet (*Solanum dulcamara*), common knapweed (*Centaurea nigra*), and creeping jenny (*Lysimachia nummularia*). The occasional local rarity may also be found, such as the discovery in 1980 of 419 spikes of green-winged orchid (*Orchis morio*) growing in a relict piece of damp meadow land in a south London cemetery.

Derelict land

Of equal interest to these seminatural urban habitats is the ability of nature to adapt to, and colonize vacant urban land and derelict sites which are awaiting the next cycle of development. The Department of the Environment's definition of derelict land, 'land so damaged by industrial or other development that it is incapable of beneficial use without treatment', conjures up a picture of biological poverty. While there are a number of industrial and mining wastes, such as colliery shale, which on initial deposit are inhospitable to plant growth (see Bradshaw, Chapter 11), there are a number of herbaceous species which, either through biochemical adaptations or the development of specific genotypes, have successfully colonized a range of polluted soils. Spring sandwort (*Minuartia verna*), a species of calcareous rocks and moors, may be found on lead-mining spoil and is not averse to exploiting suitable microhabitats; one pioneering plant of this species has been found growing on the remains of a car battery in an Oxford car breaker's yard!

Urban derelict sites range in type and size from the vacant corner plot to the large subsidence pools, open quarries and waste heaps of the west Midlands and northern conurbations. Small urban derelict plots are usually associated with the more typical urban ruderal weeds such as oxford ragwort (*Senecio squalidus*) and rosebay willowherb (*Epilobium angustifolium*). But the floral diversity of these sites may be far greater. A survey undertaken in central London between 1952 and 1955 (Jones, 1958) revealed 342 plant species associated with these derelict sites. Of this total, a number of species were garden outcasts; some reflected local industry—such as hops growing adjacent to a Whitbread brewery; while many were more typical rural species which had moved in to exploit a specific niche—for example calcicole flora including marjoram (*Origanum vulgare*), wild mignonette (*Reseda lutea*), viper's bugloss (*Echium vulgare*) and upright brome (*Bromus erectus*) growing on the lime-rich mortar of building rubble. Turning to fauna, a widely publicized event was the spread of the black redstart, a European rock-haunting species, into inner-city bomb sites during the 1940s (Fitter, 1978). Although this species remains a relatively rare British bird, London has consistently supported a high proportion of the total British breeding population.

On a larger scale, London's derelict docklands with their large areas of open grassland, hard and soft-edged docks and derelict buildings, have provided a wealth of habitats for birds. Of particular note is Surrey Docks, which between closing in 1968 and the start of redevelopment in 1976, attracted over 121 different bird species. Of the 30 or so species which nested in Surrey Docks, no less than ten provided the first ever proven breeding records for their kind in inner

London, including lapwing, ringed plover, little ringed plover and reed bunting (Alderton, 1977).

Of equal value to fauna and flora are the many wet and dry mineral workings and the subsidence flashes associated with mining. The derelict Queslett Quarry sand and gravel workings in Birmingham offer a diverse range of habitats, from flooded pits which provide nesting sites for the great crested grebe and little grebe, through small fringing pools attracting toad and common frog, to the towering artificial sand cliffs supporting one of the largest breeding sand martin colonies in the west Midlands. Perhaps less spectacular but no less important is the small marshland community which has developed in a subsidence flash at Stubbers Green in the Walsall area. Here the rich plant community includes such species as marsh arrowgrass (*Triglochin palustris*), common cottongrass (*Eriophorum angustifolium*), lesser bulrush (*Typha angustifolia*), marsh orchid (*Dactylorhiza sp.*) and *Sphagnum* mosses. Many of these species are hardly known in the West Midlands County and all are extreme rarities within the truly urban context (Teagle, 1978).

Although tipped waste material is often inhospitable, there is increasing evidence that some waste materials are providing alternative habitats for regionally rare and localized herbaceous species. Greenwood and Gemmell (1978), following a 10-year survey of predominantly urban industrial sites in west Lancashire have demonstrated that certain wastes including Leblanc process waste, pulverized fuel ash (PFA) and blast furnace slag may, in time, support herb-rich calcicolous associations. Species recorded on these wastes, which are strongly alkaline in their raw unweathered state (pH 8.0 to pH 12.7) include yellow-wort (*Blackstonia perfoliata*), early marsh orchid (*Dactylorhiza incarnata*), southern marsh orchid (*D. praetermissa*), northern marsh orchid (*D. purpurella*), common broomrape (*Orobanche minor*), and round-leaved wintergreen (*Pyrola rotundifolia*). Not only are these species rare in west Lancashire but the apparent hybridization between a number of the orchid species found on these sites led Greenwood and Gemmell to suggest that the provision of these waste tips may be contributing to the breakdown of isolating mechanisms between species.

From this brief review, it is obvious that derelict sites can make a direct contribution to the natural wealth of urban areas. And as suggested by Davies (1976) such sites may be performing one of a number of valuable functions. They may, as in the case of the Stubbers Green marsh, be replacing habitats which are rapidly declining in number in rural areas; they may, as in the case of the west Lancashire alkaline tips, be providing additional habitats for species not naturally well catered for either regionally or nationally; or they may, as in the case of urban bomb sites and the black redstart, be offering suitable habitats for new immigrant species. In addition, in an age when we are seeking guidelines on how best to recreate certain natural habitats, such sites can provide a number of pointers.

THE CONSERVATION OF URBAN WILDLIFE HABITATS

The safeguarding of sites of existing natural interest within urban areas is complex when taken in conjunction with the intricate system of urban land-use planning and consequent land values. Areas designated as public open space have no commercial value, but sites designated for office or industrial development range in value from £47 000/ha for an undeveloped site in London's derelict dockland to over two million pounds for a prime central London office location.

Many urban sites of natural interest are under threat. The threats take many forms of which the most obvious is development, including not only built development and associated drainage and servicing requirements but also all manner of transport schemes. Mineral extraction works continue, while urban water courses are used as receptacles for a wide range of pollutants and are frequently drastically modified supposedly for flood protection. Solid waste disposal is another major problem that ranges from insidious fly tipping on vacant land to full-scale landfill. Landfill still remains the most economic method of waste disposal, costing as a national average in 1977–78 some £1.87 per tonne of waste tipped compared to £11.03 for incineration (Society of County Treasurers, 1978). Because of these prices, 69 per cent of UK waste is still disposed of as crude landfill. Given high haulage costs and the total quantity of UK waste that is generated each year—estimated by the Department of the Environment in 1976 to equal some 18 million tonnes of industrial and building waste (Parry and Brummage, 1981)—it follows that large holes, such as Queslett Quarry in Birmingham, lying in or adjacent to urban areas, will be sought for landfill.

Another threat, although perhaps not so obvious, is the so-called environmental or landscape improvement scheme. Examples include the partial drainage of many of our urban heaths and commons, such as Sutton Park in Birmingham, and Wimbledon and Wandsworth Commons. Today improvement schemes are more usually associated with areas of downgraded or derelict land. Landscape improvement schemes which invest in the existing natural value of a site are of obvious and lasting benefit to the community. However, those that impose a design insensitive to the site can destroy the very amenity they seek to provide. Finally, as in rural areas, inappropriate management can greatly reduce the natural value of a site. Tidy horticultural maintenance is the hallmark of the urban park and may, on occasions, be applied inappropriately to seminatural open spaces brought within the open space network.

Turning to the mechanisms of control, few urban sites receive statutory protection under the National Parks and Access to Countryside Act (1949). There is only one truly urban national nature reserve, the Wren's Nest in Dudley, and three local nature reserves, of which two lie within the Greater London Council area. This latter statistic is appalling given that one of the major aims of the local nature reserve designation, as embodied in the 1949

Act, was to provide areas for education. However, other forms of protection exist. Certain sites of natural interest are protected from development, although not necessarily from transport proposals or inappropriate management, through designation as areas of public open space; by physical inaccessibility; or by function—the canals of Birmingham now form an integral part of the city's drainage system and cannot be filled.

But the main potential for protection lies with development control procedures. A Department of Environment circular 108/77 (Department of the Environment, 1977, paras. 3, 5 and 24) draws specific attention to the nature conservation interest of urban areas, and requests local authorities to take full account of this when considering individual planning applications. All the major forms of threat noted above require planning permission with the single exception of open space management. This contrasts with the rural scene where the major threats to wildlife—namely agricultural and forestry practices—lie outside planning controls.

In assessing planning applications, the weight given to the natural value of a site as compared with other competing land uses, will be influenced by the information made available to the planning authority concerned. Under the 1949 Act (section 23) (superseded by section 28 of the Wildlife and Countryside Act, 1981), the Nature Conservancy Council are required to notify local planning authorities of sites which they regard as being of special scientific interest (SSSI). There are a number of urban sites which qualify as SSSIs. Within London alone there are 26, covering 1779 ha or 1 per cent of the Greater London Council area (Nature Conservancy Council, 1981). However, there is a plethora of other urban sites of natural history value which do not measure up to the general criteria by which SSSIs are judged. Not surprisingly, the Nature Conservancy Council do not wish to devalue the national currency of SSSI notification by making particular exceptions for urban areas (see Goldsmith, Chapter 14). Yet the true value of 'natural' urban sites, which relates primarily to their rarity within the urban context and to the number of people which they serve either directly or indirectly, cannot be judged against the regional or national context where similar habitats may be relatively common. Indeed, in certain instances the yardstick of SSSI notification has been a positive disadvantage in fighting the case for a non-notified site. There is a temptation on the part of planning authorities to dismiss the wildlife importance of these apparent 'poor relations' in assessing planning applications. And, of course, there is no statutory requirement for planning authorities to prohibit development, even on SSSIs.

To try to provide a wider perspective on the value of urban sites, a number of initiatives have been set in train. Following the comprehensive survey of the west Midlands conurbation undertaken for the Nature Conservancy Council during 1975 (Teagle, 1978), the Council notified relevant planning authorities of sites, which although not designated as SSSIs, it felt were of significant local value. In South Yorkshire the County Council are publishing *A Review of*

Nature Conservation in South Yorkshire. This lists a large number of non-statutory sites, termed for convenience as sites of scientific interest (SSIs) which the County would wish to protect through planning and development control. The information for this publication has come from a wide range of sources including the Nature Conservancy Council, the Yorkshire Naturalists Trust, local natural history societies and the three museum databanks operating in the county.

The very fact that these particular initiatives are worthy of mention, highlights another major problem—few comprehensive surveys have been undertaken of the wildlife habitats of individual conurbations. In their absence planning control procedures do, however, alert local residents to potential threats. Unlike certain isolated rural habitats, nearly every urban site has been mentally adopted by at least one section of the local community. And it is frequently through campaigns mounted by voluntary and locally formed pressure groups, often aided by the Nature Conservancy Council, that the conservation and amenity interest of urban sites are made known—often prior to or during public enquiries. Such campaigns include the successful attempt by the Welsh Harp Conservation Group to prevent development on the fringe of the Brent Reservoir SSSI (Cole, 1980); the Moseley Bog 'Save the Bog' campaign aimed at preventing development on this valuable Birmingham bog—the inspiration for Tolkien's 'Mirkwood' and the home of the rare wood horsetail (*Equisetum sylvestris*); and the 'Save the Marshes' campaign aimed at preventing gravel extraction on the relict and biologically rich Walthamstow Marshes, London (see Goldsmith, Chapter 14).

THE CREATION OF NEW URBAN SITES OF NATURAL HISTORY VALUE

The other half of the urban nature conservation equation is the creation of new habitats of natural interest. With nature's apparent ability to colonize urban sites, albeit over some time, the purist might argue that nature should be left to heal the scars of past land misuse. Bradshaw (1979), pointing to the enormous value of derelict areas which have been left to natural colonization, has asked 'is the tidying up going too far?' But landscape schemes are a requirement of twentieth-century life where instant solutions are sought to long-term problems. Large areas of raw uncolonized derelict land degrade and dispirit, and cause much needed industry to leave urban areas for more prestigious greenfield locations. So appalling was the dereliction of Stoke-on-Trent during the early 1960s, that Crossman, in his diaries, questioned whether the city should be razed to the ground and the inhabitants rehoused. And the displays of urban ruderal weeds—the delight of both botanists and children—are to many urban residents the very emblems of past blight which they wish to dispel from their memories.

Britain has a long history of landscape design. The influence of early nineteenth-century designers such as Loudon, who chose as one of his criteria 'the display of trees and other plants individually' (quoted in Ruff, 1979, page 6) has held sway over much of twentieth-century landscape design. In the early 1970s, a few members of the landscape profession began to rebel against such formalism. In 1970, Nan Fairbrother published her much acclaimed book *New Lives, New Landscapes* which exhorted designers and planners to work with, rather than against, nature, to use ecology as the cornerstone to design. In Sweden, Germany and particularly in Holland, exciting man-created 'natural' landscapes were emerging (Cole and Keen, 1976).

Holland, with a population density twice that of Great Britain, is a country where countryside planning has always been an urgent matter of achieving maximum agricultural productivity. As an antidote to this increasingly manmade rural scene, planners, educationalists and parks departments alike began to look to the urban open spaces as one of the few remaining areas where nature could be encouraged. Starting in the 1930s with the creation of the Amsterdam Bos (some 500 ha of planted native woodland specifically designed for the urban public), through the creation of urban wild flower gardens, to the more recent development of *haemparks* (homeparks) where a range of natural habitats are accurately reconstructed, the Dutch had certainly proved that an alternative approach was feasible.

As first debated in 1974 at the Landscape Research Group Conference 'Nature in Cities' and as described by Ruff (1979), these Dutch landscapes sought to re-establish the bond between man and nature and to put the fun back into natural history. As stated by Hanke Bos (1981) Director of Municipal Parks and Open Spaces in the Hague, 'a child who has not tasted honey from a deadnettle or enjoyed the flying seeds of maple trees will never grasp the charm of real nature, even if the biology teacher is first class. We must go through life experiencing nature all round us.'

In Britain, large-scale urban dereliction offered an opportunity to test an 'ecological approach' to design—the speeding up of the natural successional process. In the late 1960s and 1970s, under a range of Government Acts, including the Local Government Act 1966, the Local Employment Acts of 1970 and 1972 and the more recent Urban Programme, a range of central government grants have been made available for the reclamation of urban derelict land to public open space.

The Stoke-on-Trent reclamation programme, which progressively covered over 600 ha of derelict mineral workings and railway lines, was one of the first major initiatives to get off the ground in the late 1960s. Inspired on the one hand by the Dutch and constrained on the other by an average approved expenditure of £4800/ha, Land Use Consultants, the designers, sought to re-establish the natural climax vegetation patterns which existed in the area prior to large-scale disturbance. In each case detailed ecological surveys were under-taken to assess the site potential and to delineate areas of natural vegetation to be conserved. Subsequently, in areas of open ground, natural plant communities were established which matched the proposed use of the site and which were adapted to the existing edaphic conditions (Figure 16.2).

Parallel with these developments in reclamation, certain new town corporations, notably Milton Keynes and also Warrington and Runcorn, started to establish natural communities common to the surrounding rural areas (Kelcey, 1977, 1978; Tregay and Moffat, 1980).

More recently, there has been detailed, practical experimental establishment of typical rural plant communities on small urban derelict plots, often on a temporary basis—the 'ecological prefabs' of tomorrow (Rose, 1981). Examples include work undertaken by the Rural Preservation Association 'Greensite 1 Project' in Liverpool and the creation in 1977 of the William Curtis Ecological Park in London. This Park, established on a 0.8 ha (2 acre) parking lot, cost a total of £2000 with the aid of voluntary help, including the British Trust for Conservation Volunteers. Today the site boasts some 300 vascular species, spread over a range of minihabitats from pond and reed bed, containing greater spearwort (*Ranunculus lingua*), bogbean (*Menyanthes trifoliata*) and Jacob's ladder (*Polemonium caeruleum*), through open meadowland to scrub woodland and sand dune. From inauspicious beginnings, the site, run by the Ecological Parks Trust, now receives 5000 schoolchildren a year and a further 2000 individual visitors (Ecological Parks Trust, 1978, 1981).

During the 10 years of practical experimentation into urban 'creative ecology', backed by research into reclamation techniques (Bradshaw and Chadwick, 1980; see also Bradshaw, Chapter 11) and practical research into the creation of herb-rich swards (Lloyd, 1976; Wells, Bell and Frost, 1981; Wathern and Gilbert, 1978; and others), techniques continue to be refined. At the same time, despite resistance from certain quarters of the landscape and planning professions, changing public tastes on the one hand and financial constraints on the other, have lent support to this approach. Natural woodland blocks and meadowland can initially be established at 25 per cent of the cost of a formal planting scheme.

Apparently similar financial arguments exist for the long-term management of these new 'natural landscapes'. Crowe (1956) estimated that the costs of maintaining Wimbledon Common were less than 3 per cent of those of Kensington Gardens, but such comparisons can be misleading. Under current methods of urban open space management, with work studies and bonus schemes on the one hand and restrictive work rotas on the other, cheap maintenance often equates with unskilled, easily programmed repetitive mechanized tasks, such as grass cutting. And often there is insufficient flexibility to allow a change to very infrequent but labour-intensive management techniques, such as coppicing, which are more appropriate to the ecological approach. However, there are a few encouraging developments. Merton Parks Department in south London is trying to improve the natural diversity of its open spaces (Berry, 1976) and has indicated that hay-making can prove cheaper than twelve to fifteen gang mower cuts a year if costs are offset against revenue from the crop—with the burgeoning

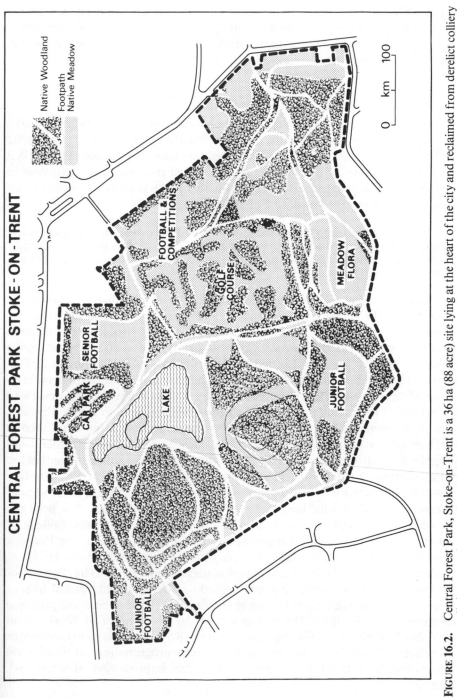

FIGURE 16.2. Central Forest Park, Stoke-on-Trent is a 36 ha (88 acre) site lying at the heart of the city and reclaimed from derelict colliery spoil tips to native meadowland and woodland between 1969 and 1974 (designed by Land Use Consultants).

city farm movement there is a ready market. But such developments remain the exception, and there is little doubt that the newly created 'natural landscapes' will have little chance of long-term survival if there is not a dramatic change in open space management policies.

INVOLVEMENT IN URBAN CONSERVATION

Urban nature conservation is not a new phenomenon. In 1771 William Curtis established one of London's first wild flower gardens on an 0.4 ha (1 acre) site in Bermondsey. Seven years later, cramped for space, Curtis moved his garden to Lambeth Marshes. But in 1789 he was forced to move again, this time to Queen's Elm, Brompton. The reason for this move is sadly recounted in Dr Thornton's memoirs (quoted in Curtis, 1941, page 84):

> I [Curtis] had an enemy to contend with in Lambeth Marshes, which neither time, nor ingenuity, nor industry, could vanquish; and that was the smoke of London; which . . . constantly enveloped my plants, and spreading its baneful influence over them destroyed many.

During this century, developments have been divided between landscape design (described previously), education, and natural history and conservation. Since the war, increasing attention has focused on environmental education and the need for practical project work as part of, and as an extension to, the school curricula (Council of Europe, 1977; Newbould, 1974; Withrington, 1977). This is an enormous subject in its own right, but a few of the outward manifestations of this movement can be noted. Glasgow Parks Department offered a lead in providing urban nature trails and nature trails appeared in London's parks during the early 1970s (Carter, 1971). At the same time a few rural studies centres were established in urban areas, such as the Botanic Gardens and Martineau Centres in Birmingham, both run by the local education authority. More recently, a number of urban schools have adopted sites as school nature reserves. But perhaps one of the most spectacular environmental improvement schemes, combined with in-school and out-of-school activities, has been the Lower Swansea Valley Project, begun in 1961 by the University College of Swansea and subsequently supported by the City Corporation, the Forestry Commission and a whole range of other funding bodies (Lavender, 1981).

In terms of natural history and nature conservation, emphasis has been, until recently, placed on monitoring and recording. Urban-based natural history societies, such as the London Natural History Society have been the principal source of recorded data. The revival in 1947 by the Minister of Works of an advisory Committee for Bird Sanctuaries in the Royal Parks is an early example of official recognition of the need for a long-term programme of ornithological observation and recording combined with modest improvement of urban park habitats.

It was not until the late 1970s that the Nature Conservancy Council began to take a positive promotional role in urban nature conservation, although as evidenced by the number of urban SSSIs designated during the 1950s and 1960s, individual Nature Conservancy Council regions had, to a greater or lesser extent, covered urban areas within their general remit. Their promotional role first came to public notice with the publication of *The Endless Village* (Teagle, 1978), a detailed record of the west Midlands conurbation. Parallel with this, between 1977 and 1979, the Nature Conservancy Council produced a range of publicity material on the subject and sponsored a detailed social survey into the community benefits of participation in urban wildlife projects (Mostyn, 1979). In 1979, urban conservation was further aided by the creation of a number of new Nature Conservancy Council assistant regional officer (ARO) posts which allowed greater coverage of urban areas, and in the case of London, provided for the first time an ARO specifically and only concerned with the GLC area.

The present impetus for inner-city nature conservation, however, rests primarily with the voluntary sector. This has involved not so much the long-established County Trusts for Nature Conservation, who have tended to concentrate their activities in rural areas, but more, the newly formed urban groups. This new movement embraces a whole range of activities.

Many groups are involved with schools in the setting up of school nature reserves and in-school and out-of-school practical project work. Examples include the work of the Environmental Resources Centre in Edinburgh and the regional offices of the British Trust for Conservation Volunteers. Many action groups have been formed to answer a threat to a local site (as described previously). Others have been more specifically concerned with adoption, either of sites of existing interest such as Highgate Cemetery (Friends of Highgate Cemetery, 1978) or of derelict sites to be planted and managed as an assemblage of seminatural habitats, to be used for children's play, outdoor studies, scientific research or a combination of all three.

More recently, a number of metropolitan groups have been established with a remit that extends from pressure group to management body. Two notable examples are the Urban Wildlife Group in Birmingham and the recently formed London Wildlife Trust. Both these groups place particular emphasis on the creation of affiliated local groups responsible for their local area.

Through the work of these groups, local authorities have become increasingly aware of the wildlife potential of urban areas. Following the Civic Trust report on vacant and dormant land (Cantell, 1977) and the grants made available under the partnership programme, a number of councils are releasing derelict plots to voluntary groups on simple, easily administered short-term licences. Other councils are seeking voluntary groups to manage ecological study areas and areas of existing ecological interest, such as small woodlands, that are in the Council's ownership. They appreciate that these areas do not fall easily within standard parks department management rotas.

Despite these encouraging developments, many aspects of urban nature conservation require review and development. Few approved comprehensive urban industrial and other developments take note of conservation objectives either in the conservation of existing resources or the creation of new habitats. Examples of good practice are required. To this end the Nature Conservancy Council, the Landscape Institute and the Council for Environmental Conservation, as a contribution to Urban Renaissance Year, are undertaking a comprehensive planning appraisal of the Blackbrook Valley 'Enterprise Zone' in Birmingham. This aims to demonstrate how comprehensive development can be integrated within a framework of nature conservation and community objectives (Baines and Barker, 1981).

In terms of development control, there is a need to move further away from 'scientific' criteria in assessing the natural value of sites and to accept public accessibility and enjoyment as a major reason for restricting development on sites of natural interest. In the field of site management, the potential for parks departments to employ unemployed labour through manpower services schemes, to undertake specific labour-intensive management operations, including haymaking and woodland work, should be investigated further. To date, union rulings have frequently prevented such outside involvement. Turning to government grant aid, there is an urgent need for long-term revenue funding for the voluntary sector. Present government grants only cover initial capital expenditure with no long-term maintenance allowance. Yet many urban groups, in taking on responsibility for a site, require a long-term management commitment, if ecological diversity of the sites is to be retained and enhanced.

These and many other problems may seem insurmountable, but given the enormous strides which have been made in urban nature conservation over the last 10 years, the omens are good.

CONCLUSION

Urban nature conservation may be despised by some conservationists as merely a 'popular' movement. This is precisely what it is, and rightly so—it seeks to bring the concept of conservation to the general populus. While some may wish to label it as a passing fad, the support which it now receives, and the continuing rise of this support in an exponential curve, suggest that this section of nature conservation is here to stay. Urban nature conservation forms an integral part of the total conservation scene. Our fast dwindling natural heritage will only be retained with popular support.

ACKNOWLEDGEMENT

I would like to thank Max Nicholson, Richard Findon and all those at the Ecological Parks Trust for all their helpful comments.

REFERENCES

Alderton, R.E. (1977). Birds at Surrey Commercial Docks, January 1973 to December 1975. *London Bird Report*, **40**, 85–90.

Baines, C. and Barker, G. (1981). The Blackbrook Valley Project. *Landscape Design*, **134**, 15–17.

Batten, L.A. (1972). Breeding bird species diversity in relation to increasing urbanisation. *Bird Study*, **3**, 157–166.

Batten, L.A. (1973). Population dynamics of suburban blackbirds. *Bird Study*, **20**, 251–258.

Berry, J.G. (1976). Countryside inside London's fringe: manager's view. *Countryside Recreation Review*, **1**, 21–25.

Bos, H. (1981). The ecological design of urban green spaces in Holland. *Landscape Research*, **3**, 19–21.

Bradshaw, A.D. (1979). Derelict land—is the tidying up going too far? *The Planner*, **3**, 85–88.

Bradshaw, A.D. and Chadwick, M.J. (1980). *The Restoration of Land: The Ecology and Reclamation of Derelict and Degraded Land*, Blackwell Scientific Publications, Oxford, 317 pp.

Burton, J.A. (1974). *The Naturalist in London*, David and Charles, Newton Abbot, 176 pp.

Burton, J.A. (1976). Fowls in foul air, *New Scientist*, **71**, 400–401.

Brain, P.J.T. (1981). *The Role of Cemeteries as Refuges for Wildlife in Metropolitan London*, City of London Polytechnic, London (unpublished), 53 pp.

Cantell, T. (1977). *Urban Wasteland*, Civic Trust, London, 56 pp.

Carter, G. (1971). Urban nature trails: pupils and parks. *Your Environment*, **2**, 66–68, 77.

Cole, L. (1980). *Wildlife in the City: A Study of Practical Conservation Projects*, Nature Conservancy Council, London, 26 pp.

Cole, L. and Keen, C. (1976). Dutch techniques for the establishment of natural plant communities in urban area. *Landscape Design*, **116**, 31–34.

Council of Europe (1977). International Study Conference on Environmental Education in a Strictly Urban Setting, Bristol 1977. *Report*, Council of Europe, Strasbourg. European Committee for the Conservation of Nature and Natural Resources, 110 pp.

Cramp, S. and Gooders, J. (1967). The return of the house martin. *London Bird Report*, **31**, 93–98.

Crowe, S. (1956). *Tomorrow's Landscape*, Architectural Press, London, 207 pp.

Curtis, W.H. (1941). *William Curtis 1746–1799*, Warren and Son, Winchester, 142 pp.

Davies, B.N.K. (1976). Wildlife, urbanisation and industry. *Biological Conservation*, **10**, 249–291.

Department of the Environment (1977). *Nature Conservation and Planning, Circular 108/77*, HMSO, London, 10 pp.

De Worms, C.J.M. (1972). A review of the macrolepidoptera of the London area for 1970 and 1971. *London Naturalist*, **51**, 28–38.

Ecological Parks Trust (EPT) (1978). *The William Curtis Ecological Park: First Annual Report 1977–78*, Ecological Parks Trust, London, 31 pp.

Ecological Parks Trust (EPT) (1981). *The William Curtis Ecological Parks: Fourth Annual Report 1980–81*, Ecological Parks Trust, London, 38 pp.

Fairbrother, N. (1972). *New Lives, New Landscapes*, Penguin, Harmondsworth, 382 pp.

Fitter, R.S.R. (1945). *London's Natural History*, Collins, London, 282 pp.

Fitter, R.S.R. (1978). Benefited from the Blitz: black redstarts in Britain. *Country Life*, **163**, 833.

Fordham, R.C. (1975). Urban land use change in the United Kingdom during the second half of the twentieth century. *Urban Studies*, **12**, 71–84.

Friends of Highgate Cemetery (1978). *Highgate Cemetery*, Friends of Highgate Cemetery, London, 48 pp.

Gilbert, O.L. (1965). Lichens as indicators of air pollution in the Tyne Valley. In G.T. Goodman, R.W. Edwards, and J.M. Lambert (Eds). *Ecology and the Industrial Society*, Blackwell, Oxford, pp. 35–47.

Gooders, J. (1968). The swift in Central London. *London Bird Report*, **32**, 93–98.

Greenwood, E.F. and Gemmell, R.P. (1978). Derelict land as a habitat for rare plants: S. Lancs (v.c.59) and W. Lancs (v.c.60) *Watsonia*, **12**, 33–40.

Harrison, J. and Grant, P. (1976). *The Thames Transformed*, André Deutsch, London, 227 pp.

Jones, A.W. (1958). The flora of the City of London's bombed sites. *London Naturalist*, **37**, 189–210.

Kelcey, J.G. (1977). Creative ecology. Part 1: selected terrestrial habitats. *Landscape Design*, **120**, 34–37.

Kelcey, J.G. (1978). Creative ecology. Part 2: selected aquatic habitats. *Landscape Design*, **121**, 36–38.

Laundon, J.R. (1970). London's lichens. *London Naturalist*, **49**, 20–69.

Lavender, S.J. (1981). *New Land for Old*, Adam Hilger, Bristol, 137 pp.

Lloyd, C. (1976). Meadow gardening, part 1. *Garden Journal of the Royal Horticultural Society*, **6**, 322–329.

Mabey, R. (1973). *The Unofficial Countryside*, Collins, London, 157 pp.

Mostyn, B.J. (1979). *Personal Benefits and Satisfactions Derived from Participation in Urban Wildlife Projects: A Qualitative Evaluation*, Nature Conservancy Council, London, 66 pp.

Nature Conservancy Council (1981). *Statutory Sites Under the National Parks and Access to the Countryside Act 1949, as Amended by the Nature Conservancy Council Act 1973. Greater London Supplement*, Nature Conservancy Council, London, 3 pp.

Newbould, P.J. (1974). Conservation in education. In A. Warren and F.B. Goldsmith (Eds), *Conservation in Practice*, Wiley, London, pp. 437–452.

Parry, G.D.R. and Brummage, M.K. (1981). Solid wastes: reclamation and management. *Landscape Research*, **3**, 15–18.

Readett, R.C. (1971). Flora of Sutton Park. *Proceedings of Birmingham Natural History Philosophical Society*, **22**, 1–88.

Rose, C.I. (1981). *Wildlife in Southwark: A Survey of the Ecological Potential of Vacant Land*, Southwark Wildlife Group, London, 53 pp.

Rose, C.I. and Hawksworth, D.L. (1981). Lichen recolonization in London's cleaner air. *Nature (London)*, **289**, 289–292.

Ruff, A.R. (1979). *Holland and the Ecological Landscapes*, Deanwater Press, Stockport, 150 pp.

Simms, E. (1975). *Birds of Town and Suburb*, Collins, London, 256 pp.

Society of County Treasurers and County Surveyors' Society (1978). *Waste Disposal Statistics*, 1977–78.

Stamp, L.D. and Hoskins, W.G. (1963). *The Common Lands of England and Wales*, Collins, London, 349 pp.

Teagle, W.G. (1978). *The Endless Village*, Nature Conservancy Council, West Midlands Region, 58 pp.

Tregay, R. and Moffat, D. (1980). An ecological approach to landscape design and management in Oakwood, Warrington. *Landscape Design*, **132**, 33–36.

Wathern, P. and Gilbert, O.L. (1978). Artificial diversification of grassland with native herbs. *Journal of Environmental Management*, 7, 29–42.

Wells, T., Bell, S., and Frost, A. (1981). *Creating Attractive Grasslands using Native Plant Species*, Nature Conservancy Council, Shrewsbury, 35 pp.

Withrington, D. (1977). *Out-of-school Youth Activities*, Department of Education and Science, London.

CHAPTER 17

Ecological Effects of Upland Afforestation

F.B. Goldsmith and J.B. Wood

There can be no doubt that professional foresters in the United Kingdom have a collective love for the countryside and a strong sense of responsibility about their stewardship of it. Nevertheless, within the forest service, employees often jokingly confess that they have no need to worry about the mistakes they make, since the consequences will not become evident until after they are dead. Although this cliché may only hold true for the silvicultural section of the profession, which nowadays is largely outnumbered by other specialist groups, it is symptomatic of the forester's universal recognition of the very long time-scale that is involved in his vocation, compared to that of most other businesses. But even foresters may be insufficiently aware of the crucial role that time plays in their management of the countryside, and few of their strongest critics or, more importantly, the politicians within whose schemes the foresters must operate, possess even the forester's sense of time-scale. In this chapter we will examine some of the consequences that this may have for our understanding and appreciation of the effects of commercial afforestation, and in particular of the ecological consequences that are the special concern of conservationists.

First, we should point out to foresters that the trees they may plant today, even the fastest-growing species on sites where rotations may be unduly short (perhaps as a consequence of the threatening effects of windblow) are unlikely to be harvested until at least the year 2020. Few of us would be willing to even guess what the political, economic and social climate of this country might be at that time. If these trees were to be left to reach their natural maturity, most would persist well into the twenty-second century. In return, we must recognize that today's plantations, whose location, structure and species-composition we may wish to question, are the products of the efforts of generations of foresters who were often working under very different guidelines and priorities to those that now prevail. Moreover, we should also understand that the policy-makers, and most of the practitioners within our forest service, received their professional training at a time when little thought was being given to the ecological consequences of afforestation, save those directly pertinent to the production of the crop itself, and were certainly trained under the guidance of people who, despite the title of the senior professional grade within the Forestry Commission (Conservator), had little or no appreciation of the term conservation, in the sense that most of us now attach to this word. Except through their amateur interests in natural hisory it is hardly surprising that few of these people have any appreciation of the peculiar values attached by conservationists to some of the elements of our native flora and fauna and particularly to certain assemblages that may occur in regions they consider suitable for afforestation, nor that, despite the pressing urgency of current economic constraints, there may be interest-groups who think in terms of time-scales far beyond even the forester's crop rotations. Equally, ecologists and conservationists often fail to appreciate the severe constraints that the economics of discounted costs and revenues, the scale of processing operations, the need to operate machinery over difficult terrain and the need to stabilize labour requirement may impose on the silviculture that the forester would otherwise wish to promulgate.

If our forest estate were ever to achieve the truly multipurpose role that most of us desire, it would first be necessary to reduce the polarization of interest-groups which now exists, and to promote the understanding by each of the other's aspirations and limitations.

To appreciate fully the circumstances that have produced the present forest estate in Britain, it would be necessary to examine in some detail our forest history over the past 200 years or more. However, there is insufficient space to do that here and, besides, this has been adequately documented elsewhere (Edlin, 1970; James, 1981; Johnston, Grayson & Bradley, 1967; Ryle, 1969). Nevertheless, we should remember that the latter half of the nineteenth century saw the nadir of forestry in the United Kingdom and only about 5 per cent of our land remained under woodland at the start of this century. The blockades of the First World War, exacerbated by those of the Second World War, provided the impetus for 35 years of massive new planting by the Forestry Commission (Figure 17.1) and not until the mid-1950s was the concept of a strategic reserve subsumed by other motives. Thus, almost all of today's productive plantations were established with the aim of growing a large volume of timber as quickly as possible. It is therefore hardly surprising that the foresters of the interwar years paid little attention to the visual effects of the alignment of forest rides and boundaries, of monocultural blankets of conifers or expressed any concern for the wildlife which the forests may have sheltered. Indeed, even now the principal constraint on forestry operations in this country is determined by the requirement of economic viability (Countryside Commission, 1972). Only since the late 1960s has the Forestry Commission officially concerned itself with the amenity or recreational impact that its forests may have (it was charged with a duty to recognize these under the 1968 Countryside Act), although it has retained the services of a qualified landscape architect since 1964. So far it has employed remarkably few ecologists and does not extensively consult that profession, although it has a good record for safeguarding statutory sites notified

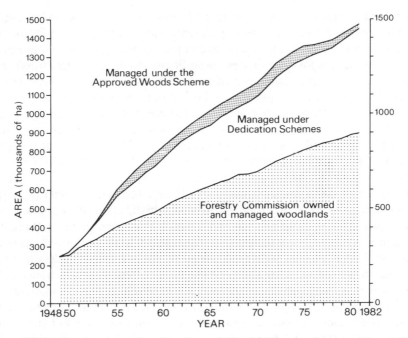

FIGURE 17.1. The increase since the Second World War in the area of woodland managed for commercial purposes with government financial support.

by the Nature Conservancy Council. The recently proposed (Centre for Agricultural Strategy, 1980) massive extensions of plantation forestry in the Scottish uplands are based entirely on economic arguments and, if they are to go ahead, will undoubtedly be designed with commercialism in mind, and will be established with the most economically efficient techniques and considerations of scale. Statutory concerns for other uses of the countryside form largely inadequate constraints upon this type of development.

It would be possible to dispute the economic arguments underlying an expansionist forestry programme, although the debate is fraught with pitfalls and complications. Essentially, the commercial exploitation of the uplands revolves around two possible uses: plantation forestry or hill sheep farming. Both are in receipt of government subsidies, show varying rates of return according to individual circumstances, have social costs and benefits, affect our national balance-of-payments and have emotive supporters and critics with differing shades of political hue. Too many issues that cannot be adequately reduced to a financial statement solely are involved in any balance sheet produced for either of these two landuses, for it ever to be possible to clearly determine which should be favoured as a commercial proposition alone. During postwar years the balance in favour of one or the other has largely been tipped by political considerations, though farming, as the existing landuse, has often had the upper hand. Rather than allowing ourselves to be drawn into this particular conflict, we ought perhaps to ask the more pertinent question from our point of view, 'would it not be better to undertake no commercial management of our uplands at all rather than support either of these financially marginal pursuits?' Even this is charged with possible misunderstanding, as there are strong emotive beliefs that all potentially productive land should be made to yield a material crop for the benefit of mankind: witness the postwar Ministry of Agriculture, Fisheries and Food (MAFF) campaign for agricultural self-sufficiency, and the strong hint of a similar line of reasoning in the foresters' proposals for expansion. At risk of also being drawn into an argument that is likely to be reduced to emotive statements of the desires of particular interest groups, and of further polarization of the issues, we would instead like to turn to a consideration of the form that may be taken by a forestry expansion in the uplands, the management practices that are likely to prevail, and the ecological consequences of these. By doing this, we do not want merely to dismiss the possibility of zero management as a non-starter. Instead we would like to promote a better understanding of why foresters produce plantations of the type currently seen in many upland areas of Britain, why substantial modifications of the current commercial practices to the possible benefit of amenity or conservation are unlikely to occur so long as the present economic constraints exist, and to point out the possible ecological disasters (or delights) that may result. Central to this theme will be the need to establish those issues for which our present knowledge is inadequate, and to indicate the investigations we would like to see undertaken so as to help resolve dilemmas.

The present-day distribution of the Forestry Commission's land holding in Great Britain is largely a consequence of the former system of land tenure in the uplands. It has been modified in postwar years as a consequence of the often vociferous opposition of farmers, most effectively by way of parliamentary protest and as a result of parliamentary support for agriculture. Local opposition from groups of people with a recreational and amenity interest in the countryside has also often been considerable. The type of plantation that has been created, its species composition and the management regime imposed on it, in turn derive largely from the type of site that has been available for afforestation, the original demand for a strategic reserve, and more recently from the consumers' demand for a specific product. These statements apply equally well to private sector afforestation, except that the often massive tax incentives and the apparently lesser concern for public opinion, may have constrained the distribution of new private plantations to a lesser extent. In contrast, much reafforestation of derelict woodland by private individuals has been of a more traditional nature and has used a wider variety of species and more diverse distribution of plantations, often with a view to the non-commercial benefits of plantations. Nevertheless, degraded woodland on the poorer sites in private hands has also undergone critical economic scrutiny, and this has often brought about change in species and management regimes at the time of restocking, to produce plantations of nearly identical form to many strictly commercial Forestry Commission woodlands. Thus, the forest estate in Britain today may be seen to derive not purely from the forester's desire to fulfil the task he has been charged with, but also to a critical extent upon the forces of competing landuse, market demand and interest rate (Price, 1976).

Right from the moment it was created, there was no question that the Forestry Commission should acquire all the land it would need of a quality suitable to enable it easily to create a massive strategic reserve of timber, no matter how urgent the initial task seemed to be. Though the transfer of some Crown woodlands and the agricultural depression of the 1920s and 1930s enabled some of the better types of site to be acquired, it was soon realised that the greater part of the Commission's holding would always consist of the poorer upland sites, and that new skills and techniques would need to be developed, and non-native species planted, if it was to stand a chance of achieving its goal. Today, upland sites with often very degraded soils are almost the only ones that are being acquired for afforestation. Most of these have not grown trees for at least several hundreds of years, and some probably grew no more than scrub, even during the equable Atlantic period. Except in Scotland the Forestry Commission has more or less abandoned any hopes of further acquisitions and the high proportion of unplantable land in their Scottish holding (Table 17.1), testifies to the problems they face in establishing trees even with their now considerable skill in preparing sites and the refinement in their choice of species and

TABLE 17.1. The state-supported Forest Estate in Scotland (at 31 March 1980) (areas in square kilometres).

	Under plantations	Retained scrub	To be planted	Other land	Total area of forest (km²)
Forestry Commission					
North Scotland	1370	6	270	1224	2870
East Scotland	947	4	63	163	1177
South Scotland	1380	2	84	334	1800
West Scotland	1280	23	195	651	2149
Private					
Basis I and II dedications	2417				
Basis III dedications	339				
Dedications in preparation	493				
Approved woodlands	74				
		Scrub			
Total (all woods)	8300 km²*	35 km²			

Source: Forestry Commission Annual Report and Accounts 1979–80.
* Excludes woodlands receiving no state aid via Forestry Commission.

provenance. Moreover, the continuing level of subsidies for hill-farming makes it difficult for them to buy land on the open market, and the majority of their new acquisitions probably come by way of estates that are offered in lieu of death duties. Even so, the Department of Agriculture and Fisheries for Scotland usually has the first choice of such land, and only releases for forestry the parts that are agriculturally unsuitable, even as hill-pasture. Faced with this problem, the Forestry Commission initially devoted extensive resources to evaluating the growth rates of a wide variety of species and developing site treatments that would promote the fastest early growth and enable successful establishment of the new plantings. Although they were included in many early trials, broad-leaved species were never really in serious contention as crop species; not only did they require better soils and grow more slowly than coniferous species, but for a long time the often mismanaged woodlands of lowland Britain produced more low-grade hardwoods than the timber trade could absorb. Largely through trial and error, the initial wide choice of coniferous species available from private collections, mostly established on country estates in the nineteenth century, was narrowed down to those that would tolerate the poorer sites and yet yield an economic harvest (Osmaston, 1968). Now, the choice lies almost exclusively between only two species, sitka spruce (*Picea sitchensis*) whenever conditions will allow and, on the highest and poorest sites afforestable, lodgepole pine (*Pinus contorta*) (Table 17.2). To a considerable extent this restricted choice of species has been reinforced by market demand, for, with its rather soft, white timber with long tracheids and relatively low resin content, sitka spruce is almost ideal for conversion to either chemical or groundwood pulp. Pulp and paper has been the biggest growth area in woodland produce, but the future is less easy to foresee.

Whilst many people regard monocultures of coniferous species as the most undesirable aspect of forestry in the uplands, probably even more ecologically undesirable is the even-aged uniformity of the plantations and the shortness of the rotation. As we will discuss below, the meagre light penetration, and the sudden structural changes imposed by clear felling practices are the two features that the woodland flora and fauna find it most difficult to survive, whilst the actual species comprising the plantation is often of lesser importance. Indeed, readers familiar with the commercial beech plantations on the South Downs will have noticed how devoid of wildlife species they too are, and how similar is the composition of the associated flora and fauna to that of conifer plantations in the same vicinity. Yet, conversely, the monocultures of European silver fir (*Abies alba*) in central Europe, are often a haven for wildlife. There the structural diversity produced by a long tradition of selection management and the relatively long life of individual trees allows a stability akin to that of a natural woodland and promotes diversity of wildlife species.

What then are the prospects for the introduction of changes in management practice in our upland plantations that might ameliorate the disruptive ecological conditions they now display? Sadly, there are very few.

TABLE 17.2. Representative yield class and rotation of trees on sites available for afforestation in Scotland.

Conservancy	Elevation zone	Soil group	Species[1]	Yield class[2]	Economic rotation
North Scotland	Lower	Podsols	SP	8	65
			LP	10	55
		Ironpan soils	SP	9	60
			LP	11	55
		peaty gleys	SS	12	55
			LP	10	50
	Upper	Podsols	SP	7	60
			LP	9	50
		Ironpan soils	SP	7	50
			LP	9	50
		Peaty gleys	SS	11	50
			LP	8	50
East Scotland	Lower	Brown earths[3]	SS	14	55
			DF	14	50
			GF	16	50
			HL	10	45
			Broadleaved	8	150
	Upper	Podsols	SS	10	55
			LP	8	55
		Peaty gleys	SS	10	45
			LP	8	45
West Scotland	Upper	Podsols	SS	12	50
		Peaty gleys	SS	12	40
South Scotland	Upper	Podsols	SS	10	55
		Peaty gleys	SS	11	45

Source: Forestry Commission record no. 97 (1974).
[1] SP = Scot's Pine, LP = Lodgepole Pine, SS = Sitka Spruce, DF = Douglas Fir, HL = Hybrid Larch
[2] Yield class measured as M^3 increment/ha/year.
[3] Not typically available sites—included for comparative purposes.

As long as afforestation in the uplands is subject to the laws of economics, consequent on the wishes of the government, rotations will be short and large-scale management will be practised. Forestry is a poor bet as a strictly commercial proposition because its products are of low value and take many years to grow. Although the costs of establishing and maintaining plantations may be minimized by the utilization of mechanical aids, massive economies of scale and the low prevailing wage for forestry workers as compared to similar grades in industry, the returns are also small. These must be discounted over the length of the rotation at a realistic rate of interest so as to be compared with costs, and hence determine the financial viability of the project. Even at extremely low rates of interest imposed by government on commercial operations in the Forestry Commission, most plantations will only show a profit over a very narrow range of rotation lengths, averaging 50–55 years for most upland spruce plantations (Figure 17.2). At interest rates approaching the prevailing bank rate, no plantations in upland areas would ever be profitable (Johnston, Grayson, and Bradley, 1967). On many sites, even short rotations must be artificially curtailed so as to avoid the disastrous effects of windblow. Moreover, any additional costs occurring near the start of the rotation impose a crippling financial burden, so that the most time-saving practices of lowest labour intensity are always sought. Thus, site amelioration by ploughing (resulting in the planting of trees in straight lines), aerial fertilizer application to maintain growth rates, line-thinning using machinery, and clear felling to minimize operational costs, are unlikely to be abandoned in favour of more in-tensive techniques that could produce more natural-looking woodlands. Selec-tion forestry is most feasible when interest rates are low, sites available with soils capable of sustaining large trees without risk of windthrow or need for fertilizers, and a premium put on the production of massive individual stems. The silvicultural management of open-grown oaks to produce timbers for a navy and merchant fleet of wooden ships probably produced a woodland of enormous conservation value, but such products may never be required again.

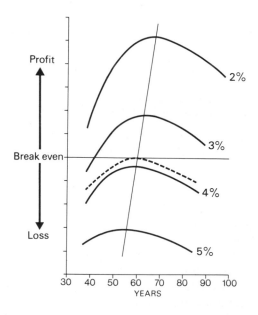

FIGURE 17.2.   The effect of interest rate on financial return. Based on data in Hart (1967).

ECOLOGICAL CONSEQUENCES

Most readers of this book will have gazed at some time into a 30-year-old stand of sitka spruce and noticed both the silence and the absence of vegetation on the floor. Except for a few strands of heather (*Calluna vulgaris*) or wavy-hair grass (*Deschampsia flexuosa*) and mosses there is no other plant growth and there is usually a thick layer of needles on the ground. The crop itself is invariably in rows and the light intensity often less than 10 per cent of that outside. It is hardly surprising that people get emotional about plans for further reafforestation of the uplands.

Let us however consider what types of ecosystem we are dealing with and what was present prior to afforestation. Natural woodlands are the most complex of all ecosystems and are dominated by a spatial and temporal mosaic of trees. Usually there are several species of various age-classes and at any one time at any spot there are interactions between canopy, soils, biotic factors and myriads of smaller plants and animals. We still know rather little about these interactions and the mechanisms involved (Bormann and Likens, 1979). On the other hand, a plantation is a simplified type of woodland which usually consists of a single species of tree with all the individuals of the same age and from the same batch of seed. In the future they might also be genetically identical. Consequently these woodlands are structurally simple and tend to have a low diversity of associated species. However they are being established in the uplands on areas that once supported other kinds of woodland but are now clothed in rough hill grassland, heather moorland or blanket bog. The degraded vegetation is the result of forest clearance, burning and repeated grazing (see Miller and Watson, Chapter 7). It is therefore a biotic or anthropomorphic subclimax in areas with a wet and cold climate, nutrient-poor soils and a fairly low potential for plant growth. For these reasons such areas have remained free from arable farming and the application of fertilizers and pesticides. Therefore, though unproductive, they have retained a degree of naturalness with long food-chains supporting large mammals (for example, red deer) and distinctive raptorial birds (for example, golden eagle, buzzard, hen harrier, merlin). People become emotional about the uplands for a variety of reasons but these inlcude the wildness of the landscape (the similarity to tundra did not escape the notice of the International Biological Programme (IBP) who included Moor House at 55°N in their tundra biome programme), the naturalness of the vegetation and the absence or low level of agricultural intensification, as well as the distinctive bird, mammal and plant species. However, plantation forests and associated habitats will support greater biomass, higher rates of primary production and possibly a higher biomass of birds and mammals than blanket bog.

The critical issues facing ecologists and conservationists are how the plantations will be managed and what proportion of the uplands will be planted.

Management involves considerations of the density of planting, the frequency and intensity of thinning, the width of the rides, the use of fertilizer and pesticides as well as the type of edge the plantation is given. The proportion of an area that is planted will vary with the terrain, the ownership of the land and the attitudes of the foresters involved. The proportion of an area afforested is sometimes obscure as official statistics are often produced on a county or regional basis but these conceal the fact that the proportion of the uplands in the county planted could be much greater. For example it may be that only 30 per cent of a region is planted whereas 70 per cent of the uplands in that area are afforested.

The ecological consequences of afforestation are two-fold. First, the *direct* effects consist of shading, litter and canopy drip (throughfall) whereas, secondly, the *indirect* effects consist of the effects of optional management activities such as the effects of fertilizer and pesticides and the acidification of streams. Fertilizers are now routinely applied from the air at frequent (often 6-year) intervals in many parts of Scotland and they affect streams, lakes, bogs and rocky outcrops as well as the plantations of spruce. Fertilization can be seen as beneficial if you perceive the uplands to be wet, nutrient-deficient, barren or as a total disaster if you value uncommon plants or animals which are restricted to these upland habitats, for example, plants of acid bogs such as the sundew (*Drosera rotundifolia*). Pesticides were hardly used up to the time of the outbreak of the pine beauty moth (*Panolis flammea*) which has always occurred at low densities in scots pine stands. However in some lodgepole pine plantations it has increased to such an extent that the Forestry Commission has resorted to an aerial spray programme using the insecticide fenitrothion. Experience in north America with the spruce budworm (*Choristoneura fumiferana*) which affects balsam fir as well as spruces suggests that the 'once-and-for-all' application of DDT in 1953 and 1954 failed to contain the outbreak and currently a vast area is still being sprayed with a variety of pesticides including fenitrothion without any hope of doing any more than marginally reducing the severity of the outbreak (Baskerville *et al.*, 1975). This has brought about serious impacts on populations of birds, bees, aquatic organisms and even the possibility of effects of human health such as the occurrence of Reyes syndrome in children (it affects the brain and is often fatal; see May, 1977). Many ecologists view the aerial application of pesticides as excessive as well as dangerous and the analogy of 'using a sledgehammer to crack a nut' is not inappropriate. Equally worrying is the fact that this kind of treatment rarely if ever stops the outbreak, it simply reduces the level of infestation for a short time.

ECOLOGY OF THE SPECIES

All the coniferous species grown in the British Isles are introductions with the exception of scots pine and even this is not native everywhere. Of our other native gymnosperms, yew is only needed for specialist purposes and juniper has little commercial value except possibly for flavouring gin or some culinary uses. Thus our plantations consist predominantly of introductions whose role in the British flora and fauna is a controversial one which has been discussed at length recently (Green, 1979; Jarvis, 1980; and Pinder, 1981). It had been thought for many years (Southwood, 1961) that introduced tree species support fewer invertebrates than our native British species (Table 17.3) although Welch (personal communication) has now shown that *Nothofagus* is very rich in invertebrates. Some foresters would argue that fewer insects suggests a reduced number of potential pests and is therefore a desirable characteristic.

TABLE 17.3. Number of insect species associated with various tree species (after Southwood, 1961; reproduced by permission of the British Ecological Society).

Native tree species	Insect species	Introduced tree species	Insect species
Oak	284	Norway spruce	37
Birch	229	Douglas fir	16
Willow	266	Larch	17
Scot's pine	91	Sycamore	15

Having introduced the ecological consequences of afforestation in general terms let us now focus on particular groups of organisms and explore the complex interactions that occur.

Plants

Probably less attention has been paid to the effects of afforestation on plants than on other groups of organisms (see, however, Hill and Jones, 1978; Hill, 1979). However plants are the basis of all foodchains, and are of interest to many specialists. The range of vegetation types in the uplands can be considerable as is shown by Figure 17.3. The bryophytes of mire communities and the insectivorous plants that grow there are especially interesting to botanists.

Each species of plantation tree produces litter of a slightly different character, the needles are shed after different intervals and at different times of the year, and they decompose at different rates (Sydes and Grime, 1981). These variables have hardly been investigated, but we know more about the shade cast by each species (Hill, 1979). They can be ranked in order of shade cast as follows:

Greatest shade   *Least shade*

Western hemlock ⟩ Sitka spruce ⟩ Norway spruce / Douglas fir ⟩ Pines ⟩ Larches / Broadleaved species

However shade is also a function of the density of planting and the thinning

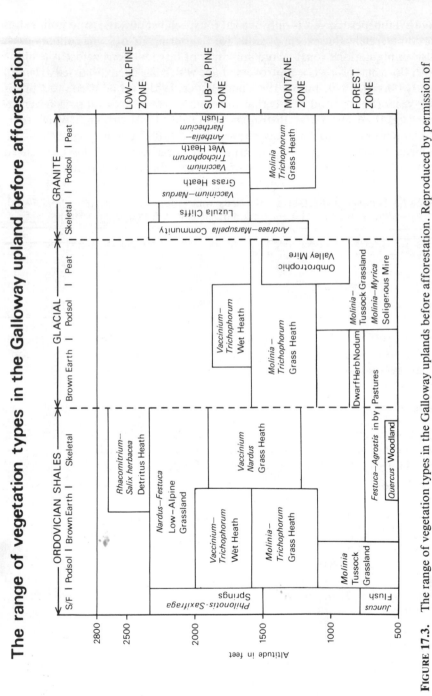

FIGURE 17.3.   The range of vegetation types in the Galloway uplands before afforestation. Reproduced by permission of Dr D.W. Shimwell from Shimwell (1971).

regime (Anderson, 1979). Indeed widely spaced conifers can have an understorey with similar species to broadleaved woodland. Larches are the only deciduous conifers that are planted in Britain, and whilst many old plantations of larch develop a fairly rich herb layer, younger plantations can be floristically poor. This is presumably because the tree canopy is complete through most of the growing season (May to September), and the high light intensities only occur when other climatic constraints prevent plant growth.

Trees can be grown on rotations of different length. For example, sitka spruce, which is planted most frequently, is often grown on quite short rotations of 45 years, and with the frequent occurrence of windblow in many areas these times may be further reduced in the future. Douglas fir, on the other hand, favours better sites with brown-earth soils and is usually managed on considerably longer rotations, often over 80 years. Environmentalists argue for longer rotations because:

(1) There is less disturbance to the site and less frequent nutrient losses after harvesting.
(2) The trees are more likely to become nesting sites for birds, including birds of prey.
(3) Thinning makes forests suitable for larger animals, such as deer.
(4) Shade-tolerant herbs are favoured.
(5) Shade-tolerant trees may regenerate naturally.
(6) The value of the product tends to be greater (larger-diameter trees) although this may not apply in terms of discounted revenues.

On the other hand, shorter rotations could resemble a sort of coppice management with frequent periods of high light intensity after harvesting and a better developed understorey. Short rotations may be more frequent in the future for there are likely to be attempts to produce 'biomass' or wood chips for energy production but we should be cautious as the supposedly high ecological interest of short rotation hardwood systems such as coppice compared with longer rotation systems such as high forest has *not* been proven. The frequent disturbance of the site at harvesting could be very damaging, especially because nutrients are removed in solution and particulates are washed into streams.

The question of the optimum rotation length can be approached from several different points of view. The forester, even if only for aesthetic reasons, favours a long rotation so long as the crop remains sound and disease-free, whilst the economist wishes to achieve a high price for the fibre as quickly as possible in order to reduce the interest notionally payable on the costs of planting and thinning. The ecologist and conservationist has only just begun to think about this question. We do not yet have the answers for deciduous woodlands, although Peterken (1977), Rackham (1980) and Rose (1976) all favour long-established woodland on sites with a continuous cover of trees as against frequent harvesting. Smart's (personal communication) studies on broad-leaved

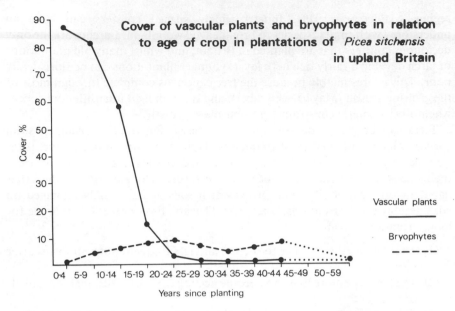

FIGURE 17.4. From Hill (1979); reproduced by permission of the Institute of Terrestrial Ecology.

plantations, for example, in the New Forest, suggest that soil type, canopy species, basal area and grazing pressure may all be more important variables than plantation age.

Hill, Evans, and Stevens (unpublished) of the Institute of Terrestrial Ecology (ITE) have, under contract to the Nature Conservancy Council, studied the ecological effects of plantations on the flora, and Figure 17.4 is taken from Hill (1979). It shows that the cover of vascular plants decreases rapidly after planting and reaches about 50 per cent by 15 years and almost nothing by 25 years. Bryophyte cover on the other hand often increases in the first 15 years, due to the removal of competition from the vascular plants of the ground layer, and from then on a cover of nearly 10 per cent is maintained. After the cover of vascular plants has decreased, some will die out and will not recover. After harvesting, some such as fireweed (*Epilobium angustifolium*) will come in as seed, whilst others will develop from the seed bank in the soil, for example, fox-glove (*Digitalis purpurea*) (Brown, 1981; Kellman, 1970). The relative proportions of these categories will vary from place to place depending upon:

(1) the original vegetation;
(2) the number of rotations that have occurred;
(3) the latitude, altitude, climate, soils, biotic influences at the site;
(4) the light intensity during the growth of the stand;

(5) the distance from sources of suitable seed;
(6) the species of tree crop grown and the chemistry of its needle-fall and canopy-drip;
(7) the type of harvesting and the degree of disturbance.

Regrettably we know rather little about many of these variables, but let us consider a few of them in greater detail. If the site were an acid brown-earth at relatively low altitude, say less than 300 m, then brambles (*Rubus fruticosus*) and bracken (*Pteridium aquilinum*) could survive a short rotation as live plants. It is even possible that in a heavily thinned fertile stand bluebells (*Hyacinthoides non-scriptus*) and anemone (*Anemone nemorosa*) will continue and recover after harvesting. On a peaty podsol, on the other hand, at an altitude near to the maximum for planting (around 650 m) the original species are more likely to be heather (*Calluna vulgaris*), purple moor-grass (*Molinia caerulea*), or deer grass (*Trichophorum caespitosum*) and cotton grass (*Eriophorum* spp.) if the drainage is impeded. These upland plantations (which are actually more common) are usually grown at high density, the site is prepared for planting by deep-ploughing and the trees eventually help to dry the site out. Bog species then disappear from the site but heather (*Calluna vulgaris*), bilberry (*Vaccinium myrtillus*) and bedstraw (*Galium saxatile*) may survive but with a substantially reduced cover. The ranking of the ground layer in order of shade tolerance enables us to predict the species which are likely to occur in plantations. It has been studied by Anderson (1979) for plantations in the lowlands and Hill (1979) in the uplands. In the uplands the tolerance is approximately as follows:

Shade tolerant				*Shade intolerant*
Calluna	*Vaccinium*	*Deschampsia*	*Rubus*	*Epilobium*
vulgaris	*myrtillus*	*flexuosa*	*fruticosus*	*angustifolium*
	Galium saxatile	*Agrostis canina*	*Pteridium*	*Betula* spp.
		Dryopteris dilatata	*aquilinum*	

It is not possible to give figures for the minimal light intensities that each species can tolerate, as they are a function of the duration as well as the intensity of the light. However, most vascular plants are shaded out in plantations where the light intensity is less than 12 per cent and most bryophytes at below 8 per cent. It is also important to note that some species such as the berry-bearing shrubs and cloudberry (*Rubus chamaemorus*) respond positively to low degrees of shading. It appears that it is only the *dense* shade of a plantation which is damaging. Shading is only one of a number of variables, all related to the character of the tree crop, that are highly correlated with each other (Grime, 1981). The others are root competition, soil moisture, the effects of coniferous leaf litter and drier soils. Nevertheless, shading is usually believed to be the most important factor.

Another question that is difficult to answer is how survival of different species in the ground layer relates to their growth rates or other aspects of their physiology. It is known for example that heather (*Calluna vulgaris*) has a low growth rate whereas cloudberry (*Rubus chamaemorus*) has a high one, yet both occur together in blanket bog. Some species have the capacity to scramble and presumably grow better over slash that is left after thinning or brashing. Examples would include brambles (*Rubus fruticosus*) and the climbing cory-dalis (*Corydalis claviculata*).

Nobody has yet suggested that any plant species will be made extinct as a result of the reafforestation of the uplands. Some will decrease, principally as the result of the draining and planting of boggy areas and these minor habitats deserve the greatest attention in the future. Some species might possibly in-crease in abundance although this appears unlikely. Some of the Breckland rarities now only survive on the rides in the pine plantations, but this is almost coincidental, because all their other habitats have gone under the plough. Anyone who has studied Scandinavian or Canadian coniferous woodlands (Goldsmith, 1980) will realize that they have quite a well developed herb/dwarf shrub layer containing several interesting species. This has led a few people in Britain to think that the new plantations may have potential for a kind of creative conservation where we might promote hitherto rare species such as the following:

Linnaea borealis	twinflower
Cornus suecica	dwarf cornel
Moneses uniflora	one-flowered wintergreen
Trientalis europaea	chickweed wintergreen

However, the woods in which these or related species occur in other countries have a very open canopy with at least 20 per cent light intensity on the ground. It is possible that they could be cultivated along the edges of rides but they are most unlikely ever to be a feature of the stands themselves.

Effects on animals

Since the biomass of animals that can be supported by a site is probably directly related to the primary productivity of the site (Coe, Cummings, and Phillipson, 1976; Newton *et al.*, 1977), upland forestry plantations almost certainly sup-port a greater number of animals than did the open moorlands that preceded them. However, it is not so much the quantity of animal life in commercial plantations that is of direct concern to conservationists, but the differences in the species composition of planted and unplanted uplands, and also the fact that animals living within dense woodlands are often extremely difficult to observe.

Open moorlands afford a rather harsh environment for most animal species, indeed their likeness to tundra has already been alluded to. In essence, this likeness to a vegetation zone that would naturally be rather rare in Britain is one possible explanation for the attractiveness of such areas. Characteristically they support a rather low diversity of animals. Probably the most numerous mammal in the uplands is the field vole (*Microtus agrestis*), which can survive beneath the predominantly short vegetation but whose populations fluctuate markedly from year to year. In some areas the mountain hare (*Lepus timidus*) is another numerous herbivore, but both of these species must compete almost everywhere with the very heavily subsidized sheep, and in the Scottish Highlands also with the red deer (*Cervus elaphus*), displaced from its natural woodland environment. On the better heather-dominated moors they compete also with the commercially exploited red grouse (*Lagopus lagopus scoticus*).

But mammals are not the most interesting faunal component of the uplands: here, pride of place must go to the birds. Since the environment is harsh, many species are large in size and consequently need large areas (see Wood, Chapter 8). Their populations may therefore be harmfully affected by any fragmentation of open moorland that would be produced in a partial afforestation programme. Ground and crag-nesting raptorial species are the most spectacular, including the hen-harrier (*Circus cyaneus*), merlin (*Falco columbarius*) and golden eagle (*Aquila chrysaëtos*), but almost equally spectacular are the species of wading birds, such as curlew (*Numenius arquata*), golden plover (*Charadrius apricarius*) and dunlin (*Calidris alpina*), which are largely summer visitors to the uplands and dependent on mainly invertebrate prey. Paradoxically, populations of some of these species are considerably enlarged during the earliest stages of afforestation (Hope Jones, 1966; Williamson, 1970, 1972). The benefits that plantations of young trees afford are two-fold: greater protection from illegal trapping and shooting and considerably enhanced populations of small-mammal prey as a consequence of the exclosure of sheep. Hen-harriers and short-eared owls (*Asio flammeus*) benefit most conspicuously from these measures. However these and other predominantly open-ground species usually disappear entirely from afforested areas when the trees start to close their canopy (Marquiss, Newton and Ratcliffe, 1978; Moss, Taylor and Easterbee, 1979; Watson, 1979).

Within commercial plantations red deer fare much better than on the open hill, and they are joined by roe deer (*Capreolus capreolus*), a more exclusively woodland species. Several of our rarer carnivores may owe their survival to the advent of extensive programmes of afforestation (Langley and Yalden, 1977), notably the pine marten (*Martes martes*), the polecat (*Mustela putorius*) and the wild cat (*Felis silvestris*).

As far as birds are concerned, there are also gains to be made. Plantations afford a safe retreat to the sparrowhawk (*Accipiter nisus*) and provide effective protection from the still damaging consequences of agricultural pesticides (Newton, 1976; Newton and Bogan, 1978), so that the breeding success of this particular bird of prey is often greatest in regions of extensive coniferous woodland. Other species that naturally associate with conifers have shown remarkable extensions of their range as more and more of our uplands have been planted, notably the redpoll (*Carduelis flammea*), and the siskin (*Carduelis spinus*), although the bird that has probably the strongest claim to being Britain's only endemic bird species, the Scottish crossbill (*Loxia scotica*) has not markedly increased, since it is primarily an inhabitant of open pine woodlands. In contrast, the continental crossbill (*Loxia curvirostra*) which specializes on spruce seeds, has become a more widespread and regular breeder throughout Britain (Nethersole-Thompson, 1975).

The real key to the diversity of the avifauna of plantations is in the structure of the vegetation (Moss, 1978a and b, 1979). Clear-fell systems and short rotations impose an unnatural uniformity which is only relieved by the interspersion of blocks of markedly different ages. Consequently the bird community is also rather poor in species and dominated by small canopy-feeding species such as goldcrest (*Regulus regulus*), coaltit (*Parus ater*), chaffinch (*Fringilla coelebs*) and willow warbler (*Phylloscopus trochilus*). As all of these are not uncommon in small woodlands or hedgerows throughout the country, their greatly increased abundance in commercial forests is not particularly appreciated by naturalists.

The promise of a greater diversity of fauna in second and third rotation crops, where structural diversity may possibly be greater because of more varied growth rates, is a hope for the future that remains as yet largely unfulfilled (Currie and Bamford, 1981). Nevertheless, increased afforestation can be seen to have produced both losses and gains as far as animals are concerned, although many conservationists will bemoan the reduction in numbers of the larger, more spectacular and rarer species, and judge these to be of far more significance than the numerical gains made by mainly smaller and less attractive species.

Effects of soils and water

Afforestation affects the soils of the uplands and water draining from them both directly through the effects of the trees and litter, and indirectly, through treatments of the stands such as the application of fertilizer and pesticides.

The leaves of conifers are poor in bases whether it is measured in the ash, by the pH of a suspension of tissue in water, by the acidity of the cell solution, or by the colloidal or organic acid content. Their high content of fulvic acid, phenols and their high lignin and cutin content result in slow rates of invertebrate comminution and bacterial decomposition. The processes are complex and incompletely understood (see, for example, Miles, 1981; Nihlgard, 1970; Pyatt and Craven, 1979) but the lower rates of decomposition, the tannins of litter and the higher acidity are known to increase the degree of soil podsolization, the levels of aluminium ions in the soil at low pH values and to reduce the levels of calcium, nitrate nitrogen and phosphorus that is available to plants. This tends to slow plant growth, including tree growth, and the growth of salmonid fish in streams and lakes (Smith, 1980). At a low pH aluminium becomes a toxic element, and affects the eggs and alevins of these fish in particular.

This problem is probably accentuated by recent increases in the acidity of rain water, thought to be due principally to the increase in sulphur dioxide emissions from power stations (Likens *et al.*, 1977). Natural rainwater has a slightly acid pH of 5.6 due to carbon dioxide in solution but this drops to around 4.0–4.3 in south-east Scotland (Harriman and Morrison, 1980). It is often difficult to separate the effects of an increase in the area planted to conifers from the increase in the acidity of the rainwater (see Figure 17.5 and Nicholson, Paterson and Last, 1980 for methods and bibliography.)

Another direct effect of afforesting a catchment is the substantial reduction in water yield (Kirby and Rodda, 1974). It has been suggested that coniferous forests reduce the albedo and increase surface air turbulence, and that the trees intercept and hold precipitation, increase evapotranspiration and draw

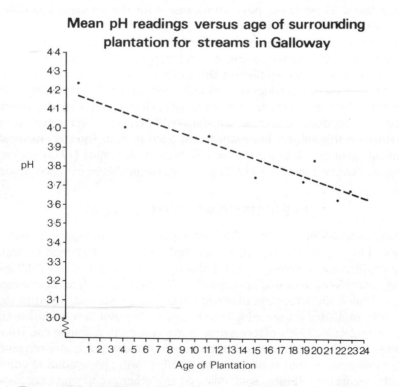

Mean pH readings versus age of surrounding plantation for streams in Galloway

FIGURE 17.5. From T. Drakeford, personal communication (1981).

moisture from greater depth in the soil (Binns, 1979). The total volume of the runoff is therefore reduced and its acidity probably increased. It is possible that there are differences in the effects of different tree species; lodgepole pine may use more water than spruce, but this subject needs further research.

Sites also suffer some indirect effects when they are afforested, for example improvement of their drainage by ploughing prior to planting and soil disturbance caused by harvesting. These activities increase the sediment yields in streams, this being four times greater in an afforested part of the Severn catchment than in a nearby unplanted catchment of the river Wye (Binns, 1979). The type of ploughing is presumably as important as the extent; downhill deep ploughing is particularly bad.

Another indirect effect of afforestation is caused by the fertilization of plots, these often being aerially sprayed at approximately 6-year intervals in some areas of Scotland. Harriman (1978) has shown that 3½ years after the application of fertilizers to forested catchments in Scotland, the losses of phosphorus were equivalent to 2 kg/ha^{-1}year^{-1}, and that this might represent 15 per cent of the amount applied. Potassium, being more soluble, represented 20 per cent of the amount applied, and nitrogen about 9 per cent, being mostly in the valuable nitrate form. These losses have implications for the biological equilibria of water courses and upland lochans.

The effects of herbicides such as 2,4,5-T (which contains dioxin as a contaminant and is known to cause teratogenic defects) and 2,4-D (which sometimes taints water supplies in Britain) are other serious indirect impacts. Similarly the use of pesticides such as DDT, malathion, fenitrothion (as well as their emulsifiers and 'vehicles' for spraying such as diesel fuel) may interact in sunlight to produce dangerous substances. There is an extensive American literature on this subject but readers will need to scan forestry, medical and chemical literature to grasp the overall picture. Kimmins (1975) reviews the herbicide literature and May (1977) discusses the problems of forest pesticides.

PRESCRIPTION FOR THE FUTURE

Certain variables in a scheme of afforestation can be manipulated more than others. The length of rotation, for example, will be determined by foresters without reference to ecologists but a mosaic of different ages in a catchment is usually considered desirable by ecologists. Fertilizer rates, thinning regimes, the width of rides, the percentage of broadleaves that are planted, and the extent of areas left unplanted are more likely to be negotiable. But how will the optimum pattern be decided? New afforestation is not subject to planning control and it is unlikely in the future that any agency, governmental or non-governmental, whose principal interest is the growing of fibre, will give ground to conservationists, environmentalists, academics, or any other group who may plead for the benefits of golden plover, greenshank, merlin, peregrine or golden eagle.

Scientists will have to harden their case by collecting more data and by conducting more appropriate experiments whilst the Royal Society for the Protection of Birds (RSPB), the Ramblers, the Council for the Protection of Rural England (CPRE) and the Nature Conservancy Council will have to keep reminding the Forestry Commission that one of their stated objectives is to protect and enhance the environment. The Commission and large private forestry organizations such as the Economic Forestry Group ought to employ more ecologists, just as they have employed landscape designers, to advise on which areas should be left unplanted and how the planted areas should be treated. However, the ecologists should be financially independent of the Commission or tenured in some way so that they can speak with authority and without fear of recrimination.

It is impossible to formulate a single prescription that will cover all aspects of upland afforestation but the following short list of topics merits consideration.

Areas to be left unplanted

Prior to planting, a detailed ecological survey should be conducted to indicate the most important areas from the nature conservation viewpoint. The best of these should then be set aside and not planted though possibly managed in some other appropriate manner. The Conservation Course (1981) indicated how this might be carried out in Galloway, for example. The Forestry Commission has recently prepared new guidelines for the management of streams (Mills, 1980). A generous width on either side of steams and around lakes should be left unplanted for ecological reasons. This will only marginally reduce the fibre yield for the whole estate for such areas are difficult to plant and sometimes almost impossible to harvest. Drains should be discontinued well before they reach a stream or a lake margin in order to reduce the sediment load entering the water.

Bogs and marshes have a higher ecological interest than the surrounding areas and, moreover, if planted, the trees on them are particularly vulnerable to windblow. It therefore makes good commercial and ecological sense to leave them unplanted. Similarly, rocky outcrops fail to produce good tree growth because of seasonal drought or nutrient deficiency and should not be planted. Finally any pockets of broadleaves such as birch, alder or rowan should be left, as their seeds, berries and associated insects are important to birds and their litter ameliorates the soil (Miles, 1981). Being deeper rooted than conifers they act as 'nutrient pumps' and some, such as alder, have nitrogen-fixing bacteria associated with their roots.

Rides, roads, wayleaves and firebreaks

These are an essential part of the commercial forest but they also form the ecologically richest part of the plantations. The number of plant species can be at a maximum (for a plantation) along roads because competition is reduced and seeds are introduced by people and vehicles. Rides, wayleaves and firebreaks may be more important to animals because they are less disturbed there than on roads. It has not been established that wider roads and those with scalloped edges are more interesting than narrow ones, but it appears probable. The orientation probably does not matter in upland Britain but in climates with hotter, drier summers an east–west orientation is favoured to maximize shading.

The management of stands

Lower planting densities are now being favoured by foresters to reduce their planting costs, but this often necessitates the use of herbicides such as 2,4-D and 2,4,5-T to control woody growth. Lower densities of planting and early thinning both reduce the intensity of shading and this is ecologically desirable. For the same reason, repeated thinnings can be advocated. Selection systems of silviculture are preferred by ecologists but are seldom practised in Britain, even with broadleaves, let alone conifers. Clear felling is invariably the rule, but the longer the rotation the better ecologically. Moreover only small parts of a catchment should be harvested at a time to reduce sediment losses (Bormann and Likens, 1979).

The edges of plantations

In British commercial afforestation stands have traditionally been given sharp, fenced edges. In the Black Forest (*Schwarz walden*) of Germany, however, chevron-shapes have been used to break the wind, especially on slopes, and such indented edges are likely to be important as shelter for birds and the larger mammals. An alternative would be to decrease the density of trees towards the edge of a stand.

Percentage of broadleaves

Foresters find that stands of a single species are easier to manage and harvest at maturity, but there are strong ecological arguments for an admixture of a small proportion of broadleaves in a plantation. They discourage the spread of fire, speed up the recycling of nutrients, reduce the possibility of pest outbreaks, attract the attention of deer and possibly divert their attention from the coniferous crop. There should be some research in this area to evaluate the potential of broadleaves in mixtures with conifers.

The need for research

There are many aspects of the growth of the trees and selection of provenance that require the attention of foresters, but ecological considerations also merit more research. The Forestry Commission's research budget is about £5 million, and although this sounds considerable, if it is compared with the annual estimates of losses to windthrow of £30 million it is very small. We need to know more about the flora and fauna of our new plantations, how their abundances and behaviour change, how soil processes (nitrogen fixation, mineralization) are affected by the trees (their throughfall and litter), the role that broadleaves might play in mixtures, the possibility of natural regeneration in some areas in the future, the ecological consequences of indented edges, as well as the processes of acidification.

REFERENCES

Anderson, M. (1979). The development of plant habitats under exotic forest crops. In S.E. Wright and G.P. Buckley (Eds) *Ecology and Design in Amenity Land Management*, Wye College/RERG, pp. 87–108.

Baskerville, G.L., Belyea, R.N., Kettela, E.G., Varty, I.W. and Bradford, K. Marshall (1975). The spruce budworm. *Forestry Chronicle*, **51(4)**, 1–26.

Binns, W.O. (1979). The hydrological impact of afforestation in Great Britain. In G.E. Hollis (Ed) *Man's Impact on the Hydrological Cycle in the UK.*, Geo. Abstracts, Norwich, pp. 55–69.

Bormann, F.H. and Likens, G.E. (1979). *Pattern and Process in a Forested Ecosystem*, Springer-Verlag, New York, 253 pp.

Brown, A.H.F. (1981). The recovery of ground vegetation in coppice wood: the significance of buried seed. In F.T. Last and A.S. Gardiner (Eds) *Forest and Woodland Ecology*, Grange-over-Sands, pp. 41–44.

Centre for Agricultural Strategy (CAS) (1980). *Strategy for the UK Forest Industry*. Report 6, Centre for Agricultural Strategy, Reading, 347 pp.

Coe, M.J., Cummings, D.H., and Phillipson, J. (1976). Biomass and production of large African herbivores in relation to rainfall and primary production. *Oecologia*, **22**, 341–54.

Conservation Course (1981). The afforestation of the uplands: the botanical interest of areas left unplanted. *Discussion Paper in Conservation*, **35**, 88 pp.

Countryside Commission (1972). *Note of the Countryside Commission's views on the Consultative Document on Forestry Policy published 28 June 1972*, Cheltenham, Glos., 6 pp.

Currie, F.A. and Bamford, R. (1981). Bird populations of sample pre-thicket forest plantations. *Quarterly Journal of Forestry*, **75**, 75–82.

Edlin, H.L. (1970). *Trees, Woods and Man*, London, Collins New Naturalist (revised edition), 272 pp.

Ford, E.D., Malcolm, D.C. and Atterson, J. (Eds) (1979). *The Ecology of Even-aged Forest Plantations*, Institute of Terrestrial Ecology, Cambridge, 582 pp.

Goldsmith, F.B. (1980). The evaluation of a forest resource: a case study from Nova Scotia. *Journal of Environmental Management*, **10**, 83–100.

Green, B.H. (Ed) (1979). *Wildlife Introductions to Great Britain*, Nature Conservancy Council, London, 32 pp.

Grime, J.P. (1981), Plant strategies in shade. In H. Smith (Ed) *Plants and the Daylight Spectrum, British Photobiology Society Symposium*, Leicester, Academic Press, 508 pp.

Harriman, R. (1978), Nutrient leaching from fertilized forest watersheds in Scotland. *Journal of Applied Ecology*, **15**, 933–942.

Harriman, R. and Morrison, B.R.S. (1980). Forestry, fisheries and acid rain in Scotland. *Scottish Forestry*, **35(2)**, 89–95.

Hart, C.E. (1967). *Practical Forestry for the Agent and Surveyor*, Estates Gazette Ltd., London, 440 pp.

Hill, M.D. (1979). The development of a flora in even-aged plantations. In E.D. Ford, D.C. Malcolm and J. Atterson, *The Ecology of Even-Aged Forest Plantations*, Institute of Terrestrial Ecology, Cambridge, pp. 175–192.

Hill, M.D. and Jones, E.W. (1978). Vegetation changes resulting from afforestation of rough grazing in Caeo Forest, South Wales. *Journal of Ecology*, **66**, 433–456.

Hope Jones, P. (1966). The bird population succession at Newborough Warren. *British Birds*, **54**, 180–189.

James, N.D.G. (1981). *A History of English Forestry*, Blackwell, Oxford, 339 pp.

Jarvis, P.J. (1980). *The Biogeography and Ecology of Introduced Species*, Working Paper 1 (Geography), Birmingham, 40 pp.

Johnston, D.R., Grayson, A.J., and Bradley, R.T. (1967). *Forest Planning*, Faber, London, 541 pp.

Kellman, M.C. (1970). Preliminary seed budgets for two plant communities in coastal British Columbia. *Canadian Journal of Botany*, **48**, 1383–1385.

Kimmins, J.P. (1975). *Review of the Ecological Effects of Herbicide Usage in Forestry*, Canadian Forestry Service Information Report BC-X-139, 44 pp.

Kirby, C. and Rodda, J.C. (1974). Managing the hydrological cycle. In A. Warren and F.B. Goldsmith (Eds), *Conservation in Practice*, Wiley, London, pp. 73–84.

Langley, P.J.W. and Yalden, D.W. (1977). The decline of the rarer carnivores in Great Britain during the nineteenth century. *Mammal Review*, **7**, 95–116.

Likens, G.E., Bormann, F.H., Pierce, R.S., Eaton, J.S., Johnson, N.M. (1977). *Biogeochemistry of a Forested Ecosystem*, Springer-Verlag, New York, 146 pp.

Marquiss, M., Newton, I., and Ratcliffe, D.A. (1978). The decline of the raven (*Corvus corax*) in relation to afforestation in southern Scotland and northern England. *Journal of Applied Ecology*, **15**, 129–144.

May, E.E. (1977). Canada's moth war. *Environment*, **19(6)**, 16–24.

Miles, J. (1981). *Effect of Birch on Moorland*, Institute of Terrestrial Ecology, Banchory, 18 pp.

Mills, D.H. (1980). *The Management of Forest Streams*, Forestry Commission Leaflet 78, HMSO, London.

Moss, D. (1978a). Song-bird populations in forestry plantations. *Quarterly Journal of Forestry*, **72**, 5–13.

Moss, D. (1978b). Diversity of woodland songbird populations. *Journal Animal Ecology*, **47**, 521–527.

Moss, D. (1979). Even-aged plantations as a habitat for birds, pp. 413–427. In E.D. Ford D.C. Malcolm and J. Atterson (Eds), *The Ecology of Even-aged Forest Plantations*, Institute of Terrestrial Ecology, Cambridge, 582 pp.

Moss, D., Taylor, P.N., and Easterbee, N. (1979). The effects on song-bird populations of upland afforestation with spruce. *Forestry*, **52**, 129–147.

Nethersole-Thompson, D. (1975). *Pine Crossbills*, T. and A.D. Poyser, Berkhamsted.

Newton, I. (1976). Breeding of sparrowhawks (*Accipiter nisus*) in different environments. *Journal of Animal Ecology*, 45, 831–849.

Newton, I. and Bogan, J. (1978). The role of different organochlorine compounds in the breeding of British sparrowhawks. *Journal of Applied Ecology*, 15, 105–116.

Newton, I., Marquiss, M., Weir, D.N., and Moss, D. (1977). Spacing of sparrowhawk nesting territories. *Journal of Animal Ecology*, 46, 425–441.

Nicholson, I.A., Paterson, I.S. and Last, F.T. (1980). *Acid Precipitation in Forest Ecosystems*, Institute of Terrestrial Ecology, Cambridge, 36 pp.

Nihlgard, B. (1970). Precipitation, its chemical composition and effect on soil water in a beech and a spruce forest in south Sweden. *Oikos*, 21, 208–217.

Osmaston, F.C. (1968). *The Management of Forests*, George Allen and Unwin, London, 384 pp.

Peterken, G.F. (1977). Habitat conservation priorities in British and European woodlands. *Biological Conservation*, 11, 223–236.

Pinder, N. (Ed) (1981). Conservation and introduced species. *Discussion Paper in Conservation*, 30, University College London, 63 pp.

Price, C. (1976). Forestry. In M. MacEwen (Ed), *Future Landscapes*, Chatto and Windus, London, 224 pp.

Pyatt, D.G. and Craven, M.M. (1979). Soil changes under even-aged plantations. In E.D. Ford, D.C. Malcolm and J. Atterson (Eds), *The Ecology of Even-aged Forest Plantations*, Institute of Terrestrial Ecology, Cambridge, pp. 369–386.

Rackham, O. (1980). *Ancient Woodland*, Arnold, London, 402 pp.

Rose, F.H. (1976). Lichenological indicators of age and environmental continuity in woodlands. In D.H. Brown, D.L. Hawksworth and R.H. Bailey (Eds), *Lichenology— Progress and Problems*, Academic Press, London, pp. 279–307.

Ryle, G. (1969). *Forest Service: The First Forty-five Years of the Forestry Commission of Great Britain*, David and Charles, Newton Abbot, 340 pp.

Shimwell, D.W. (1971). *Description and Classification of Vegetation*, Sidgwick and Jackson, London, 322 pp.

Smith, B.D. (1980). The effects of afforestation on the trout of a small stream in southern Scotland. *Fisheries Management*, 11(2), 39–58.

Southwood, T.R.E. (1961). The number of species of insect associated with various trees. *Journal of Animal Ecology*, 30, 1–8.

Sydes, C. and Grime, J.P. (1981). Effects of tree litter on herbaceous vegetation in deciduous woodland, I. Field investigations; II. An experimental investigation. *Journal of Ecology*, 69, 237–248, 249–262.

Watson, J. (1979). Food of merlins in young conifer forest. *Bird Study*, 26, 253–258.

Williamson, K. (1970). Birds and modern forestry. *Quarterly Journal of Forestry*, 64, 346–355.

Williamson, K. (1972). The conservation of birdlife in the new coniferous forests. *Forestry*, 45, 87–100.

PART III
ORGANIZING CONSERVATION

Introduction

Our third section deals with the issues that confront conservation organizations and agencies: the behaviour of competitors for rural land, how voluntary groups exert pressure, the response of other government agencies, how national park and local authorities protect the areas for which they are responsible.

The section starts with two chapters (18 and 19) that are historical perspectives, the first showing how the use of the land in the past has affected the patterns we find in nature today, and the second analysing the evolution of the agencies involved in nature protection. Chapters 20 and 21 document the more recent changes in land use that have had such a vast impact on our landscape and its wildlife: Chapter 20 traces the intensification of modern farming and Chapter 21, in its analysis of the controversies over land drainage, shows just how flimsy are the economic arguments that are so often advanced to justify such activities. Controversy over the use of the land is most intense in our ten national parks (the topic of Chapter 22). They are usually thought to have a more recreational than nature conservation function, but are shown to have a role in nature conservation in the broader sense. Local authorities also have a responsibility to mediate in the conflicts, and in Chapter 23 the ecologist employed by our largest new city presents his views on ecology and conservation at this level of government. Chapter 24 is a review of the voluntary sector which has taken a keen interest in the debate and which, it is claimed, now represents about 3 million conservationists in Britain. Finally in Chapter 25 we broaden our horizons to some of the ways in which other countries have met these problems.

Conservation in Perspective
Edited by A. Warren and F.B. Goldsmith
© 1983 John Wiley & Sons Ltd.

CHAPTER 18

The Historical Perspective

JOHN SHEAIL

In selecting wildlife communities for conservation and in deciding on management policies for them, the conservationist needs to make close reference to the past in order to understand the character and composition of those communities and wilderness areas today. When were they established, how did they develop, and what has been the relative importance of natural processes and human influence? The answers to these questions about the past should help to place in perspective the problems of managing the natural environment of today for conservation and other purposes.

Never before has so much information suddenly become available to the historian. His ability to collate, digest and analyse the information has been severely tested. This explosion of data reflects not only the unparalleled opportunities to gather material, and the introduction of an increasing array of methods and techniques, but also the unprecedented interest being taken in the past. Among ecologists the term 'historical ecology' has been coined to describe their particular approach to the past. It is this recent conjunction of so many academic interests that has provided much of the stimulation to historical research.

Ellenberg (1979) recounted how many ecologists left Europe and other densely settled regions in search of 'real nature', only to discover the effects of human influence in even the most remote highlands and lonely valleys. The mosaics, or mixtures, of ecosystems encountered by the ecologist represent 'different stages of landscape history and different levels of human influence'. Another kind of historian has been involved in such ventures as the multivolume *Agrarian History of England and Wales* (Piggott, 1981), but an interest in the man−environment relationship is also prominent. The first volume begins with a reconstruction of the natural environment of early man, on the premise that man can only be understood 'as a piece of natural history himself'. The first human settlements in Britain were 'a natural consequence of the northward expansion of plant and animal species to fill new biological niches' (Piggott, 1981).

A man−environment relationship does not imply that human behaviour is determined by environmental factors, or that the environment is moulded by man in some preset pattern. Although the environment is to all intents and purposes permanent, the means and will to use it in given ways are not (Jones, 1979). Tubbs (1968) explored the extent and variety of this man–environment relationship in his ecological history of the New Forest. He described how the history of human land use, management and exploitation, and the economic factors governing them, formed a coherent framework within which to describe the development of the range of forest habitats. Rackham (1980) investigated the relationship in the context of a particular habitat. In his pioneer study, *Ancient Woodland*, he related 'the complex variety of native woodlands' both to 'the biology of the woods' and to the part that they have played in human affairs. In doing so, he was able to portray 'the richness of detail to be found within many apparently ordinary woods'.

ANCIENT AND MODERN

The making of the landscape has been much more complicated and drawnout than once supposed: the time-perspective of the historian has been greatly extended by recent research. In no way did Domesday England of 1086 represent the start in the development of the present environment. According to Taylor (1980), the countryside of 1000 years BC had much more arable and less woodland than the countryside visited by the Norman clerks during the compilation of the Domesday Book in the eleventh century AD. Far from early man ignoring the uplands, these areas may have been among the first to experience the impoverishing effects of man's exploitation of natural resources. The essential features of the higher parts of the North Yorkshire Moors may be largely prehistoric in origin (Spratt and Simmons, 1976). Intense grazing and the cultivation of crops accentuated the leaching of plant nutrients from the soils and the natural tendency of the lime-poor soils towards podsolization (Ball, 1975). By the first century BC, much of the higher ground was covered in blanket bog and, elsewhere a heather moor was evolving, representing in aggregate the transition from a brown earth soil–forest ecosystem to a more impoverished podsol soil–heathland system. The forest never returned.

In view of this widespread degradation of vegetation and soils in ecological terms, examples of ancient wood and grassland have a particular significance as relicts of the past. Dimbleby (1974) emphasized how their relative freedom from human interference made them a unique standard by which to measure changes that have occurred in the neighbouring countryside. So far, however, the use of primary woodland sites as a measure of change has been only rudimentary, and the sites themselves continue to be destroyed by changes in their use and management (Ball and Stevens, 1981).

It is extremely unlikely that even these long-established wood and grassland sites are the result of an entirely static management system in the past. There is, for example, evidence that farmers began to manipulate the natural succession in their meadows and pastures much more consciously and competently during the sixteenth and seventeenth centuries (Lane, 1980). In the *Surveiors Dialogue* by John Norden (1610), the Surveyor speaks of how:

> even the best meadows will become ragged, and full of unprofitable weedes, if it bee not cut and eaten; some will become too moist, and so growe to bogges—some too dry, and so to a hungrie mosse . . .

Such grasslands had to be ploughed and sown with 'the seed of the claber grasse, of the grasse hony-suckle, and other seed that fall out of the finest and purest hay'. As the Surveyor remarked (Norden, 1610), if the land was:

> not fed with nutriture, and comforted and adorned with the most expedient commodities, it would pine away, and become forlorne, as the mind that hath no rest or recreation, waxeth lumpish and heavy.

Much more needs to be known about the management of these habitats in the past.

At the other end of the time-spectrum, the ecologist's knowledge of recent changes in the natural environment is often deficient. This is in spite of the large-scale changes that have taken place in the countryside. As early as the 1920s, de Lotbiniere (1928–29) warned of the impact of afforestation on Breckland. The lack of sun and food in the new plantations would lead to an impoverishment of wildlife, compared with that of the open environment of the former brecks. Only when the conifer trees had reached full maturity would they offer much in the way of amenity to the sightseer and rambler. Despite these premonitions of change, the chronology of events in these areas was rarely recorded consciously. The widely quoted survey of the Dorset Heaths by Moore (1962) was exceptional for the way it portrayed, in a comparatively precise and systematic manner, the decline and fragmentation of a habitat in the earlier part of the century.

The main impetus for such studies has come only recently, as a result of the need for the conservationist to substantiate his claims of unprecedented losses of wildlife habitat. It may be no coincidence that the preliminary results of some of these historical studies appeared at about the same time as parliament debated the Wildlife and Countryside Bill of 1980–81. A paper was published in *New Scientist* (Goode, 1981) showing how almost 90 per cent of the lowland 'raised bog' systems in four parts of Scotland and northern England had been destroyed, according to a comparison of Ordnance Survey maps compiled about 1850 and 1900, and a series of more recent aerial photographs. Half the remainder had been reduced to fragments of less than 10 ha. In another comparative study, Parry, Bruce, and Harkness (1981) estimated on the basis of changes in four national park areas (Figure 18.1) that 5000 ha of moorland in England

Reclamation of Moorland

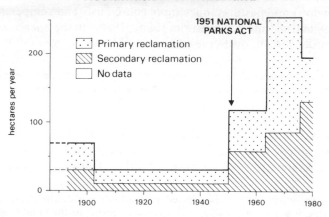

FIGURE 18.1. The reclamation of moorland area in the North Yorkshire Moors, Dartmoor, Brecon Beacons and norther Snowdonia National Park areas, as indicated by a comparison of aerial photographs and Ordnance Survey maps. From Parry, Bruce, and Harkness (1981); reproduced with permission.

and Wales were being enclosed and reclaimed each year, and that a significant proportion of this moorland had remained unploughed since at least pre-Roman times (primary reclamation). Such studies highlight the value of detailed appraisals of the recent past as part of the background briefing for those whose decisions will affect contemporary and future trends in the environment.

THE NATURE OF THE EVIDENCE

The extent to which the past can be reconstructed must depend on the period and subject under study. For those early periods for which archaeological data only are available, one is more than usually dependent on the potential evidence having escaped destruction or concealment by modern land-use change. Even then the archaeologist can never, in practice, give a complete and accurate impression of what was taking place in the environment. Whilst there is considerable scope for model-making, the lack of opportunity to test the validity of any patterns and activities discerned will be a severe limiting factor.

There may, nevertheless, be more scope for studying the past where archaeological and ecological expertise collaborate in reconstructing land-use systems and their impact on the environment (Widgren, 1979). Sufficient data can be found on the environment and human occupation for the island of Oland in the Baltic to provide a detailed and plausible model of ecological instability during the Iron Age, as expressed in terms of ecosystem complexity and the flow of energy, matter and information. In response to the increasing demands from the Roman Army for animal products, the limestone plains of the Great Alvar were overgrazed; output greatly exceeded (energetical) input, particularly on the more marginal soils (Enckell, Konigsson, and Konigsson, 1979).

It is not enough to know what happened and when; the mechanisms by which these changes came about have to be reconstructed, and their implications understood. Considerable interest has been focused, for example, on the efficiency of fire, grazing and cutting as management tools in a variety of habitats. By reference to the archaeological record, ethnographic literature and research carried out in connection with wildlife conservation programmes in North American forests, Mellars (1976) has speculated on the ecological relationships that might have prevailed between fire, animal populations, and human groups in prehistory. The improvements brought about by fire in at least some forest environments might have increased not only the total carrying capacity, but also the relative growth and reproductive rates, of the animals. So striking was the overall increase in productivity that hunting and gathering populations could have been encouraged to adopt systematic policies of forest burning. This in turn might have led to the emergence of more complex patterns of man–animal relationships that were more closely similar to those of traditional 'herding' and 'pastoralist' economies.

The most serious limitations of any hypothesis concerning the early development of the environment/man relationship is the possibility that the mechanisms for change did not leave any discernible evidence in the archaeological record. Where the significance of these mechanisms is so great, the search for the evidence must nevertheless be made. One such case is the decline in elm (*Ulmus*), which seems to have taken place so dramatically between 3400 and 2800 BC over a wide range of localities in north-west Europe (Godwin, 1975). Some research-workers have suggested that man must have discovered a mechanism that was capable of inflicting tremendous damage on the forests. Was such a mechanism available, and could its presence be identified by the palynologist? In order to find the answer, Garbett (1981) constructed a high-resolution pollen diagram by freezing the peat core and taking samples at 0.2 cm intervals at Ellerside Moss in Cumbria, a site where the elm decline had already been recorded in earlier work. The diagram suggested that elm, oak and lime were exploited for leaf-fodder in the initial stage of the elm decline, followed by a period when elm only was taken. This seems to have led to the near-destruction of the elm, the partial clearance of elm-dominated areas, and ultimately the production of small temporary clearances throughout the forest. The transition in forest ecology over the course of 100 years may have implied not so much a simple change from one type of economy to another, but a series of steps in the development from a hunter-gatherer economy to one based on domesticated animals.

No individual historical technique will reveal the truth, the whole truth, and nothing but the truth about the past. By perfecting existing techniques, and devising further means of investigation, there will, however, be greater opportunities to check and cross-check the validity of results obtained from each. If a consensus emerges from an array of independent approaches, hypotheses may become less speculative and the risks of circular-argument may diminish. It is for these reasons that strides in such techniques as dendrochronology and soil-phosphate analysis have such great significance, not only within their respective specialisms, but for the environmental-historian generally. In pollen analysis, particular attention has been given to places and periods for which documentary, place-name and cartographic evidence might also be available. In Lancashire, the different forms of evidence have provided further confirmation of a late Romano-British farming maximum, a phase of Norse forest-clearance and pastoral farming, and a pre-Cistercian, later Norman phase in which farming retreated and the forest regenerated (Oldfield, 1969).

The evidence of old maps, books and papers is rarely straightforward. All is not what it seems! It is always important to discover the purpose for which the evidence was compiled, the competence of the author, and the nature of any constraints imposed on him by his publisher or patron. When using large-scale Ordnance Survey maps to reconstruct the former incidence of rough pasture, for example, it is pertinent to discover how far it was within the terms of reference of the surveyors to record such features (Harley, 1979). To some extent, all documents have to be treated as if they were encoded. Whilst the obviously technical, obsolete and foreign terms have to be 'decoded', the trickiest terms may be those with which we seem to be most familiar. It is however always worth asking what 'ploughing', 'mowing' or 'ditching' would have meant within the times and places described in the text.

Like the anthropologist, the historian has to bear in mind the circumstances and perceptions of those whose records are now being used in a reconstruction of the past. The preoccupations of writers with the rare and unusual often meant that the commonplace features passed unrecorded, unless they were noted by a stranger. Strangers brought their own, and different, sets of prejudices to their observations (Sheail, 1980). The records of topographers and explorers have been intensively used by the historian, and McArthur (1973) has demonstrated their value and limitations in his attempts to reconstruct the vegetation of parts of arid Australia in the early nineteenth century, and the likely impact of the aborigines' use of fire on the natural environment. In giving advice on the use of fire as a management tool in the present-day management of forest and scrub, the records left by the 'meticulous recorder and describer of vegetative associations and new species' on an expedition of 1817 are of more than academic interest.

Much of the stimulus for local history comes from the realization that the 'richest historical record' of all is 'the English landscape itself' (Hoskins, 1970; Meinig, 1979). As a major component of the landscape, plant and animal life form part of that study. Even to the untrained eye, there are patterns in the number and mixture of species present in hedgerows, that seem to reflect their age and function (Hoskins, 1975). Empirical observations may be reinforced by the dating-techniques of the ecologist, based on the relationship between the number of shrub-species present and the period of time available for colonization (Pollard, Moore, and Hooper, 1974). Not only has this relationship been examined in different parts of the country, but the opportunity has been taken to assess the influence of the origins and age of hedgerows on the presence of animals with poor powers of dispersal. Cameron, Down, and Pannett (1980) found that hedges originating near, or in, woodland had a richer snail fauna than those planted in open fields, and that some snail species could be used as indicators of hedges of woodland origin. Snail diversity increased with the age of the hedge, but the effect became slight in hedges of over 100 years in age.

A wide range of phenomena has been investigated in the search for ways of estimating the age and history of a site. One of the more recent to emerge in the grassland habitat has been the use of anthills of the species *Lasius flavus* (L.) as a dating-technique. King (1981) used an index based on anthill size to estimate the dates at which thirteen chalk grasslands in Wiltshire were last ploughed. The mean volumes of the five largest anthills out of 1000 increased linearly with the actual age of the grassland, in fields last ploughed between 50 and 165 years ago (Figure 18.2). As a rule of thumb, the index of grassland age may help to corroborate or supplement other forms of field and documentary evidence.

Perhaps the most remarkable form of evidence is a combination of documentary and field evidence in the form of old photographs, taken from the air or ground. Where the location of these photographs can be identified, they provide a unique type of baseline from which to measure subsequent change. This has been graphically illustrated in a study of changes in Wistman's Wood on Dartmoor, where a comparison of photographs taken from the late nineteenth century onwards indicates the survival of many individual trees, a general rise in the canopy, change in tree-growth form, and a marginal expansion of the wood to nearly twice its former area. Not only do these observed changes form a basis for hypotheses on the effects of grazing pressure and climatic change, but they provide an invaluable context for deciding how the site might be managed in the future (Proctor, Spooner, and Spooner, 1980).

The changing status of species may provide some index of environmental change. Victorian naturalists attributed the loss of lichens and mosses from Epping Forest to the increase of smoke pollution from London, and a survey of 1970 indicated that 129 species of lichen had disappeared since 1800 from an area within 16 km of Trafalgar Square (Rose and Hawksworth, 1981). Since 1960, however, the sulphur dioxide content of the air in towns and cities has been halved, and surveys of selected sites have shown that not only have some lichens extended their range, but some species last recorded in London in 1800 have now been rediscovered. Wherever possible, any correlation between the incidence of species and the character of their habitat should be based on the

Development of Anthills in Ageing Grassland

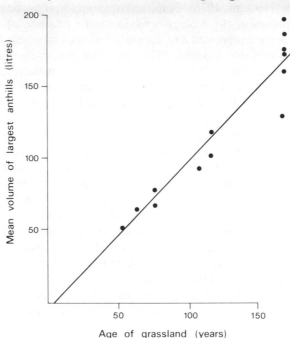

FIGURE 18.2. Mean volumes of the five largest anthills out of a thousand, plotted against actual grassland age for 13 Wiltshire sites. In most cases the actual year in which the field was last ploughed is not known with certainty, and the midpoint in the range is taken. For grasslands ploughed over 130 years ago, the arbitrary figure of 165 years old was used since most grasslands were probably ploughed during the Napoleonic wars, which ended in 1815. The regression line was plotted through the six means of grassland age. Its equation is $y = 1.04, x - 5.23$, where y is the mean volume in litres of the five largest anthills out of a thousand, and x is the actual age of the grassland in years. From King (1981); reproduced by permission of Blackwell Scientific Publications.

changing status of a group as opposed to a single species. Even then, its validity may remain in doubt owing to a lack of knowledge of the general distribution of the species, their habitat requirements and the interaction with other species (Hawksworth, 1974).

In some instances, a human health hazard has been used by the historian to prove some wider trend in the environment. All too often, these attempts have been based on an oversimplified understanding of cause and effect, and they have, in practice, underlined both the complexity and far-reaching ramifications of any change in the natural environment. Reference may be made to the correlations that have been drawn between the extinction of bubonic plagues in seventeenth-century England and changes in building materials, personal hygiene, burial practices and other aspects of the urban environment. There is no evidence to indicate that any of these hypotheses identify the primary reason why the rat—flea—man chain of infection was completely broken. It may have been an environmental change at the microscale rather than the macroscale that broke the chain. There may, for example, have been an increase in the population of insects that preyed upon the eggs and larval stages of the flea, coinciding with the aftermath of an epidemic when the rat and adult-flea population would have been severely reduced and more vulnerable to extinction (Bayliss, 1980). There may similarly be little substance in the simple correlation between the demise of marsh-fever in England and the drainage of the marshes, the traditional home of the ague. The recession of malaria was a long process. Although the number of infective cases decreased, the geographical distribution remained the same. It was probably a combination of factors that reduced the frequency and severity of malaria in England in the eighteenth and nineteenth centuries— including not only changes in the use and management of the marshlands but also changes in the biology of the mosquito and the receptivity of the local human population (Dobson, 1980).

THE HISTORICAL CONTENT OF ECOLOGY

Right from the start, the science of ecology had a distinctive perspective on events in the past. Whereas the geologist was primarily interested in large-scale changes, extending far back in time, the field-ecologist was much more concerned with the 'minor changes and dimensions' that were crucial to an understanding of the lives of organisms and to the control of plant and animal communities (Godwin, 1978). In the East Anglian fens, for example, the subtle interplay of seasonal, episodic and Quaternary changes in water regime, and the repercussions of man's attempts to drain and farm the soils, were quickly recognized by the ecologist as being essential to any understanding of the incidence and character of the organisms found on such nature reserves as Wicken Fen, or more generally in the peat profiles.

Writing in a woodland context, Peterken (1979) identified six basic questions which the ecologist was likely to ask of his historical sources:

(1) Did the wood exist, and if not what was the land used for?
(2) How extensive was the wood, and where were the boundaries?
(3) How was the wood managed, and what was the structure of the stand?
(4) What species of tree and wildlife were present?
(5) What forms of disturbance occurred?
(6) What was the use and vegetation of the adjacent land?

Most studies tend to be a synthesis of these types of questions in various combinations and relating to various dates, spans of time and geographical areas.

The ecologist's concern for, and knowledge of, past events have become more discriminating and precise. The character and variety of the chalk grasslands of many parts of central England were attributed at first to their management over many centuries as sheepwalk, but this traditional view has been challenged by the realization that over half of Salisbury Plain, for example, was under the plough in the nineteenth century. More particularly, the sharp linear boundaries that separate the vegetation-types today often correspond with the edge of cultivation, as recorded on old maps and aerial photographs. In a study of chalk grassland on the Porton Ranges, Wells *et al.* (1976) distinguished four age-classes of grassland from documentary evidence; groups of plants were identified as 'indicator species' of each of the four classes. There were also differences in soil chemistry, which appeared to reflect the influence of grassland succession, postcultivation management, and soil faunal activity on the build-up of soil organic matter and other soil properties, rather than being a primary cause of the differences in vegetation. An annual rate of accretion of organic matter content was observed of about 0.08 per cent dry weight over the 0–10 cm soil depth. This was comparable with rates reported from other humose soils of high biological activity.

Very rarely does the ecologist encounter a simple, undisturbed succession of plants—even on the embankments and cuttings of operational railway lines. Although these earthworks were built over a comparatively short space of time, and were the responsibility of a relatively small number of agencies, almost every bank and cutting has been subjected to a wide variety of disturbance and modification arising out of improvements to the track and management of the earthworks. This has not only been on a piecemeal basis, but has left little in the way of written evidence. The ecologist is never quite sure what he is studying, or how representative it may be of railway earthworks more generally (Sheail, 1979). Similar difficulties may be encountered when tracing the speed and direction of changes arising from the reversion of upland pasture to moorland. Ball *et al.* (1982) have recorded wide variations in the long time-sequence involved in the transition through rough pasture, grass heath to shrubby heath. This may reflect differences in postreversion grazing regimes, and the use of fire as a management tool.

Considerable interest is taken in the biogeography and ecology of introduced species. As Jarvis (1981) has explained, status is a dynamic attribute; the status of a species will change in relation to the vicissitudes of its establishment and competition with other species. The ecological and environmental consequences of the appearance of alien species may vary in significance, according to time and place. Ouren (1980) has used a wide range of plant records to illustrate the part played by nineteenth-century shipyards in the introduction of aliens to Norway. The plants were introduced in the ballast, or cargo, thrown onto the shore as the boats were prepared for careening operations. The introduction of the rabbit to Britain has been much more significant. Over 900 years it has become one of the most abundant and ubiquitous animals. Introduced to different areas for its fur and meat, and for sport, the animal only gradually came to be regarded as a pest. The postmedieval period may have been the turning-point, when large numbers were kept in an increasing number of commercial warrens, and when the destruction of predators and increase in winter crop-cover may have enabled feral colonies to survive in larger numbers (Sheail, 1978). One way of assessing the competence of farmers in their choice of sites, crops and husbandry practices might be to investigate the incidence and significance of damage caused by alien species. Bowen (1980) has recounted how the invasion of jackrabbits was just as important as drought in causing the demise of dry-farming in parts of Nevada in the early years of this century.

The historical interest of many conservationists is encapsulated in the study of the relationship between the distribution and breeding of the red kite (*Milvus milvus*) and changes in land use in Wales, as reported by Newton, Davis, and Moss (1981). Because of the rarity of the species, the study sought to identify the types of local landscape in which the species flourished, and to obtain a basis for predicting the effects of further changes in land use. Over the previous 33 years, there had been a large-scale extension of forestry, reservoir development and tourism. Although the increase in population of the kite had coincided with these changes, there was little evidence of a causal relationship. This rise in recent years might be due to a decrease in human persecution as a result of a decline in keepering and in the territory over which poison-baits were used by shepherds. Looking to the future, the effects of further changes in habitat depended on the actual location and scale of disturbance. Some districts might already have more trees than were good for the birds, whereas other, more open areas, distant from nest sites could probably take more trees without inflicting harm to the kite.

The experiences of the past can never provide a comprehensive and certain guide as to what must happen in the future. So many changing parameters have to be taken into account, and the ecologist's knowledge of the incidence and significance of each remains very fragmentary.

A DISTANT MIRROR

Writers of both fact and fiction have looked to the past for precursors of the trials and tribulations that afflict the modern age. In her best-selling book, *A Distant Mirror*, Barbara Tuchman (1980) found the fourteenth century to be 'a violent, tormented, bewildered, suffering and disintegrating age'. In view of the 'unusual discomfort' of our own times, it was reassuring to know that humans had lived through much worse. By comparing experiences over time, it may be possible to develop new approaches and secure greater objectivity in looking at human and environmental crises, whether in the past, present or future.

In English history, the man—land relationship has excited particular interest in the medieval period. In their recent study of rural society and economic changes in the years 1086—1348, Miller and Hatcher (1978) devote considerable attention to the consequences of the growth in population, land colonisation, and the competition for land and employment. There is ample evidence of low corn yields in that period. Not only may there have been soil exhaustion on a significant scale, but Harwood Long (1979) has suggested that productivity was further reduced by the farmer's inability to suppress the growth of weeds among his crops of corn.

For the historian, the major difficulty arises from the lack of documentary evidence for such hypotheses. As White (1962) remarked, 'the peasant has seldom been literate'. This has encouraged some historians to apply modern ecological theory and the results of experimental land-husbandry in order to analyse what little is known of the fortunes of agriculture in individual localities. According to Cooter (1978), an awareness of even some of the 'central concepts of modern ecological science' would make the historian even more pessimistic as to the levels of productivity in medieval agriculture. Except for self-limiting bursts, the peasant had little or no scope for raising output.

The inclusion of environmental themes in historical studies continues to become more evident. In his study of the American Indian as an ecologist, Jacobs (1980) wrote of how environmental history can act as 'a window to a clearer image of the past and can offer us unique perspectives on generally accepted historical concepts of unlimited growth, frontier expansion, and the rapid use of nonrenewable natural resources'. In another American publication *Ecology in Ancient Civilizations*, Hughes (1975) focuses on the relationship of the Greek, Roman, Jewish and early Christian civilizations to the Mediterranean environment. In a third volume, *Historical Ecology: Essays on Environment and Social Change* (Bilsky, 1980), contributors set out to analyse ecological crises in the ancient world, the European middle ages, and the more recent past. Clearly, there is much to gain from looking at even the well-worked themes of biblical Palestine, the fall of the Roman Empire, and the Black Death from a new environmental perspective, but a concentration on human ineptness in the name of 'historical ecology' runs the risk of introducing further false perspectives. Studies of the past produce precedents not only of human folly but also of man's greatest achievements, both in the environment and elsewhere.

REFERENCES

Ball, D.F. (1975). Processes of soil degradation. In J.G. Evans, S. Limbrey and H. Cleere (Eds), *The Effects of Man on the Landscape—The Highland Zone*, Council for British Archaeology, London, pp. 20—27.

Ball, D.F., Dale, J., Sheail, J., and Heal, O.W. (1982). *Vegetation Change in Upland Landscapes*, Institute of Terrestrial Ecology, Cambridge, 45 pp.

Ball, D.F. and Stevens, P.A. (1981). The role of 'ancient' woodland in conserving 'undisturbed' soils in Britain. *Biological Conservation*, **19**, 163—176.

Bayliss, J.H. (1980). The extinction of bubonic plague in Britain. *Endeavour*, 4, 58–66.
Bilsky, L.J. (Ed) (1980). *Historical Ecology: Essays on Environment and Social Change*, Kennikat Press, New York, 195 pp.
Bowen, M. (1980). Jackrabbit invasion of a Nevada agricultural community. *Ecumene*, 12, 6–16.
Cameron, R.A.D., Down, K., and Pannett, D.J. (1980). Historical and environmental influences on hedgerow snail faunas. *Biological Journal of the Linnean Society*, 13, 75–87.
Cooter, W. (1978). Ecological dimensions of medieval agrarian systems. *Agricultural History*, 52, 458–477.
Dimbleby, G.W. (1974). The legacy of prehistoric man. In A. Warren and F.B. Goldsmith (Eds) *Conservation in Practice*, Wiley, London, 279–289.
Dobson, M. (1980). 'Marsh-fever'—the geography of malaria in England. *Journal of Historical Geography*, 6, 357–389.
Ellenberg, H. (1979). Man's influence on tropical mountain ecosystems in South America. *Journal of Ecology*, 67, 401–416.
Enckell, P.H., Konigsson, E.S., and Konigsson, L.K. (1979). Ecological instability of a Roman Iron Age human community. *Oikos*, 33, 328–349.
Garbett, G.G. (1981). The elm decline: depletion of a resource. *New Phytologist*, 88, 573–585.
Godwin, H. (1975). *The History of the British Flora*, Cambridge University Press, Cambridge, pp. 243–247.
Godwin, H. (1978). *Fenland: Its Ancient Past and Uncertain Future*, Cambridge University Press, Cambridge, 196 pp.
Goode, D. (1981). The threat to wildlife habitats. *New Scientist*, 89, 219–223.
Harley, J.B. (1979). The Ordnance Survey and land-use mapping. *Historical Geography Research Series*, Geo Books, 2, 58 pp.
Harwood Long, W. (1979). The low yields of corn in medieval England. *Economic History Review*, 32, 459–469.
Hawksworth, D.L. (1974). Man's impact on the British flora and fauna. *Outlook on Agriculture*, 8, 23–28.
Hoskins, W.G. (1970). Editor's introduction. In C. Taylor (Ed), *Dorset*, Hodder and Stoughton, London, pp. 15–19.
Hoskins, W.G. (1975, reprint). Historical sources for hedge dating. In *Hedges and Local History*, edited for Standing Conference for Local History and the Botanical Society of the British Isles, NCSS/Bedford Square Press, London, pp. 14–19.
Hughes, J.D. (1975). *Ecology in Ancient Civilisations*, New Mexico University Press, Albuquerque, 181 pp.
Jacobs, W.R. (1980). Indians as ecologists. In C. Vecsey and R.W. Venables (Eds), *American Indian Environments: Ecological Issues in Native American History*, Syracuse University Press, Syracuse, New York State, pp. 46–64.
Jarvis, P.J. (1981). Aspects of the biogeography and ecology of introduced species. In N. Pinder (Ed), *Conservation and introduced species, Discussion Papers in Conservation 30*, University College, London, pp. 1–23.
Jones, E. (1979). The environment and the economy. In P. Burke (Ed), *The New Cambridge Modern History XIII. Companion Volume*, Cambridge University Press, Cambridge, pp. 15–42.
King, T.J. (1981). Ant-hills and grassland history. *Journal of Biogeography*, 8, 329–334.
Lane, C. (1980). The development of pastures and meadows during the sixteenth and seventeenth centuries. *Agricultural History Review*, 28, 18–30.
de Lotbiniere, H.G.J. (1928–29). Afforestation of Breckland. *Transactions of the Norfolk and Norwich, Naturalists' Society*, 12, 673–677.

McArthur, A.G. (1973). Plotting ecological change. In D. Dufty *et al.* (Eds), *Recreating the Past*, Hicks Smith, Sydney, pp. 27–48.

Meinig, D.W. (1979). Reading the landscape. In D.W. Meinig (Ed), *The Interpretation of Ordinary Landscapes*, Oxford University Press, Oxford, pp. 195–242.

Mellars, P. (1976). Fire ecology, animal populations and man: a study of some ecological relationships in prehistory. *Proceedings of the Prehistoric Society*, **42**, 15–45.

Miller, E. and Hatcher, J. (1978). *Medieval England*, Longman, London, pp. 27–63.

Moore, N.W. (1962). The heaths of Dorset and their conservation. *Journal of Ecology*, **50**, 369–391.

Newton, I., Davis, P.E., and Moss, D. (1981). Distribution and breeding of red kites in relation to land-use in Wales. *Journal of Applied Ecology*, **18**, 173–186.

Norden, J. (1610). *The Surveiors Dialogue*, Busby, London, pp. 163–167.

Oldfield, F. (1969). Pollen analysis and the history of land use. *Advancement of Science*, **25**, 298–320.

Ouren, T. (1980). The impact of the old shipyards on the invasion of alien plants to Norway. *Norsk Geografisk Tidsskrift*, **34**, 145–152.

Parry, M., Bruce, A., and Harkness, C. (1981). The plight of British moorland. *New Scientist*, **90**, 550–551.

Peterken, G.F. (1979). The use of records in woodland ecology. *Archives*, **14**, 81–87.

Piggott, S. (1981). Early prehistory. In S. Piggott (Ed), *The Agrarian History of England and Wales, Vol. I.I: Prehistory*, Cambridge University Press, Cambridge, pp. 3–23.

Pollard, E., Hooper, M.D., and Moore, N.W. (1974). *Hedges*, Collins, London, 256 pp.

Proctor, M.C.F., Spooner, G.M., and Spooner, M.F. (1980). Changes in Wistman's Wood, Dartmoor: photographic and other evidence. *Transactions of the Devon Association of the Advancement of Science*, **112**, 43–79.

Rackham, O. (1980). *Ancient Woodland: Its History, Vegetation and Uses in England*, Arnold, London, 402 pp.

Rose, C.I. and Hawksworth, D.L. (1981). Lichen recolonisation in London's cleaner air. *Nature (London)*, **289**, 289–292.

Sheail, J. (1978). Rabbits and agriculture in post-medieval England. *Journal of Historical Geography*, **4**, 343–355.

Sheail, J. (1979). British Rail land—biological survey. Interim report. *The history of railway formations, CST Report no. 276*, Nature Conservancy Council, Banbury.

Sheail, J. (1980). *Historical Ecology: The Documentary Evidence*, Institute of Terrestrial Ecology, Cambridge, 21 pp.

Spratt, D.A. and Simmons, I.G. (1976). Prehistoric activity and environment of the North York Moors. *Journal of Archaeological Science*, **3**, 193–210.

Taylor, C.C. (1980). The making of the English landscape. *Local Historian*, **14**, 195–201.

Tubbs, C.R. (1968). *The New Forest: An Ecological History*, David and Charles, Newton Abbot, 248 pp.

Tuchman, B.W. (1980). *A Distant Mirror*, Penguin, Harmondsworth, 677 pp.

Wells, T.C.E., Sheail, J., Ball, D.F., and Ward, L.K. (1976). Ecological studies on the Porton Ranges: relationships between vegetation, soils and land-use history. *Journal of Ecology*, **64**, 589–626.

White, L. (1962). *Medieval Technology and Social Change*, Oxford University Press, Oxford, p. 39.

Widgren, M. (1979). A simulation model of farming systems and land use in Sweden during the early Iron Age, *c.*500 BC–AD 550. *Journal of Historical Geography*, **5**, 21–32.

Conservation in Perspective
Edited by A. Warren and F.B. Goldsmith
© 1983 John Wiley & Sons Ltd.

CHAPTER 19

Values and Institutions in the History of British Nature Conservation

P.D. LOWE

Never before has there been so much contention about the purposes of nature conservation and the means of achieving them. What is becoming clear is that conservation is as much about difficult value judgements as it is about scientific facts. An historical perspective can help to clarify exactly what is at stake in contemporary issues. Many values now implicit in an activity or institution were once explicit. Much as one might peel away the layers of an onion, historical study can reveal the accretions of values and symbolic meanings that an activity has acquired over the years.

In search of these meanings I want to examine four historical periods in the evolution of nature conservation in Britain:

The natural history/humanitarian period	1830–90
The preservationist period	1870–1940
The scientific period	1910–70
The popular/political period	1960–present day

The dominant ideas within each of these periods were not discrete, but each had a distinct value orientation, which was superseded by and subsumed in the next.

THE NATURAL HISTORY/HUMANITARIAN PERIOD, 1830–90

As an organized movement of opinion, concern for wildlife conservation in Britain dates from the latter part of the nineteenth century, contemporary with similar movements in Europe and the United States (Conwentz, 1909). It was rooted in two other intellectual currents deriving from the end of the previous century: the strong enthusiasm for natural history; and the crusade against cruelty to animals. It was from the latter source that the initial impetus for wildlife protection arose. Cruelty to domestic animals was one of the major humanitarian preoccupations of the nineteenth century. The Society for the Prevention of Cruelty to Animals, founded in 1824 and given a royal charter in 1840, became the largest and perhaps the most influential voluntary organization in Britain in the second half of the nineteenth century (Harrison, 1973). The strength of its hold on the Victorian middle and upper classes arose, in part, from the growing recognition of the kinship of people and animals. From the late eighteenth century the concept of man as a supranatural being, both different from and superior to other living beings, was steadily eroded. Evidence of the biological affinity of human beings and the higher animals accumulated from studies in comparative anatomy, physiology, evolution and animal psychology. This new sense of animal kinship fostered empathy for the sufferings of beasts and inspired more considerate treatment of man's 'dumb relatives'.

It had an even more profound effect on the image Victorians had of themselves and their fellow human beings (Burrow, 1966). They became obsessed by the threat of human animality to the dignity and uniqueness of man and to the maintenance of morality and civilization (Pearsall, 1969; Turner, 1980). Cruelty to animals was so disturbing, not only because of what it did to the victims, but also because of what it implied about human nature. Conversely, kindness to animals seemed a sure refutation of man's bestial savagery. In rescuing animals from cruelty, animal protectionists could believe that they were helping to preserve the very fabric of society (Turner, 1980). As well as being the object of such high-mindedness, animals were made the receptacle of the gushing sentimentalism which was such a feature of the age.

Through the promotion of legislation, individual prosecutions and humane education, the RSPCA applied itself successively to the treatment of cattle and horses, the suppression of animal baiting and blood sports, the care of stray cats and dogs, the regulation of slaughterhouses and the evils of vivisection (Fairholme and Pain, 1924). It was only a matter of time before wild creatures attracted its zealous attention. The two ingredients necessary to elicit humanitarian indignation were conspicuous suffering by animals with which people could readily identify, associated with blatant human barbarity and callous disregard for the suffering that was inflicted. Not surprisingly, birds were the first form of wildlife to arouse concern (Turner, 1964).

The first successful campaign for bird protection was against the slaughter of seabirds by sportsmen during the breeding season, when it was easy to kill nesting gulls, guillemots and kittiwakes in very large numbers. Charles Waterton, whose Yorkshire estate was a sanctuary for birds, condemned one of these hunts in terms which could have been calculated to shock humanitarian sensibilities (Waterton, 1838):

> no profit attends the carnage; the poor unfortunate birds serve merely as marks to aim at, and they are generally left where they fall. Did these heartless gunmen reflect, but for one moment, how many innocent birds their shot destroys; how

many fall disabled on the waves, there to linger for hours, perhaps for days, in torture and in anguish; did they but consider how many helpless young ones will never see again their parents coming to the rock with food; they would, methinks, adopt some other plan to try their skill, or cheat the lingering hour.

A flurry of correspondence provoked a *Times* leader in October 1865 to bemoan the folly of slaughtering small birds. In 1867 the Reverend Francis Orpen Morris, vicar of a large rural Yorkshire parish and a leading ornithologist and advocate of animal protection, petitioned Parliament for protection of wild birds (Morris, 1897). He failed, but the next year his friend, the Reverend H.F. Barnes, vicar of Bridlington, convened a meeting of local clergy to discuss ways of stopping the annual hunts at nearby Flamborough Head. The outcome was the formation of the East Riding Association for the Protection of Sea Birds—possibly the first wildlife preservation society in the world. In 1869 the local MP was pressed to promote a Sea Birds Protection Bill, which, with the support of the RSPCA, was passed, thereby securing a close-time for sea birds during the breeding season.

Though the reform was prompted by humanitarian motives, the debate which surrounded it also raised the issue of the impact of shooting on bird populations. Much was made, for example, of the birds' role in preserving the balance of nature and in keeping down harmful insects. This new concern derived from scientific interest in the subject. In 1868, Alfred Newton, Professor of Comparative Anatomy at Cambridge and a leading member of the British Ornithologists' Union, warned the British Association that the continued slaughter of wild birds during the breeding season would lead to their extinction. The British Association responded by setting up a committee to consider the 'possibility of establishing a close-time for the protection of indigenous animals'. This committee along with the RSPCA and the Association for the Protection of British Birds (established in 1870 by Morris) promoted additional legislation in 1872 and 1876 to protect wildfowl, as well as the Wild Birds Protection Act of 1880 which consolidated the previous legislation and extended it to other species of birds (Russell, 1897).

By the 1890s only pets aroused more intense interest than wild birds among animal lovers. The issue which solidified the bird protection front was the plumage trade (Doughty, 1975). That the demands of fashion accounted for the annual destruction of thousands of egrets, herons and birds of paradise was a particularly sensitive point for the predominantly female movement against cruelty to animals. In the 1880s, a sharp increase in the import of tropical bird skins for the millinery trade aroused considerable concern. In 1885, the Plumage League and the Selborne League were both founded and, shortly after, joined forces as the Selborne Society. Another amalgamation (in 1891) between the Manchester-based Society for the Protection of Birds (formed in 1889) and the Fur, Fin and Feather Folk of Croydon (formed in 1889) produced

FIGURE 19.1. The story of the egret: sandwichmen employed by the Royal Society for the Protection of Birds to patrol London West End streets, July 1911. Reproduced with the permission of the RSPB.

what is now the Royal Society for the Protection of Birds (RSPB)—the Royal prefix was added in 1904 (Figure 19.1).

Both groupings sought to discourage the wearing of feathers (except for ostrich plumes, which are plucked almost painlessly from the live birds) through propaganda and moral pressure. Significantly, they also had broader aims, which is fortunate as it was not until 1921 that the Importation of Plumage (Prohibition) Act was passed, banning all plumage imports, except of the African ostrich and the eider duck. Besides opposition to the plumage fashion, the RSPB had as its objective the general protection of wild birds, whereas the Selborne Society was committed 'to preserve from unnecessary destruction such wild birds, animals, and plants, as are harmless, beautiful, or rare . . . and to protect places and objects of interest or natural beauty from ill-treatment or destruction'. It should be added that one of the reasons that legislation took so long was that the two societies became publicly opposed over the matter. The RSPB was much more militant in its opposition to the plumage trade than was the Selborne Society which preferred public persuasion and education. In 1913, after the failure of a number of Plumage Bills through staunch opposition from the millinery trade, the Selborne Society joined with representatives of the feather merchants to promote a study of the status of the exploited species and to devise means by which the trade, through self-regulation, could avoid endangering any species and thereby obviate legal prohibition. This infuriated the RSPB which had just succeeded in persuading the government to introduce its own Plumage Bill to ban the trade. The rift effectively sealed the Bill's fate and delayed the enactment of such a measure until after the First World War (Bensusan, 1913; Hornaday and Lemon, 1914).

The other wellspring of wildlife concern in the second half of the nineteenth century was the strong enthusiasm for natural history. Most Victorian naturalists were specifically motivated not by a wish to protect nature, but by a desire to collect and study it. Nevertheless from their observations came the clear evidence of a declining flora and fauna, and from their ranks came the first people to voice concern. The enormous popularity of natural history had a number of causes (Allen, 1976; Barber, 1980). It was a byproduct of the new prosperity of industrial Britain, of expanded opportunities for education and leisure. It provided an outlet for the contemporary obsession with travel and self-improvement. For devout Victorians, it was one of a restricted range of morally accepted pastimes (Figure 19.2). In the words of the President of the Cotteswold Naturalists' Club, 'We seek a healthy, a most fascinating, far more than either, a most holy study. For what is the study of Natural History, but an approach to the Creator through his works' (Lloyd Baker, 1849). Nature was a revelation of God's order and purpose, of His Design—the so-called 'open book of Nature'. To study it, therefore, was itself a devout act, as expressed in the recurring phrase 'through Nature up to Nature's God'.

As a scientific subject, natural history was passing through a period of great intellectual excitement, yet still remained accessible to the educated layman. From the 1830s onwards, a flood of natural history books and cheap manuals appeared, reflecting and stimulating the growing popularity of the subject. Natural history societies and field clubs proliferated. By the 1880s, there were several hundred of them across the country with a combined membership of around 100 000 (Lowe, 1978). Thus, when the time came, here was a large public susceptible to an appeal to the intrinsic significance of wildlife. Indeed, as groups such as the Commons Preservation Society, the Selborne Society, the Society for the Protection of Birds and the National Trust appeared they were able to mobilize this extensive latent concern. Many field clubs and natural history societies began to assume the role of local environmental watchdogs for the preservation societies, which thereby were able to establish themselves quickly as national pressure groups.

Even so, Victorian naturalists occupy an ambivalent position in the annals of conservation. They were avid collectors. There seemed nothing especially culpable about collecting, as long as it was still believed that 'The vast domain of nature can never be fully explored, her attractive resources being infinite and inexhaustible' (Sim, 1864). Yet as the number of naturalists increased and their equipment improved, collecting began to take its toll, particularly on many

FIGURE 19.2.   Victorian naturalists on a field outing: the Belfast Naturalists' Field Club at the Giants Causeway, 1868. Reproduced with the permission of the Ulster Museum.

rarer species, although this seldom arose from wanton destructiveness. In part, the collecting zeal is intrinsic to natural history. Specimens have to be taken for the purposes of identification, as well as to bear witness to a new record. In part, also, the collecting instinct of Victorian naturalists reflected the acquisitiveness of their society. It was this which drove natural history collecting to excess—as people competed to build up bigger and bigger private collections, as exchange clubs were established to barter rare and critical specimens, and as the professional collector arrived on the scene to fulfil the rising commercial demand for natural curiosities.

Implicit in natural history, however, is a regard for the objects studied and an interest in preserving them, if only for the purpose of study. One of the first local field clubs, the Tyneside Naturalists' Field Club, set up in 1846, included the following amongst its original rules: that the Club should 'discourage the practice of removing rare plants from those localities of which they may be characteristic' and 'the extermination of rare or interesting birds'; that members should 'use their influence with landowners and others for the protection of the characteristic birds of the country'; and that they should not seek to enrich their private collections 'at the expense of nature's great museum out of doors'. Though such admirable regulations were not copied by any other society until the 1870s, the period from the 1840s onwards saw growing disquiet amongst naturalists about the excesses of collecting (Allen, 1980). The first group to become generally aware of the dangers were botanists. They were shocked out of their complacency by the extreme depredations of the fern craze of the 1850s and 1860s, when whole districts were stripped of their ferns by commercial collectors to grace fashionable drawingrooms (Allen, 1969).

Perhaps the most important contribution of Victorian naturalists to the cause of wildlife conservation was the accumulation of detailed botanical and zoological records, county by county, which provided a benchmark against which any loss could be gauged. Thus, Sir William Jardine, Scottish landowner, ornithologist and the great impresario of early Victorian natural history, commended local field clubs in the following terms (Jardine, 1858):

These Clubs are of much importance. The preservation of the condition of the present physical characters of our country will be far more dependent on them than at first appears. The last fifty years have made a great change in the surface of the country; population has increased; so have agricultural improvement, plantations, drainage, enclosure of waste lands, in short artificial works of every kind. These have often completely altered the nature and aspect of the country, and in consequence the productions, both animal and vegetable . . . It will be to these Clubs that we shall be indebted for a record of what in their days did exist . . . there is nothing that should prevent an active Club to fill up in a few years a list of the productions within their beat, and so lead into a complete and accurate Fauna and Flora of our own time and age; and generations succeeding would be able not only to mark the changes of the productions, but to judge and reason upon the effects which these now so-called improvements have produced on the climate and soil, and the fertility and increase of the latter.

THE PRESERVATIONIST PERIOD, 1870–1940

The late nineteenth and early twentieth century saw the formation of a spate of societies devoted to preserving open land and its associated wildlife:

Commons Preservation Society	1865
Selborne Society for the Protection of Birds, Plants and Pleasant Places	1885
Society for the Protection of Birds	1889
National Trust for Places of Historic Interest and Natural Beauty	1894
Society for the Preservation of the Wild Fauna of the Empire	1903
British Empire Naturalists' Association	1905
Society for the Promotion of Nature Reserves	1912
Council for the Preservation of Rural England	1926

A crucial factor in this movement was a gathering sense of the vulnerability and loss of wildlife, demanding urgent countermeasures. Acceptance of Darwinian evolutionary theory and evidence from the fossil record of extinct animals laid to rest the belief in a natural plenitude. *The Origin of Species* presented a different vision—of an uncaring nature, mindlessly murderous, with death and wastage on an enormous scale, whole species formed and then blindly squandered (Fleming, 1961; Houghton, 1957). Equally disturbing was a growing sense of man's own destructive power. A number of species had already been extinguished by over-hunting and human persecution—including the blauwbok, the quagga, the sea mink, the dodo, the Antarctic wolf, the great auk, the marno, the moa and the passenger pigeon. Many more were threatened, especially the large mammals of Africa, the fur-bearing animals of North America and Siberia, and the colourful birds of the tropics. In Britain, several bird species had become locally extinct, including the osprey, the avocet, the black-tailed godwit, the bustard and the capercaillie.

It was not simply a matter of the reckless greed of the hunter or the wanton destruction of the sportsman. The causes were more complex and pervasive. The avocet and the black-tailed godwit, for example, were lost during the first half of the nineteenth century, largely due to drainage of their breeding haunts in eastern England. In the words of Charles Rothschild, the founder of the Society for the Promotion of Nature Reserves, 'civilization and progress are gradually but surely forcing back and destroying the beauties of wild life' (Rothschild, 1914). Evidence that many plant species were becoming locally rare or extinct induced the Selborne Society to set up a Plant Protection Section whose recorder gave the following causes for the decline or extermination of wild plants (Horwood, 1913):

Smoke; atmospheric abnormalities; drainage; cutting down of woods; desiccation; drought; cultivation; building operations; sport; hawking and collecting; professional collecting; nature-study operations.

The once-stalwart objects of nature now seemed vulnerable and fragile. This was reflected in the changing concept of the balance of nature (Egerton, 1973). In the eighteenth century, the term had implied a robust, preordained system of checks and balances which ensured permanency and continuity in nature. By the end of the nineteenth century it conveyed the notion of a delicate and intricate equilibrium, easily disrupted and highly sensitive to human interference. Thus, according to Raphael Meldola, an organic chemist and leading amateur entomologist who led the campaign to retain Epping Forest as a 'biological preserve' (Meldola, 1883): 'The workings of nature are connected and bound up in such endless and unsuspected ways that any interference on the part of man may unknowingly upset the adjustments that have taken ages for their perfection' (Meldola, 1880). Sir Ray Lankester, the eminent zoologist, flatly declared, 'Once man is present in the neighbourhood, even at long distance, he upsets the "balance of nature"' (Lankester, 1915).

Though Darwin had placed man firmly in nature, the Industrial Revolution had irrevocably broken man's bondage to the natural world. Human beings no longer lived in a primitive state. Freed from nature's control, they now saw themselves running amok, reeking havoc and destruction with a vengeance. The relentless advance of civilization seemed inimical to other living things, and a new image began to emerge of man as a cosmic freak in the evolutionary process. Such awesome power as man increasingly possessed over nature needed to be exercised with proper restraint and care and in full consciousness of the terrible responsibility that man now bore for the fate of the natural world. As Peter Chalmers Mitchell, the Secretary of the Zoological Society, declared in a strong appeal to the British Association for the organized preservation of the world's fauna, 'Each generation is the guardian of the existing resources of the world; it has come into a great inheritance, but only as a trustee' (Mitchell, 1912).

As we have seen, concern at the loss of valued natural features was sharpened by a growing sense of their rarity and vulnerability. Yet this still does not explain why changes which in the past had been considered generally advantageous, now aroused passionate opposition, at least amongst an influential minority of upper-class and intellectual people. What was different was a new evaluation of the features being obliterated and a new orientation towards the forces and motives which wrought these changes. In particular there was a reversal of the rationalist, progressivist outlook deriving from the Enlightenment which, with its confidence in the perfectibility of all things, had looked always to the improvement of nature and society through the exercise of human reason. Victorian and Edwardian preservationists rejected the imperative to improve— whether it be the enclosure of common wastes for agricultural or building purposes, the control of 'vermin' on country estates, the clearance of wood-land, or the drainage of old meadows and marshes. On the contrary, they saw in these acts deformation and vandalism.

It seems that preservationist concern was an integral part of the late Victorian intellectual reaction to many of the tenets of economic liberalism. It is no coincidence, for example, that some of the social philosophers and writers in the vanguard of this reaction, such as John Stuart Mill, John Ruskin, Lord Avebury and William Henry Hudson were founder members of preservationist groups. This profound shift of opinion arose from a reassessment of the social and economic changes of the nineteenth century. The optimism and belief in boundless prosperity that had characterized mid-century Britain was replaced by pessimism about the prospects for social and economic advance. The Victorians' earlier self-confidence was sapped by the Great Depression of the 1880s, and by the intellectual crisis of the post-Darwinian years which cast doubt on the nature of the human condition and the possibility of its improvement (Burrow, 1966). Britain's increasingly disappointing industrial performance in the final decades of the century was matched by a growing equivocation towards industrialism itself; the source of the nation's economic and political power was coming to be seen as destructive of the moral and social order, human health, traditional values, the physical environment and natural beauty. This growing antipathy to the industrial spirit in late nineteenth-century Britain reflected the absorption of the urban bourgeoisie into the upper reaches of British society and its genteel value system—a value system which disdained trade and industry, which stressed the civilized enjoyment, rather than the accumulation of wealth, and which preferred social stability to enterprise (Wiener, 1981).

Nature offered an alternative value system to that of work and trade, one which expressed social pessimism and other-worldliness. 'It is always pleasant and often wise,' advised the President of the Essex Field Club, 'to turn aside for awhile from the busy and relentless human world, with its ceaseless anxieties, worries, sorrows, labours and cares to contemplate the silent and wonderful economy of that other world of nature, with which we are so closely related, and on which we are so dependent' (Fitch, 1889). Charles Kingsley (1880) informed the young clerks and apprentices who comprised the Chester Society of Natural Science that he had known many men 'who in the midst of smoky cities have kept their bodies, their minds and their hearts healthy and pure by going out into the country at odd hours', to contemplate nature. Shaw-Lefevre, the founder of the Commons Preservation Society wrote in the following terms of the importance of preserving commons on the edge of expanding cities (Shaw-Lefevre, 1894):

> they form . . . oases of nature, in striking contrast to their surroundings . . . They are natural parks, over which everyone can roam freely . . . They are reservoirs of fresh air and health, whence fresh breezes blow into the adjoining town. They bring home to the poorest something of the sense and beauty of nature.

Common to all the preservation groups of the period was a moral and aesthetic revulsion to the contemporary industrial city. They hoped to preserve things and places that had not yet been corrupted by urban and industrial expansion. Their task became more widespread as it became more urgent, because the city seemed to be bursting its bounds. In the words of a *Times* leader (18 December 1912), welcoming the formation of the Society for the Promotion of Nature Reserves:

> During the last century, while the great towns enormously increased, they focused within themselves the forces of national growth, and there was no change at all comparable (elsewhere). Now not only does the tide of building and industrial development still steadily move forward, but it promises to press far more strongly than before on country areas.

Previous equilibria, for example, between change and continuity, or between the city and the country, were thrown out of balance. The forces of urbanism and industrialism seemed ubiquitous, sweeping all before them.

As one foreign observer commented (Conwentz, 1909):

> Anyone travelling in England . . . may see for himself that the constant cultivation of the land and the growth of industrial undertakings have threatened, and in many places considerably damaged, interesting tracts of country as well as natural monuments. From the economic aspect . . . it would even be justifiable . . . if man were to bring under his control almost the whole realm of nature. But, on the other hand, from the scientific and aesthetic standpoints, it is much to be regretted that so many types of scenery and of the vegetable and animal worlds should pass away irrevocably.

'The replacement of the natural by the artificial proceeds apace', complained the ecologically minded geologist C.B. Crampton (1913), 'one may now travel miles and fail to detect a single spot left as virgin ground . . . Surroundings where the naturalist feels confidence in a truly aboriginal interrelation of the various forms of life, such as results from centuries of unrestricted competition and selection, are . . . becoming scarcer every year.'

Tangible relics of the natural world had a symbolic function. They provided a truthful record, however incomplete, of what 'progress' had destroyed. Nature, untamed and primitive, also stood in stark contrast to the rigid and stultifying social conventions of Victorian society. This was why it was important not to 'reduce the wilderness to vulgar conventionality' (Lankester, 1915). Otherwise, the progress of civilization threatened a bland, ersatz uniformity, impoverishing the sources of human experience. Sir H.H. Johnston, Governor of Uganda and founder of the Society for the Preservation of the Wild Fauna of the Empire, warning of excessive game hunting, commented 'The world will become very uninteresting if man and his few domestic animals, together with the rat, mouse, and sparrow are its only inhabitants' (Johnston, 1906). Similarly, Sir

Ray Lankester (1915) suggested 'we shall . . . be blessed by future generations of men for having saved something of Britain's ancient nature, when all else, which is not city, will have become manure, shooting greens, and pleasure gardens'.

In addition, the natural vegetation and the associated animal life of an area contributed to its distinct character and thereby provided a visible guarantee of historical identity, to be preserved from the arbitrary standardization that industrialism seemed to threaten. With the gathering tide of nationalism at the end of the nineteenth century, this last factor became increasingly important. As the golden eagle came close to being exterminated, the ornithologist, Henry Seebohm (1897), appealed 'Before it is too late, Scotchmen, protect your national bird, the eagle of your ancestors'. Similarly, Octavia Hill (1899), said of the property of the National Trust that it was 'a bit of England belonging to the English in a very special way'. The Society for the Promotion of Nature Reserves—established in the jingoistic atmosphere prior to the First World War —had as its objective 'to preserve for posterity as a national possession some part of our native land, its fauna, flora and geological features'. Playing on Anglo-German rivalry, these early conservationists repeatedly pointed to the achievements of the German state in preserving its natural heritage (Horwood, 1913; Rothschild, 1914, for example). As the Prussian State Commissioner for the Care of Natural Monuments reminded the British Association, 'the care of natural monuments is not only of scientific and public interest, but it also possesses a patriotic value; for, by these undertakings, parts of the country at home become better known and more fully appreciated. In this way it is that true patriotism—the love of one's homeland—is increasingly promoted' (Conwentz, 1909).

THE SCIENTIFIC PERIOD, 1910–70

Systematic ecological research began in Britain around the turn of the century, and in 1913 the British Ecological Society was formed. Eventually, ecologists were to have a profound impact on the conservation movement, though initially their influence was modest. Ecologists soon appreciated the importance of nature reserves for their research, as ecology moved from its initial concern with extensive phytogeographical surveys to intensive studies of the dynamics of ecosystems, often involving detailed and protracted field observation and measurement. When the continuity of Francis Oliver's investigations of Blakeney Point was threatened by the death of the owner, an anonymous donation from Charles Rothschild helped secure it for the National Trust. Oliver, who was Professor of Botany at University College London, was coopted onto the Council of the Society for the Promotion of Nature Reserves (SPNR) in 1913, as was his ex-assistant Arthur Tansley, the first President of the British Ecological Society (BES). In a talk to the Society, Oliver drew attention to the absence of a 'considered policy' for the establishment of reserves and urged that 'it (is) important for ecologists to consider whether nature reserves should be formed, and if so, how the formation of such areas (is) to be promoted' (British Ecological Society, 1914). By 1918, the SPNR was summarizing the purpose of reserves as 'for the enjoyment of lovers of wild nature, the pursuit of scientific knowledge, and the well-being of the community in general' (quoted in Sheail, 1976, page 62).

English ecologists did not necessarily share the preservationist convictions of the SPNR's nature lovers. Indeed, in attempting to win government backing for ecology, they stressed its potential contribution to the economic exploitation of marginal land, such as the afforestation of uplands or the reclamation of marsh and heathlands—just the sort of land-use changes which were anathema to the early conservationists.

The interest of ecologists in the causal relationship between habitats and plant and animal communities brought an appreciation of the need to protect habitats to preserve species. Another inference, which slowly dawned, was that habitats could be manipulated and controlled to maintain particular species. Nevertheless the notion that nature might need to be managed was quite foreign to the wildlife preservationists. As the *Times* (18 December 1912) proclaimed, 'The only effective method of protecting nature is to interfere with it as little as possible'. Thus there were divergent views on the function of nature reserves. Rothschild (1914) defined a reserve as 'an area, perhaps small, perhaps large, possibly only a single tree, which is specifically kept in its wild state'. Its function was to safeguard a valued place from all human interference except quiet contemplation. By comparison, the physiologically minded ecologist, W.B. Crump (1913) evidently contemplated extensive interference when characterizing a reserve as 'an outdoor workshop for the study of plants and animals in and in relation to their natural habitats; a twentieth century instrument of research as indispensable for biological progress as a laboratory or an experimental station'.

Finally, the two groups attached quite different significance to the pheno-menon of rarity. Many early conservationists were old-fashioned naturalists seeking a new outlet for their collecting and labelling instincts. Nature reserves were the rich man's answer to the odium into which natural history collecting was falling. If it was no longer socially acceptable to collect specimens of rare species, then why not collect the sites on which they still survived? In contrast, ecology had arisen as a discipline in conscious reaction to the obsession of systematics with novelty and rarity (Lowe, 1976). Ecologists were concerned with the typical not the novel, with the common species not the rare. Many of them regarded the activities and mentality of collectors with open distaste.

The interwar years were a hiatus for ecology in Britain. There were no posts specifically for ecologists either in the universities or in government. The maintenance of the discipline depended largely upon the personal commitment of individual academics. Gradually, courses in ecology were introduced into university botany degrees, but without specific posts there was no incentive to train in ecology; indeed, Arthur Tansley (1939) actively discouraged students from specializing in the subject.

During this period, conservation was also in the doldrums. Charles Rothschild's death in 1923 was a severe financial and inspirational loss to the SPNR (Rothschild, 1979). Agricultural depression did remove some of the threats to wildlife, although the countryside faced growing pressures from urban sprawl. Popular concern for rural preservation flowed into the amenity movement, while public interest in natural history reached its lowest ebb for many years.

Somewhat unexpectedly, the start of the Second World War created a more favourable context for conservation. Preparations for postwar reconstruction provided unprecedented opportunities to influence the formulation of government policy. In 1941, the SPNR convened a conference on the preservation of wildlife after the war. The conference's recommendations, including proposals for nature reserves, received favourable publicity and aroused attention in official circles. Within government, there was already a growing commitment to a comprehensive land-use policy. Official interest in wildlife protection followed as a corollary of the general responsibility, embraced at least in principle, for preserving the British countryside (Cherry, 1975). In 1942 at the invitation of Sir William Jowitt, Paymaster-General and Chairman of the Committee on Reconstruction Problems, a Nature Reserves Investigation Committee (NRIC) was appointed under the auspices of the SPNR to develop the case for nature conservation and draw up a list of proposed reserves. There followed a period of intensive investigation and lobbying which resulted eventually in the creation of the Nature Conservancy in 1949 (Sheail, 1976).

During this period, ecologists gradually assumed the leadership of the conservation movement. They were anxious to institutionalize their discipline; and nature conservation seemed a potential vehicle for achieving government recognition. Individual ecologists such as Edward Salisbury had maintained a personal interest in nature conservation, but ecologists as a group now began to commit their professional fortunes to the cause (Duff and Lowe, 1981). The British Ecological Society established its own committee in 1943, under Tansley's chairmanship, to investigate the need for nature reserves and nature conservation. The committee, in its report, reasoned that the formulation and implementation of nature conservation policies must be based on sound ecological advice. Therefore, it recommended the formation of an Ecological Research Council which would establish a national system of *habitat* reserves, but would also undertake biological surveys and conduct fundamental ecological research.

These and other suggestions were considered by two official committees: one for England and Wales, under the chairmanship initially of Julian Huxley with Tansley as his deputy, which reported in 1947; and the other for Scotland, under James Ritchie, which reported in 1949. These committees drew up lists of proposed nature reserves. They stressed the importance of selecting and managing reserves on scientific principles and recommended for the purpose the setting up of an official 'Biological Service'. The government acted on their recommendations by establishing, in 1949, the Nature Conservancy as a chartered research council. This new government agency combined the functions of conducting and sponsoring ecological research, of giving advice and information on nature conservation, and of acquiring and managing nature reserves. Justifications for conservation swung from the moral-aesthetic standpoint of earlier in the century to arguments of public benefit through amenity and scientific study. In the words of the British Ecological Society Committee (1944), 'the ecological interest almost exactly coincides with the aim of preserving the characteristic charm of British scenery'. It became a convenient belief that nature conservation could and should provide opportunities for both scientific study and the enjoyment of natural surroundings.

Ecological concepts and terminology were assimilated into conservation strategies and in the process, 'science for conservation' was subtly transmuted into 'conservation for science' (Adams and Lowe, 1981). The NRIC claimed to have used 'primarily scientific criteria' in drawing up its list of proposed nature reserves; sites were included whose destruction 'would be a serious loss to science' (Nature Reserves Investigation Committee, 1945). Similarly, the Huxley Committee argued that reserves should be chosen for their scientific value; and that the selection of sites should 'provide a foundation for a sound ecological study of wildlife conditions' (Wildlife Conservation Special Committee, 1947). For its part, the British Ecological Society took exception to the notion of reserves for species that had been favoured by the wildlife preservationists. It argued:

> Rare species, as such, are not [the ecologist's] primary interest . . . For this reason ecologists could not possibly be content with limiting reserves to 'sanctuaries' for the preservation of rare species . . . The stress is therefore to be laid on preserving the community: if the samples preserved contain rare species their interest is increased, but the choice of reserves cannot be limited on that ground (British Ecological Society, 1944).

Cyril Diver, who was to become the first director-general of the Nature Conservancy, pursued this point within the NRIC. His reasoning is interesting: 'Every real naturalist likes to see rare species, and to secure their continued existence if possible, but the fundamental work of science . . . is carried out on the commoner forms'. This consideration, he argued, should dictate the selection of reserves though he fully recognized that 'rare species have a wide emotional appeal, and there is no reason why we should voluntarily forego the benefits of this fact' (quoted in Sheail, 1976).

In the postwar period, British ecologists were able to exploit the Conservancy's unique institutional structure to establish their profession. The Conservancy's combination of research and practical conservation facilitated the application of ecological knowledge and techniques, and the development of applied ecology. The fusion of pure and applied science enabled ecology to break out, much more quickly than other disciplines, from the ivory tower mentality that had characterized academic biology before 1939. Scientific criteria and justifications inevitably dominated the practice of conservation. Central to the Conservancy's philosophy was a conception of nature conservation as applied ecology, and it set about the task of developing the scientific management of wildlife along ecological lines.

The Nature Conservancy had to create a conservation profession *sui generis*. Before its establishment there were very few trained ecologists and their background was university research not practical conservation. As part of its support for academic research, the Conservancy began awarding studentships for postgraduate training in ecology—initially about 10 to 12 per annum. This was the first systematic provision for the training of ecologists in Britain.

The expansion of the Nature Conservancy provided posts for some—its scientific personnel rose from nine in 1950 to 61 in 1958, and twelve of the latter were former holders of studentships. Later, the Conservancy encouraged the development of postgraduate training in conservation, the first such course being that established at University College London in 1960. By the mid-1960s it was estimated that the output of postgraduates from the new courses was running at about 25 per annum (Lambert, 1967). Thus conservation had emerged as a profession (Figure 19.3).

THE POPULAR/POLITICAL PERIOD, 1960–PRESENT DAY

By the end of the 1950s, with its primary tasks of research and reserve acquisition in hand, the Conservancy was able to turn its attention outwards, to develop its advisory, information and educational services, and its public relations. Throughout the 1960s, it spearheaded a concerted and highly successful campaign to arouse greater national commitment to conservation. Having established its expertise in applying ecology to the maintenance of nature reserves, the Conservancy was well placed to foster the extension of ecological knowledge and conservation techniques to other types of land and resource use. A major vehicle for this was the 'Countryside in 1970' Conferences, held in 1963, 1965 and 1970, under the presidency of the Duke of Edinburgh. The Conferences involved the leaders of nearly all national environmental groups, representatives of farming, forestry and landowning interests, and key industrialists and government officials. A major theme was that, with the industrialization of agriculture and the increasing recreational use of the countryside, measures to conserve wildlife populations could no longer be confined to nature reserves.

In seeking to build a strong constituency for official conservation policies, Max Nicholson, the Conservancy's director-general between 1952 and 1966, became increasingly impatient at being unable to count on the active assistance of the voluntary wildlife organizations, some of which were slow to grasp the changed context and new opportunities presented by the existence of the

FIGURE 19.3. Science and professionalisation: postgraduates of University College London, on Britain's first conservation course, attending a field trip at the Old Lifeboat House, Blakeney Point Nature Reserve, 1960.

Conservancy. Some years later he described the state of the conservation movement during the 1950s as 'one of low morale, weak leadership, elderly and largely passenger memberships, feeble finances and ignorance of the mounting threats to the biosphere' (Nicholson, 1976). The Conservancy sought to reform the voluntary conservation movement as an effective political lobby. In 1958, with encouragement from the Conservancy, the Council for Nature was established to represent all wildlife and natural history organizations in public affairs and to publicize and win support for nature conservation. The early initiatives of the new Council included the establishment of the Conservation Corps (now the British Trust for Conservation Volunteers) and an Intelligence Unit. The latter, which was supported by a grant from the BBC, did much to publicize nature conservation issues, encourage natural history film-making, and cement the relationship between conservationists and the small but growing band of wildlife programme makers (the BBC natural history unit had been set up in 1957— Parsons, 1982). Through television, wildlife has become a central feature of popular culture in Britain.

In the late 1950s, a new source of initiative in nature conservation, arising from the local level, began to expand rapidly and this also was encouraged by the Conservancy. The county trusts for nature conservation which now cover the whole of the United Kingdom emerged as a coordinated movement then, although three had been formed earlier: the Norfolk Naturalists' Trust in 1926, the Yorkshire Trust in 1946 and the Lincolnshire Trust in 1948. Their objective was to acquire and manage nature reserves and arouse local interest in nature conservation. A.E. Smith, the energetic secretary of the Lincolnshire Trust, and Christopher Cadbury, a longstanding Council member of the Norfolk Trust, perceived the potential of local activity and support for conservation, and together they initiated a drive for a national network of county trusts. This led to the setting up of the Leicestershire Trust in 1955, followed by the Cambridgeshire Trust in 1956. The movement quickly gathered momentum—by 1964 there were 36 county trusts in existence, with a combined membership of 17 700 (see Perring, Chapter 24).

The existing wildlife groups were not left untouched by the growing interest in nature conservation. The Society for the Promotion of Nature Reserves found a new role as the national coordinating body for the county trusts (Lowe and Goyder, 1983, Chapter 9). In 1958, the RSPB started a national press advertising campaign to boost its membership which, at 8000, had remained stagnant since the end of the war, but which attained the undreamt of figure of 29 000 by the end of 1965. The elitism of preserving wild plants and animals for the sole amusement of naturalists had to give way to a new emphasis on presenting and interpreting nature to an eager public (Figure 19.4).

Popular interest in conservation and media attention gathered momentum through the 1960s (Brookes *et al.*, 1976). Public concern was heightened by a number of spectacular pollution incidents. For instance, heavy spring mortalities

FIGURE 19.4. Open day, 1981, at Old Warden Tunnel, a Nature Reserve run by the Bedfordshire and Huntingdonshire Naturalists' Trust. Reproduced with the permission of Mr. T.J. Thomas.

of birds and mammals occurred in 1959, 1960 and 1961 where seed corn had been dressed with organochloride pesticides (Moore, Chapter 10). Prompted by the RSPB and the British Trust for Ornithology, the Nature Conservancy took up the issue and, in negotiations with the manufacturers of pesticides, agreed on voluntary restrictions on their use. A programme of research on the effects of toxic chemicals on wildlife was initiated at the Conservancy's new experimental station at Monks Wood, near Huntingdon, and the results led to further controls. Other disasters which also projected conservationists into the news included major oil spillages, such as the Torrey Canyon incident of 1967, and the Irish Sea bird wreck of 1969 (probably caused by pollution from polychlorinated biphenyls).

Something of the same dramatic media impact was achieved by alarmist warnings of *impending* disasters. Mancur Olson (1973) coined the term 'the profits of doom' to characterize the undoubted achievements of this particular brand of journalism, which became so prevalent in the early 1970s. Amidst a gathering sense of an enveloping environmental crisis with a number of pundits predicting imminent ecological collapse, many environmental groups sprang up to question the direction that society was taking, the foremost being Friends of the Earth and the Conservation Society. These new affiliations were much more radical and international in their outlook and more explicitly political than the established conservation and preservation groups. They tended to view individual environmental problems as interrelated, identifying the profligate use of the earth's non-renewable resources as the most serious problem, and economic and population growth as the fundamental cause of the crisis.

The preservationist concern of the turn-of-the-century had returned with a vengeance. Previously people had been alarmed at the prospect that human profligacy and industrial power might eliminate the natural world. Now it seemed that these forces even threatened human survival. Anxiety over this prospect temporarily overshadowed the traditional concerns of conservationists, though eventually it gave them a new significance. The conservation of wildlife came to be seen as an integral part of the task of conserving the human habitat and husbanding the world's natural resources. The penalties of neglect were not only a loss of wild animals and plants but could hazard man's future.

Thrust to the centre of national and international politics, conservation could not remain a scientific preserve. In 1973, as part of the reorganization of civil science, following the proposals of Lord Rothschild, the research and conservation functions of the Nature Conservancy were separated. A new statutory body, the Nature Conservancy Council, was established, responsible for nature conservation directly to the Secretary of State for the Environment (a post which was itself a political response to the contemporary environmental concern). The research stations that had formed part of the Conservancy were reconstituted as the Institute of Terrestrial Ecology, remaining within the Natural Environment Research Council (NERC)—which on its formation in 1965 had been made responsible for the Nature Conservancy.

Though at the time the split caused considerable contention, it served the interests of both the ecologists and the conservationists. For the former, conservation had served its purpose as a vehicle to professionalization. Ecology had become a well-established discipline with secure sources of support. The future expansion of its role in British society depended upon the exploitation of new outlets for ecologists' skills and research findings, apart from nature conservation, for example in such fields as agriculture, fisheries, forestry, pollution control, water management and land use planning. In this endeavour, they were finding themselves hampered by being stereotyped, through too exclusive an identification in the public mind with a specific set of conservationist attitudes. Ecologists became particularly concerned for the integrity and credibility of their professional expertise as various pundits, pressure groups and even a political party (the Ecology Party, established in 1973) claimed the authority of their discipline to support radical pronouncements and social programmes. For their part, the reorganization of the Nature Conservancy gave conservationists direct access to the centre of political power. This promised more resources and greater political leverage in confronting the development-oriented interests represented by other major government departments.

The moves threatened to leave nature conservation ideologically naked—deprived of the 'smokescreen' of scientific justifications it had previously enjoyed (Green, 1975). On the other hand conservationists were freed to exploit the diverse social support for conservation that had emerged from the developments of the 1960s. Slowly, some have groped towards a new philosophy and a new set of justifications for the conservation of wildlife, emphasizing its role as an essential component of the relationship between man and environment (Mabey, 1980).

Popular interest in conservation had continued to grow through the 1970s. By the end of the decade, the combined membership of wildlife conservation groups was several hundred thousand, compared with just a few thousand in the 1950s. The conservation lobby has always been articulate and influential. The addition of such numerical support has made it a powerful political force, capable of challenging major economic interests. Conservation has become a matter of growing political contention in the face of continued technological threats to wildlife, particularly through agricultural intensification. By the early 1980s, in the debate surrounding the Wildlife and Countryside Bill (enacted in 1981), conservation had become a matter of party politics, with both the Labour and Liberal Parties dissenting from the stance of the Conservative Government that constraints on agricultural change were unnecessary or inappropriate to safeguard wildlife. It is clear that this issue will remain contentious, and lead to deepening divisions between the parties over conservation and even greater political prominence for the cause.

CONCLUSIONS

Each of the four periods discussed above has contributed certain values to the conservation movement. From the Victorian period comes a passion for collecting, a taste for rarities, a humanitarian regard for man's fellow creatures, and a fundamental reverence for nature. From the turn-of-the-century period comes an aesthetic and spiritual identity with the wild, strong anti-urban and anti-industrial sentiments, and a sense of stewardship, associated on the one hand with an appreciation of the web of life and its fragile balance, and on the other hand with a patriotic attachment to the indigenous flora and fauna. From the period of the postwar years comes the potent values of professionalism, including a managerial ethic keen to regulate habitats and species, and a scientific understanding of ecological relationships. Finally, in recent years, wildlife has become an aspect of popular culture, and the protection of wildlife, as part of a general effort to conserve the human environment, is now a central feature of domestic and international politics.

Values from earlier periods are embodied in the laws and institutions created during those periods. Thus, the contemporary conservation movement includes a plurality of values, and not a little ambivalence. The concern to preserve nature from human interference, for example, is at odds with the interventionist tendencies of the managerial ethic. Similarly, the professional exclusivity fostered by the notion of conservation as applied science, is at odds with efforts to make nature more popularly accessible. Equally, despite the international orientation expressed in *The World Conservation Strategy* (International Union for the Conservation of Nature, 1980) and the recognition that environmental problems, as well as animal and plant species, are no respecters of national boundaries, there still remain distinct nationalistic and nativistic sentiments, as in the opposition to 'alien' conifers and the efforts to preserve species rare in Britain though abundant elsewhere, such as the East Anglian avocet. Moreover, the humanitarian strand in conservation ensures that disproportionate effort goes into protecting the members of a few species that arouse strong human sympathies, irrespective of their ecological significance.

In the main, however, the diversity of values in the conservation movement is one of its great strengths. It provides conservationists with a vast armoury of justifications to suit any particular circumstance. More important, it reflects the centrality that conservation has achieved as a major cultural force.

ACKNOWLEDGEMENTS

I wish to thank David Allen and the editors for their constructive comments on a draft of this chapter.

REFERENCES

Adams, W. and Lowe, P.D. (1981). Continuity and change: science and values in nature conservation strategies. In C. Rose (Ed), *Values and Evaluation, Discussion Papers in Conservation 36*, University College London.
Allen, D.E. (1969). *The Victorian Fern Craze: A History of Pteridomania*, Hutchinson, London.
Allen, D.E. (1976). *The Naturalist in Britain*, Allen Lane, London.
Allen, D.E. (1980). The early history of plant conservation in Britain. *Transactions of the Leicester Literary and Philosophical Society*, **72**, 35–50.
Barber, L. (1980). *The Heyday of Natural History*, Jonathan Cape, London.
Bensusan, S.L. (1913). Birds and the plumage trade. *The Nineteenth Century*, **74**, 1067–1080.
British Ecological Society (1914). Report of the general meeting December 1913: F.W. Oliver on nature reserves. *Journal of Ecology*, **2**, 55–56.
British Ecological Society (1944). Nature conservation and nature reserves: a report by a special committee of the British Ecological Society. *Journal of Ecology*, **32**, 45–82.
Brookes, S.K. *et al.* (1976). The growth of the environment as a political issue in Britain. *British Journal of Political Science*, **6**, 245–255.
Burrow, J.W. (1966). *Evolution and Society*, Cambridge University Press, Cambridge.
Cherry, G.E. (1975). *Environmental Planning 1939–1969. Volume II. National Parks and Recreation in the Countryside*, HMSO, London.

Conwentz, H. (1909). *The Care of Natural Monuments*, Cambridge University Press, Cambridge.

Crampton, C.B. (1913). Ecology, the best method of studying the distribution of species in Great Britain. *Proceedings of the Royal Physical Society of Edinburgh*, **19**, 22–36.

Crump, W.B. (1913). Two nature reserves. *Country Life*, (**10 May**), 678–679.

Doughty, R.W. (1975). *Feather Fashions and Bird Preservation*, University of California Press, Berkeley.

Duff, A.G. and Lowe, P.D. (1981). Great Britain. In E.J. Kormondy and J.F. McCormick (Eds), *Handbook of Contemporary Developments in World Ecology*, Greenwood Press, Westport, Connecticut.

Egerton, F.N. (1973). Changing concepts of the balance of nature. *Quarterly Review of Biology*, **48**, 322–350.

Fairholme, E.G. and Pain, W. (1924). *A Century of Work for Animals: The History of the RSPCA, 1824–1924*, Royal Society for the Prevention of Cruelty to Animals, London.

Fitch, E.A. (1889). Presidential address. *Essex Naturalist*, **3**, 95–110.

Fleming, D. (1961). Charles Darwin, the anaesthetic man. *Victorian Studies*, **4**, 219–236.

Green, B. (1975). The future of the British countryside. *Landscape Planning*, **2**, 179–195.

Harrison, B. (1973). Animals and the state in nineteenth-century England. *English Historical Review*, **88**, 786–820.

Hill, O. (1899). The open spaces of the future. *The Nineteenth Century*, **46**, 26–35.

Hornaday, W.T. and Lemon, F.E. (1914). England's duty toward wild birds. *The Nineteenth Century*, **75**, 355–364.

Horwood, A.R. (1913). The state protection of wild plants. *Science Progress*, **7**, 629–637.

Houghton, W. (1957). *The Victorian Frame of Mind*, Yale University Press, New Haven.

International Union for the Conservation of Nature (1980). *World Conservation Strategy*, International Union for the Conservation of Nature, Geneva.

Jardine, W. (1858). *Memoirs of Hugh Strickland*, Van Voorst, London.

Johnston, H.H. (1906). Introduction to *With Flashlight and Rifle* by C.G. Schillings, Hutchinson, London.

Kingsley, C. (1880). *Scientific Lectures and Essays*, Macmillan, London.

Lambert, J.M. (Ed) (1967). *The Teaching of Ecology*, Blackwell, Oxford.

Lankester, R. (1915). *Diversions of a Naturalist*, Methuen, London.

Lloyd Baker, T.B. (1849). Address. *Proceedings of the Cotteswold Naturalists' Club*, **1**, 9–14.

Lowe, P.D. (1976). Amateurs and professionals: the institutional emergence of British plant ecology. *Journal of the Society for the Bibliography of Natural History*, **7**, 517–535.

Lowe, P.D. (1978). *Locals and cosmopolitans: a model for the social organisation of provincial science in the nineteenth century*, unpublished MPhil thesis, University of Sussex.

Lowe, P.D. and Goyder, J.M. (1983). *Environmental Groups in Politics*, Goerge Allen and Unwin, London.

Mabey, R. (1980). *The Common Ground*, Hutchinson, London.

Meldola, R. (1880). *An Inaugural Address Delivered to the Epping Forest and County of Essex Naturalists' Field Club, 28 February 1880*, Essex Naturalists' Field Club, Buckhurst Hill.

Meldola, R. (1883). The conservation of Epping Forest from the naturalists' standpoint. *Nature (London)*, **27**, 447–449.

Mitchell, P.C. (1912). Zoological gardens and the preservation of fauna. *Annual Report of the British Association*, 478–487.

Morris, M.C.F. (1897). *Francis Orpen Morris, A Memoir*, J.C. Nimmo, London.
Nature Reserves Investigation Committee (1945). *National Nature Reserves and Conservation Areas in England and Wales*, Society for the Promotion of Nature Reserves, London.
Nicholson, M. (1976). The ecological breakthrough. *New Scientist*, 72, 460–463.
Olson, M. (1973). Epilogue: the no-growth society. *Daedalus*, 102, 229–241.
Parsons, C. (1982). *True to Nature*, Patrick Stephens, Cambridge.
Pearsall, R. (1969). *The Worm in the Bud: the World of Victorian Sexuality*, Wiedenfeld and Nicholson, London.
Rothschild, M. (1979). *Nathaniel Charles Rothschild 1877–1923*, privately published by the author, printed by Cambridge University Press, Cambridge.
Rothschild, N.C. (1914). The preservation of nature. *Country-side*, 416–418.
Russell, H. (1897). The protection of wild birds. *The Nineteenth Century*, 42, 614–622.
Seebohm, H. (1897). The extermination of the Golden Eagle. *Natural Science*, 10, 303–304.
Shaw-Lefevre, G.J. (1894). *English Commons and Forests*, Cassell, London.
Sheail, J. (1976). *Nature in Trust: the History of Nature Conservation in Britain*, Blackie, Glasgow.
Sim, J. (1864). Natural history. *The Naturalist*, 1, 3–6.
Tansley, A.G. (1939). British ecology during the past quarter-century. *Journal of Ecology*, 27, 513–530.
Turner, E.S. (1964). *All Heaven in a Rage*, Michael Joseph, London.
Turner, J. (1980). *Reckoning with the Beast: Animals, Pain and Humanity in the Victorian Mind*, John Hopkins University Press, Baltimore.
Waterton, C. (1838). *Essays on Natural History*, Longman, London.
Wiener, M.J. (1981). *English Culture and the Decline of the Industrial Spirit 1850–1980*, Cambridge University Press, Cambridge.
Wildlife Conservation Special Committee (1947). *Conservation of Nature in England and Wales*, Cmnd 7122, HMSO, London.

Conservation in Perspective
Edited by A. Warren and F.B. Goldsmith
© 1983 John Wiley & Sons Ltd.

CHAPTER 20

Agriculture and Conservation: What Room for Compromise?

RICHARD MUNTON

There is no longer any serious debate as to whether modern farming practices conflict with amenity and wildlife, even if there are still wide differences of opinion as to the gravity of the conflict. Instead, the focus of discussion has moved on to how the conflict might be resolved. Progress will undoubtedly be slow. Tackling the real bases of the conflict will mean: (a) challenging well-established positions over the nature of financial support to agriculture; (b) broadening the outlook of the Ministry of Agriculture, Fisheries and Food (MAFF) away from an overriding concern with food production; and (c) achieving increased clarity and consistency on the part of the environmental movement in its definition of its own objectives for the countryside. All these depend on information about the extent and nature of the effects of farming on wildlife and this is still very patchy.

That this conflict should attract considerable public attention is hardly surprising. Farmers manage more than 85 per cent of rural land in Great Britain and mostly maintain a developmental or improving outlook on the use of their land. They do this in response to the government's longstanding goals of increasing national self-sufficiency in temperate food production and of promoting farm business efficiency (Ministry of Agriculture, Fisheries and Food (MAFF), 1975; 1979). Nevertheless, it is instructive to enquire a little more deeply as to why this conflict is so widely discussed today. After all, today's farm policies are not new. Agricultural advisory officers have sought to improve farm productivity and profitability for many years (Dexter and Barber, 1967), and in this have only been putting into effect what government intended (Beresford, 1975; Bowler, 1979). Moreover, the effects of this policy on the nature of farming practice, and in turn on the environment, have not been difficult to anticipate. Indeed, some of its consequences for the environment were noted, measured and researched during the 1960s (see, for example,

Blackwood and Tubbs, 1970; Brett, 1965; Hooper and Holdgate, 1968; Ministry of Agriculture, Fisheries and Food, 1970; Moore, 1966; Weller, 1967). So what has happened during the 1970s to raise the level of public awareness and the political temperature?

A CHANGING CONTEXT

In seeking answers one must begin with changes in agriculture itself. These are discussed in greater detail later in the chapter and it is sufficient to record here that during the 1970s farmers expanded production, modernized their farms and in some respects increased their inputs (in volume or real terms). But with the exception of unprecedented rises in land prices and interest rates, which together have contributed to the increasing wealth *and* indebtedness of farmers (Reid, 1981), these changes represent continuations of trends firmly established during the whole postwar period. They are, therefore, insufficient to account for the great increase in public awareness and political discussion. A more plausible explanation begins by accepting that there is always a time-lag between academic research and the development of public interest. Until the latter happens the subject will not reach the political agenda. Even then government may be reluctant to acknowledge the importance of the issue and in this particular case the key government department, MAFF, is not known for its responsiveness to developments that challenge the hegemony of agriculture in the countryside. By this argument the agricultural events of the 1960s would not be noticed until the 1970s. Furthermore, evidence has to be communicated in a forthright, uncomplicated and non-scientific way and this happened increasingly during the 1970s (see below).

The 1970s was also the decade during which there was a growing conviction among conservationists that wildlife resources could not be effectively conserved solely on the land that had been designated for that purpose. Not only are national nature reserves and SSSIs poorly protected but together they constitute less than 7 per cent of the land area of Great Britain (Nature Conservancy Council, 1982), and they have become increasingly vulnerable to the loss of species as a result of the general reduction in the area of wildlife habitat in the rest of the countryside (Green, 1981). Even more important, and not unrelated to the last point, there has been a change of emphasis within the conservation movement away from a particularist, scientific concern for species on nature reserves towards the conservation of whole habitats and their amenity for the benefit of the public at large (Mabey, 1980). One consequence is that the established rationale for farming has been examined anew right across the countryside by the rapidly growing memberships of amenity bodies and by those that simply enjoy visiting the countryside.

SOME NEW EVIDENCE

An enormous amount of literature was published in the 1970s on the relations between conservation and agriculture, but probably the most influential study of change in the rural landscape was *New Agricultural Landscapes* (Westmacott and Worthington, 1974; see also Countryside Commission, 1977). The study was neither comprehensive (it covered only 8320 ha in seven representative farming areas of England and Wales or only 0.07 per cent of the total agricultural area) nor scientific in that the authors accepted from the outset that their judgements on landscape quality were both subjective and personal. However, they presented their findings to great effect and by suggesting that the farming reasons for the losses would continue and that existing controls were inadequate to prevent this they were able to draw widespread public and professional attention to the issue. This attention was in no way diminished by their claim that farmers were divided in their views as to whether they, as the main occupiers of rural land, held any responsibility for the conservation of the landscape. Whether they felt they had any responsibility or not, all the farmers interviewed by the authors argued that they should be fully compensated for any financial losses they incurred for being conservators, and the Ministry of Agriculture's own study of the same question led to a similar finding (Ministry of Agriculture, Fisheries and Food, 1976).

Since the publication of *New Agricultural Landscapes* other studies have confirmed the rapid rate of landscape change (for example, Davidson and Lloyd, 1977; Hooper, 1977). Blacksell and Gilg (1981) produced data for the period 1950–76 for several parts of Devon. Their most important discovery was that changes in land use and in the landscape were occurring equally quickly on the urban edge, in the farmed countryside and within areas protected by amenity designations. Even in this western county there was a demonstrable shift towards monocultural farming practices, whilst agricultural improvement had increasingly penetrated marginal farming areas within the Exmoor and Dartmoor National Parks. Indeed, the reclamation of moorland on Exmoor was one of the bitterest disputes of the 1970s (Porchester, 1977; see also MacEwen and MacEwan, 1982 and Chapter 22 of this volume), and diverted attention from the scale of moorland loss in other national parks (Countryside Commission, 1981; Parry, Bruce, and Harkness, 1981).

Knowledge of the loss of wildlife habitat as a result of agricultural improvement, whether this takes the form of ploughing, spraying, fertilizing, draining or reseeding, has come from many sources (King and Conroy, 1980). These sources include academic research, surveys by the Nature Conservancy Council and voluntary organizations, and planning and other types of public inquiry. For example, the inquiry in 1978 into the grant-aiding by MAFF of a drainage scheme at Amberley Wild Brooks in Sussex led it to be the first such scheme to be refused a grant by the Minister for Agriculture on environmental grounds

(Parker and Penning-Rowsell, 1980, pp. 229–234); the furore over the loss of moorland on Exmoor led to the first public inquiry to be jointly sponsored by MAFF and the Department of the Environment (Porchester, 1977); and the well-publicized dispute over the drainage of the Halvergate Marshes in Norfolk has yet to be resolved (O'Riordan, 1981).

An attempt is made in Table 20.1 to bring together the information that is available on habitat loss resulting from agricultural improvement. The evidence

TABLE 20.1   Habitat loss in Great Britain: some examples attributable to agricultural improvement.

Habitat	Location	Per cent Loss	Period
Permanent grass/ rough grazings	Devon	35	1905–1977
	South Scotland	35	1946–1973
	Powys	7	1971–1977
Floodplain meadow	Oxfordshire	16	1978–1980
Chalk downland	Dorset ⎫	72	1815–1980
	Dorset ⎭	37	1967–1980
	Wiltshire	47	1937–1971
	Hampshire	20	1966–1980
	Sussex	20	1966–1980
	Isle of Wight	17½	1966–1980
Heathland	*Dorset ⎫	85	1750–1978
	Dorset ⎬	66	1811–1960
	Dorset ⎭	50	1960–1980
	North Hampshire	19	1966–1980
	South Scotland	61	1946–1973
Moorland	Dartmoor	20	1945–1980
	Exmoor	21	1947–1976
	North Yorkshire Moors	20	1945–1980
	Powys	7	1971–1977
Wetlands	South Scotland	10	1946–1973
	*Scotland and North England	87	1850–1978
Ancient woodland	*Britain	30–50	1947–1981
Deciduous woodland	Suffolk	50	1837–1970
	Scotland	56	1947–1980
Hedgerows	England and Wales	25	1946–1977
	South Scotland	25	1946–1973
	Norfolk	45	1946–1970

Sources: Goode (1981); MacEwen and MacEwen (1982); Parry *et al.* (1981); Rackham (1971); Shoard (1980); Webb and Haskins (1980).

Notes: Bracketed examples give figures for the same habitat but over different periods.

* Includes losses to urban development and coniferous afforestation.

demonstrates quite clearly that all the major semi-natural habitats in Britain have experienced *substantial diminution in their areas*. The data are by no means comprehensive but no amount of quibbling about their reliability can deny this basic fact. Moreover, loss is continuing. What the information in the table cannot do, however, is to indicate whether the *rate* of loss increased during the 1970s. Loss has been more rapid in the postwar period *as a whole* than in former times but the data cannot inform about the 1970s specifically.

The data in the table present other problems of interpretation. Because absolute figures are not always cited these have been omitted, and therefore the percentage figures may in some cases relate to small areas, and the sources of evidence and means of measuring change vary between the different investigations. Some studies employ only aerial photographs (as with the south of Scotland data), some use Ordnance Survey maps and field surveys (as in the Devon study), and others Ordnance Survey maps and aerial photographs (Parry, Bruce, and Harkness, 1981). Furthermore, the information presented here does not detail the spatial pattern of the remaining habitats, although all the studies that do, record increasing fragmentation as well as loss of area (Jones, 1973; MacEwen and MacEwen, 1982; Webb and Haskins, 1980). Neither does the information note the 'kind' of habitat loss, such as revealed by Parry, Bruce, and Harkness' study of moorland reclamation in the national parks. This particular investigation distinguishes between the loss of 'primary' moorland (moorland that, as far as can be established, has never been ploughed) and 'secondary' moorland (moorland that has invaded abandoned farmland). The authors maintain that up to and including the Second World War only 'secondary' moorland was reclaimed but as much as 60 per cent of the reclamation in the postwar period in the Brecon Beacons, Dartmoor and the North Yorkshire Moors national parks has been of 'primary' moorland.

Some research indicates a slowing down of the rate of habitat loss in recent years (Webb and Haskins, 1980), and no doubt farming interests would claim this as a measure of success for the consensus approach to limiting the environmental effects of farming activities. For a number of reasons such a view would be misplaced. First, there is now much less semi-natural habitat left to lose than 30 years ago. What remains is fragmented, and the marginal value of every additional hectare that is now lost is greater than the last. Second, it may be that the increased public concern expressed in the 1970s over habitat loss itself actually helped to encourage the decline in the rate of loss of habitat (if it occurred). Third, as noted in the first paragraph of this chapter, little progress has yet been made on tackling the *causes* of the losses; and fourth, an increasing proportion of the area of the remaining semi-natural habitat lies within designated areas where loss might be expected to be extremely limited. That this last assumption cannot be taken for granted has been a cause of even greater concern to wildlife conservationists. In preparing evidence for the debates on the Wildlife and Countryside Bill, Goode (1981) reported that SSSIs were being damaged or destroyed at a rate of 4 per cent per annum (100 out of 3900). Goode does not make clear the precise meanings, or significance in individual cases, of the terms 'damaged' and 'destroyed' but it must be hoped that procedures now available under the terms of the Wildlife and Countryside Act, unsatisfactory though they are (see Chapters 19 and 26 of this volume), will at least slow down this rate of damage.

Much less has been heard in recent years about the impact of agricultural pollution on the environment: the effects of pesticides, fertilizer residues and farm wastes. Although the Royal Commission on Environmental Pollution in its report *Agriculture and Pollution* (1979) suggested all was not well, the technical nature of its report and the fact that it was unable to demonstrate any dramatic worsening of the situation, meant that its conclusions were not especially newsworthy. They are, nevertheless, important. The Royal Commission welcomed the introduction of the Pesticides Safety Precautions Scheme (see Moore, Chapter 10) as a means of reducing the risks of persistence and low selectivity but it warned, as others have done (Perring and Mellanby, 1977), that there is no safe level of use. It was especially concerned at the growth of aerial spraying, estimated at 400 000 ha in 1976. Moreover, although the Royal Commission felt that both manufacturers and scientists had improved their understanding of the secondary consequences on the environment of excessive applications, it was not persuaded that farmers were well-informed or that they could be relied upon to self-regulate the quantities they added in the absence of fuller advice.

A CHANGING DEBATE?

The accumulation of evidence, the increased lobbying of central government by amenity groups, and the considerable amount of parliamentary time spent on the Wildlife and Countryside Bill have undoubtedly sharpened the debate. The traditional view that an efficient, prosperous agriculture could form the basis of a thriving rural economy, whilst also providing a custodial role for other countryside users, is now largely discredited. In response to this, central government and its agencies have remained largely committed to a consensus approach to resolving the conflict between modern farming and the rural environment involving education, information, persuasion, financial incentive and voluntary management agreements (Countryside Review Committee (CRC), 1978; Advisory Council for Agriculture and Horticulture (ACAH), 1978). Much is to be said for this approach, at least until it is shown to be unworkable, and it forms the basis of the Nature Conservancy Council's thinking in its report *Nature Conservation and Agriculture* (Nature Conservancy Council, 1977). It is also central to the Countryside Commission's attitude towards countryside management (Feist, 1978) and it is endorsed by MAFF through their support for the Farming and Wildlife Advisory Groups (Keenleyside, 1977).

Except amongst those bodies representing farming and landowning interests (see, for example, National Farmers' Union (NFU)/Country Landowners' Association (CLA), 1977), which themselves have been forced onto the defensive by the vehemence of environmental protest and the provocative behaviour of a small minority of their members, views outside government have become much more critical. Faith in consensus has diminished. At one level, farmers as individuals are seen as unsympathetic to conservation. Leonard and Stokes (1977) attack what they regard as a number of popular fallacies promoted by the farming community. These include the assumptions that the farmer knows best: in other words, he knows that society really wants cheap food at any environmental price; that if farmers could earn a decent living they would make fewer changes to the landscape; and that problems can be solved by voluntary action. At another level, it is the nature of agricultural policy and the role of MAFF in promoting it that has been most seriously questioned. Conservationists' goals are seen as unrealizable in the face of agricultural policies that encourage or even oblige farmers to adopt new and more sophisticated technologies (Dexter, 1977) especially when these are backed by enormous infusions of public money (see below; Shoard, 1980). To reverse this process, the goals of agricultural policy would have to be broadened to incorporate aims other than farming efficiency. These would then have to be supported by the redirection of public funds and, if necessary, by compulsory management agreements and planning controls. To achieve this, the remit of MAFF might have to be expanded (see the arguments for this put forward by the Advisory Council for Agriculture and Horticulture (ACAH), 1978; Countryside Review Committee, 1978; Royal Commission on Environmental Pollution, 1979; Wibberley, 1976); but clearly it should not be the kind of Ministry in which pollution 'problems were regarded as secondary in importance and as unavoidable concomitants of food supply' (Royal Commission on Environmental Pollution, 1979, page 3); or about which 'we have formed the view that the MAFF approach to pollution questions has been unduly defensive and protective towards agricultural interests' (Royal Commission on Environmental Pollution, 1979, page 311).

AGRICULTURAL CHANGE DURING THE 1970s

Some factual evidence on how farming has changed and what pressures have been responsible for these changes is essential if views are to be formed about whether there is much room for compromise.

Production patterns

The agricultural industry continued to grow at a substantial rate between 1970 and 1980 although its contribution to the gross domestic product fell from 2.8 per cent to 2.1 per cent over the same period (MAFF, 1982). The provisional figures for 1980 indicate that home production, including the value of food exports but excluding net imports of agricultural inputs, was equivalent to 74.8 per cent of indigenous-type food consumed in the United Kingdom and 60.5 per cent of all food. The comparable figures for 1969–71 were 58.6 per cent and 45.9 per cent respectively. (Footnote 1, end of chapter.) This increase occurred despite a rise in living standards and a reduction of 2.3 per cent in the amount of agricultural land.

Changes in agricultural land use and the numbers of livestock in the United Kingdom between 1955 and 1980 are shown in Table 20.2. The area of tillage has steadily increased, largely at the expense of temporary grass, whilst the extent of rough grazings has declined significantly during the 1970s, largely the result of its afforestation and its improvement. Cattle numbers have increased steadily and the size of the national sheep flock has expanded substantially. The quantity of purchased animal feeds has not increased since 1969 and the amount of cereals fed to livestock has fallen slightly, implying a significant net rise in the stocking rate of grassland. Increases in yields and output continue to be marked (Table 20.3). Only figures for four of the major enterprises are shown but they each record a higher rate of yield increase in the 1970s than in the previous period. This result may have been inflated by the good harvests of the late 1970s, so the figures should be treated with caution, but there is little evidence to support the notion prevalent earlier in the decade that yield increases could be expected to slow down (Centre for Agricultural Strategy (CAS), 1976).

TABLE 20.2 Agricultural land use ('000 ha) and livestock numbers (000) in the United Kingdom 1955, 1969 and 1980.

				Annual change[3]			
	1955[1]	1969	1980	1955–69		1969–80	
('000 ha)				ha	%	ha	%
Tillage	4573	4940	5031	+ 26.2	+ 0.57	+ 8.3	+ 0.18
Temporary grass[2]	2484	2307	1965	− 12.6	− 0.51	− 31.1	− 1.48
Permanent grass	5477	4997	5140	− 34.3	− 0.63	+ 13.0	+ 0.29
Rough grazings	6829	6849	6333	+ 1.4	+ 0.02	− 46.9	− 0.75
('000)[4]				numbers	%	numbers	%
Total cattle	10 668	12 374	13 426	+ 121.9	+ 1.14	+ 95.6	+ 0.85
Total sheep/lambs	22 949	26 604	31 446	+ 261.1	+ 1.14	+ 440.0	+ 1.82
Total pigs	5843	7783	7815	+ 138.6	+ 2.37	+ 2.9	+ 0.04
Total poultry	86 857	126 515	135 105	+ 2832.0	+ 3.26	+ 780.9	+ 0.62

Source: *Annual Reviews of Agriculture*, MAFF, HMSO, London.

Notes: [1] 1955 and 1969 were the years used in *Conservation in Practice* (Munton, 1974, p. 329) to allow for comparison.
[2] Some small changes have been made in Agricultural Census during the period, including the definition of temporary grass in 1959.
[3] Over the 1955–80 period the area recorded in the Agricultural Census fell by 486 000 ha.
[4] Total figures are used as the disaggregated data do not suggest any major change in the proportions of young and adult animals over the period.

TABLE 20.3 Estimated yields and total production levels for wheat, barley, potatoes and milk in the United Kingdom 1955–80.

	1954–56	1968–70	1978–80	% increase per annum	
				1954/56–1968/70	1968/70–1978/80
Yield					
Wheat (tonnes/ha)	3.07	3.92	5.38	+1.98	+3.72
Barley (tonnes/ha)	2.93	3.46	4.24	+1.29	+2.25
Potatoes (tonnes/ha)	19.61	25.78	33.45	+2.25	+2.98
Milk (litres/cow)	3187	3741	4653	+1.24	+2.44
Total output					
Wheat ('000 tonnes)	2725	3689	7311	+2.53	+9.82
Barley ('000 tonnes)	2608	8155	9937	+15.19	+2.19
Potatoes ('000 tonnes)	7405	6856	6931	−0.53	+0.11
Milk (million litres)	7569	12153	15278	+4.67	+2.50

Source: Annual Reviews of Agriculture, MAFF, HMSO, London.

TABLE 20.4 Production of the major cereal crops in England 1955, 1969 and 1979.

	1955				1969				1979			
	Wheat	Barley	Oats	Total	Wheat	Barley	Oats	Total	Wheat	Barley	Oats	Total
Area ('000 ha)	760	844	537	2141	788	2021	228	3037	1340	1786	88	3214
Per cent crops and grass	8.6	9.5	6.1	24.2	9.1	23.4	2.6	35.1	15.9	21.2	0.8	37.6
Production (million tonnes)	2.57	2.72	1.42	6.71	3.16	7.21	0.78	11.15	6.99	7.30	0.35	14.64

Source: United Kingdom Agricultural Statistics (appropriate years), MAFF, HMSO, London.

The rise in cereal production in particular raises questions of concern for conservationists, associated as it is with large fields, mechanized systems of harvesting, handling and storage, and monocultural systems of production. Although the staggering increase in barley output between 1955 and 1969 has not been sustained, wheat production has increased substantially instead (Tables 20.3 and 20.4).

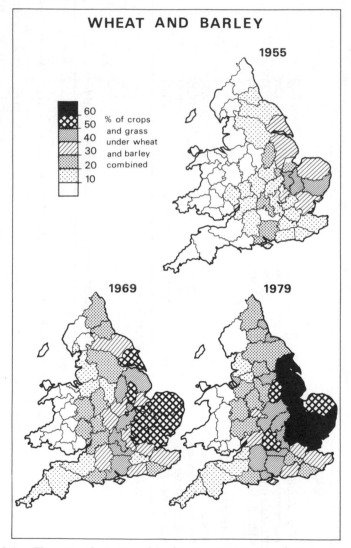

FIGURE 20.1. The area of wheat and barley as a percentage of total crops and grass acreage 1955, 1969 and 1979. Source: Ministry of Agriculture, Fisheries and Food agricultural statistics.

The pattern by county of the area under wheat and barley combined, as a proportion of all crops and grass, for 1955, 1969 and 1979, is shown in Figure 20.1. Direct comparisons are impossible following the realignment of county boundaries in 1974 but the overall picture is clear enough. The counties in eastern England that traditionally grew large quantities of cereals have retained their pre-eminence despite the westward expansion of cereal growing. By 1979 every county in England bar Cumbria had more than 10 per cent of its crops and grass in wheat and barley, and six eastern counties had figures in excess of 60 per cent. The raw data show that the really big increases in wheat production in the 1970s occurred in the cereal 'heartland' whilst the peripheral cereal growing areas, such as Devon, Cornwall, Shropshire and Northumberland went against the national trend (Table 20.4) by growing a larger area of barley. Cereal monoculture often brings problems of weed and disease control and physical damage to the soil, but Moore (1977, p. 42) has noted that: 'Most of the species of wildlife on farmland are dependent on the small woods, hedges, ditches and ponds and not the arable land itself, and so alterations in farming practices are generally less useful in conserving wildlife than not farming parts of the land'. These agricultural statistics cannot elucidate this particular point but general experience argues that cereal monoculture has also been associated with the more intensive use of all available land.

Farming inputs

The farming industry is proud of these gains in productivity. They have been achieved by increasing some but not all inputs to the industry. For example, the numbers employed fell by more than 10 per cent between 1970 and 1980 and although the industry has modernized its systems of production there was no increase in the *volume* of investment during the decade. This general situation is repeated in the specific case of field drainage. Between 1945 and 1970 the area of farmland which was drained, or redrained, rose each year. But since the early 1970s the area drained in England and Wales has been steady at about 100 000 ha per annum (Armstrong, 1981). Most draining has continued to occur on heavy arable land in eastern England and whilst these areas do not have a greater physical need for drainage than others in western England they command greater financial returns from the higher yielding arable crops that can be grown as a result. Based on a study of land drained between 1971 and 1980 Armstrong (1981, p. 10) concluded that

The majority of land (59%) does not change its land use at all when drained. However, of the land that does change its use, by far the biggest proportion is associated with the improvement of permanent pasture to include a proportion of arable crops. Hence 28% of the land originally in intensive grassland, and 38% of the land in extensive grassland moves towards a mixed farming regime with the inclusion of some arable crops, and a further 9% of the intensively used grass and

20% of the extensively used grass goes into full arable production. Hence we can estimate that a total of 15% of the area drained (a total of 116 500 ha) has been converted from permanent pasture to a cultivated regime . . . .

For some other inputs there is evidence of an increase in volume. In the case of fertilizer use, the application of nitrogen (N), phosphorus (P_2O_5) and potassium (K_2O) increased until the mid-1960s. Since then only additions of nitrogen have grown (Table 20.5). The Royal Commission on Environmental Pollution (1979) concluded that fertilizer additions to arable crops are now close to their optimum although it confidently expects further additions of nitrogen to grass where this is intensively managed. In the case of pesticides (including herbicides and fungicides) a substantial increase in use occurred in the 1970s both as a labour-saving device in the control of weeds and as an essential byproduct of monocultural systems of arable farming.

TABLE 20.5   Fertilizer additions on farms in England and Wales, 1970 and 1979 (kg/ha).

	1970				1979			
	Tillage	Leys[1]	Permanent grass	All crops and grass	Tillage	Leys	Permanent grass	All crops and grass
Nitrogen (N)	88	108	53	79	112	160	87	114
Phosphorus (P_2O_5)	56	43	24	42	49	32	21	36
Potassium (K_2O)	61	31	18	40	53	36	19	38

Source: ADAS/Rothamsted Experimental Station/Fertilizer Manufacturers' Association, 1980.

Note: [1]Seeded down 7 years or less.

Finally, reference must be made to changing land prices and rents (see Figure 20.2). Land prices rose steadily during the 1960s, doubling in current prices and increasing by 37 per cent in real terms. But these rises were small compared to the immense price changes, both up and down, that took place in the 1970s. The changes were so dramatic as to lead to the setting up of a Committee of Inquiry (see Northfield Committee Report, 1979). Land prices rose from £544/ha in 1971 to £3162/ha in 1980 but are now falling in real terms. The high rates of inflation of the period meant that whilst changes in real terms were substantial (the 1980 price was 74 per cent above the 1970 level) they were nothing like as large as current price changes. Farm rents have also increased sharply during the 1970s in current price terms but more steadily than land prices (Figure 20.3). More interestingly, based on a 1970 constant, rents rose by about 25 per cent in real terms in the 1960s but by only 7 per cent in the 1970s, and this small increase occurred at the end of the decade as farming incomes were beginning to fall.

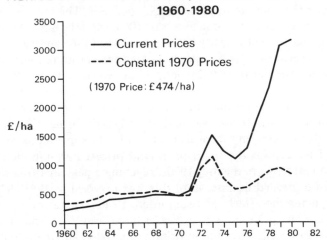

AGRICULTURAL LAND PRICES, ENGLAND AND WALES
1960-1980

— Current Prices
--- Constant 1970 Prices

(1970 Price: £474/ha)

£/ha

FIGURE 20.2.   Source: Ministry of Agriculture, Fisheries and Food/Inland Revenue Land Price Series (adjusted for 9-month time-lag).

FARM RENTS IN ENGLAND AND WALES 1961-1981

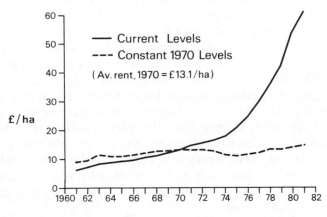

— Current Levels
--- Constant 1970 Levels

(Av. rent, 1970 = £13.1/ha)

£/ha

FIGURE 20.3.   Source: *Annual Rent Survey* (Ministry of Agriculture, Fisheries and Food).

Farms and farming income

Throughout the postwar period the number of farms has fallen and their average size increased.[2] Between 1968 and 1975 the number of holdings fell by 75 000 to 272 000 (MAFF, 1977), but over 40 per cent of this reduction was accounted for by changes in the Census, leaving a true fall of about 1.8 per cent per annum. This rate increased to 2.3 per cent per annum in the second half of the 1970s

(MAFF, 1982). Between 1970 and 1980 the average size of a full-time farm rose from 92.1 ha to 115.8 ha, an increase of 25.7 per cent. The only size group to have increased in numbers is that in excess of 100 ha and the land area occupied by this category increased by nearly one-third between 1965 and 1975. Not only have farms been getting bigger but they have also become more specialized. The number of enterprises per farm fell from 3.18 in 1968 to 2.85 in 1974 (Britton, 1977).

In real terms average net farming income per hectare has not increased substantially throughout the whole postwar period, although farmers' incomes have risen as a result of increasing farm size. The industry's overall prosperity has also increased because of the real rise in land prices. Farming income[3] rose measurably in real terms in the mid-1970s, reaching a peak in 1976. Since then there has been a marked decline, until this was reversed in 1981−82, the industry experiencing the effects of rising interest rates and a severe cost-price squeeze. Community price support increases have been lower than the rate of inflation in the EEC generally, and in Britain especially, and the fall of sterling against the European Unit of Account has reduced even further the benefit to the British farmer of community price rises. Farmers in Britain are now pressing for substantial increases in support prices to cushion themselves against this decline. The effect of this fall on farmers' incomes has been lessened by the increase in the size of farms and excellent harvests in 1981 and 1982, but most farmers regard their economic future as uncertain at best and as bleak at worst. Making long-run comparisons between different sectors of the farming industry is difficult because of changes in the way farm incomes have been compiled by MAFF and because of substantial year-to-year fluctuations in income at the individual farm level. These qualifications aside, average returns on tenant's capital, for example, suggest better returns for arable, especially cereal, producers than other kinds of farmer, and on bigger rather than smaller farms. Farmers in eastern England have, therefore, fared very much better than livestock producers in western Britain since entry to the EEC and it would be wrong to treat the aggregate trends as if they applied equally to all kinds of farmer. There may well have been an increasing difference, proportionately, between the profitability of different farming systems, but this is difficult to prove from published figures.

THE IMPERATIVES

The previous discussion shows that the 1970s were characterized by a further intensification of agricultural land use, including the removal of hedges and woods, the drainage of wetlands and the reclamation of rough grazings, but not by real increases in all kinds of farming inputs. The latter reflects in part the increasingly uncertain economic climate and declining real incomes of the latter part of the decade. In the mid-1970s the Common Agricultural Policy (CAP) seemed to promise a major rise in farm incomes but this has not been sustained because inflation (and thus cost increases) in Britain outstripped increases in support prices. Does this situation eliminate any chance of compromise with nature conservation in the near future?

The question can be approached through an examination of at least three important issues.

First, it is often suggested that it is essential to increase output on the remaining area of farmland simply to offset the transfer of agricultural land to forestry and urban uses (Coleman, 1976). This begs the basic question of what level of self-sufficiency in temperate foodstuffs we wish to sustain now that we are members of the EEC which produces relatively riskless supplies of food, albeit at a price. The case for any further expansion in food production beyond the 75 per cent level of self-sufficiency in temperate foodstuffs already achieved is extremely thin (Best, 1981; Ritson, 1980). The key point, however, is that all the recent studies of future land budgets for the United Kingdom indicate that the loss of farmland is not the critical determinant of future levels of self-sufficiency whilst a high input—high output industry is sustained (Centre for Agricultural Strategy, 1976; Edwards and Wibberley, 1971). Output is reduced to a small extent by such losses but in terms of *maintaining* current levels of food supply 'the intensification of agricultural practices that might be necessitated by land lost by year 2000 is negligible' (Royal Commission on Environmental Pollution, 1979, p. 30). Naturally, at the level of the individual farm that has lost some land to development, intensification of use on the remaining area might be essential to its economic survival (Hearne *et al.*, 1977).

Second, farmers operate in a policy context that encourages them to expand output whilst placing a premium on technological sophistication. Grant aid is given to assist capital investment, not to employ more workers. Farmers are encouraged to produce more by an open-ended system of price support, and whether large or small, needy or prosperous, receive support in proportion to the amount they produce. As Shoard says (1980, p. 24): 'In a world where any amount of production is saleable at an attractive price, every unfelled wood or undrained wetland represents forfeited profit'. However, some effort can be expected in Europe during the 1980s to reorient the Common Agricultural Policy towards an income-supplementation approach, via a system of quotas and levies for individual member states of the Community, leading eventually to a system of quotas for individual farmers that favours small producers (Commission of the European Community, 1981). This would not be especially welcomed in Britain where most farmers have larger businesses than their European counterparts, and any such move would be bound to end in compromise. Not too much should be expected of it in terms of changed farming practice, especially as over 75 per cent of all public support for our farming industry comes from UK sources, primarily as grant aid for improvements, tax reliefs and other forms of indirect support, such as the Agricultural Development and

Advisory Service (ADAS). Total direct public expenditure (that is, money in the farmer's pocket) amounted to £1012.2 million in 1980–81 and is estimated at £1038.9 million for 1981–82 (MAFF, 1982).

There are two main points here: The first is that £190.1 million was spent on farm improvements grants of various kinds in 1980–81 and £167.5 million is estimated for 1981–82 (MAFF, 1982). These monies have been disbursed on the assumption that they would increase farming efficiency and have paid little regard to their environmental consequences. To conservationists this is a contentious matter in itself. But this procedure has been called further into question by a recent study of farming efficiency in Britain (Centre for Agricultural Strategy, 1980). The study showed that British farming was less efficient on all the criteria used than that in Holland and Belgium, and in some respects less efficient than that in Denmark, France and Ireland too. The Centre suggests that a major reason for this rather mediocre performance is the 'excessive' capital investment made by many farmers for *fiscal* rather than for reasons of efficiency or productivity.

The second point is that tax reliefs, such as the agricultural and business reliefs afforded to working farmers under Capital Transfer Tax and Capital Gains Tax (Northfield Committee Report, 1979, pp. 71–90), and price supports, are to a large degree capitalized back into higher land prices. As farmers are easily the largest single group of purchasers of land (Northfield Committee Report, 1979, pp. 63–67) their complaints over the high price of land and the size of their mortgage repayments, and thus their demands for further reliefs or even higher support prices, should be treated with scepticism. Farmers almost to a man want the retention of a free market in land and therefore they should accept the exigencies as well as the benefits that the market disposes. It follows that their argument that land must be intensively farmed because of its high price is open to question. But breaking into this circular argument will not be easy because of high interest rates, which are not of the farmers' own choosing, and because tenants are currently experiencing real increases in farm rents at a time of falling incomes. Nevertheless, high rates of inflation rapidly reduce the real burden of long-term borrowing and only a small proportion of farmers have bought much land in the latter part of the 1970s (only about $1\frac{1}{2}$ per cent of the agricultural land area is sold each year). But most important, farmers *believe* they need to farm more intensively to obtain a higher standard of living and tend to regard the good years of the mid-1970s as the norm against which to assess their present difficulties.

Third, the previous discussion indicates why farmers believe they must expand their businesses, for to stand still is to go backwards. Greater size is thought to bring internal economies of scale and the present tax arrangements encourage farmers to accumulate capital in their businesses rather than to extract it as personal income (Northfield Committee Report, 1979, pp. 37–40). Recent research indicates that greater farming efficiency is usually derived from enlarging small farms but this is not so on farms employing more than two or three people (Britton and Hill, 1975; Northfield Committee Report, pp. 35–37), or precisely that category of large farm that is growing most rapidly. But because the evidence reveals greater variability in efficiency within size groups than between them, and because in terms of farm area the size of farm needed to realize all internal economies of scale is continually rising, the Northfield Committee (1979, pp. 239–246) was not prepared to recommend size limits for different types of farm as these limits could inhibit the growth of efficient farming businesses. Nevertheless, once again the farmer's argument that he must be allowed to expand production and the size of his farm under all circumstances in the interests of farming efficiency is open to question.

ASSESSMENT

The preceding discussion suggests that the farmer's case for expanding his business is neither watertight nor always in the public interest, even in terms of raising agricultural efficiency. The case also ignores the external disbenefits imposed on the environment by the farmer in his attempts to farm more intensively; and the earlier part of the chapter, despite the problems of measuring and evaluating the significance of habitat loss within and without designated areas, can leave no-one in any doubt about the extent of his impact.

Nevertheless, in the present economic and farm policy circumstances it is often difficult to persuade the farmer that this is so. Research into his attitudes towards wildlife and landscape conservation indicate that his views do vary (farmers should not be treated as having only a single view) but these are at best ambivalent and at worst hostile to conservation interests (MAFF, 1976; Newby *et al.*, 1977). MAFF in its report says that 13 per cent of the 305 farmers interviewed intended to remove at least part of the semi-natural habitats remaining on their land although an equal number planned to improve existing or to create new habitats; and whilst the survey uncovered a latent demand for conservation advice, positive conservation measures were largely restricted to those with big farms or who were members of conservation bodies. The study in East Anglia by Newby *et al.* (1977) led them to conclude that the 'new' owners of land, primarily large, owner-occupied agribusinesses and the financial institutions, were much less sympathetic to the needs of conservation than either gentleman farmers or small family farmers.

This attribution of differing attitudes may be broadly correct—although Newby *et al.* do not provide actual management evidence in support of it—but for a number of reasons a clear-cut distinction such as this between types of landowners may not materialize. First, the wealth of the traditional private landowner is being more effectively taxed today than hitherto, following replacement of estate duty by capital transfer tax. Even allowing for the substantial tax concessions now being offered to the owner, the owner may now feel less obliged or less able to adopt an altruistic position to environmental enhancement. Second, the current cost-price squeeze in farming will certainly not leave the management of small farms unscathed; and third the financial institutions do at least have the cash to make environmental improvements and are conscious of their public image, matters that Worthington (1979) feels can be turned to good effect. In any event, owners and occupiers will expect to be compensated in full both for expenditure incurred in positive conservation management to meet the public interest and for income lost from not being allowed to 'develop' their land. Payment in the first context represents a reasonable, or at least an arguable, call on public funds. In the latter it depends on the circumstances, and these are broadly defined by what role we want farming to play in different parts of the countryside, but there is insufficient space to develop this theme here. Equally important, despite its present financial difficulties, the case for *additional* public funding for agriculture is difficult to sustain and so it is appropriate to argue that the cash for 'environmental' spending in this area should come out of MAFF and not DOE monies, either directly or via a subvention to the Nature Conservancy Council and the Countryside Commission. Given the huge amounts of public money going into agriculture by comparison with those spent on conservation, there is clearly room for compromise here.

There is also room for compromise over the amount of food we produce, and therefore the amount of agricultural land put to agricultural use. As Ritson (1980) argues, the case for seeking a level of self-sufficiency much beyond that of meeting our basic dietary needs in an emergency is highly debatable now that we belong to the EEC. The European Community ensures an almost riskless source of food. Equating our basic dietary needs with what we currently produce is very difficult, not least because we eat much more protein than we need, but it is quite clear that we produce much more food today than would be required to meet our basic dietary needs (Blaxter, 1975; Mellanby, 1975). Some resources in the agricultural budget could, therefore, be transferred to promoting social and environmental improvements in the countryside, including the encouragement of much less intensive production systems, without seriously prejudicing our food supplies. But more generally, it is not only possible but very important that a compromise be struck. On the one hand, although many people value our rural heritage highly there is little evidence that this represents their first priority or that they would pay dearly for it by way of much higher food prices. And on the other, absence of compromise would only further polarize the debate over the introduction of compulsory management agreements and planning controls. There is no opportunity here to discuss the potential efficacy of the widespread use of such measures. However, they should certainly be on the statute book as measures of *last resort* to protect the best examples of our rural heritage and to control the behaviour of that minority who wish, almost as a matter of principle, to fly in the face of the interests of others. But the continued search for consensus has much to commend it, for who wants a countryside swarming with members of local amenity bodies spying on farmers who then respond by proliferating the number of 'keep out' notices?

ACKNOWLEDGEMENT

I would like to thank Paul Littlewood for helping to compile Table 20.1.

NOTES

1. The basis of calculating these figures was changed in 1975, accounting for about one-fifth of the increase of 15–16 per cent recorded between 1969–71 and 1980.
2. Temporal trends are difficult to establish in detail as several times MAFF has raised the minimum size of unit returned in the Agricultural Census. The Census also records holdings and not farms, a holding being a piece of land for which a Census return is made. Several returns may be made for one farm and this means that the average size of farms is underestimated (see Northfield Committee Report, 1979, pp. 32–35).
3. Aggregate pre-tax income for the industry after allowing for depreciation and interest repayments but excluding stock appreciation (see MAFF, 1982, Table 21).

REFERENCES

Advisory Council for Agriculture and Horticulture (ACAH) (1978). *Agriculture and the Countryside*, Robendene, Chesham.
Agricultural Development and Advisory Service, Rothamsted Experimental Station and Fertiliser Manufacturers' Association (1980). *Survey of Fertiliser Practice*, ADAS, London.
Armstrong, A.C. (1981). *MAFF Drainage Statistics 1978–80*, Technical Report 80/1, Field Drainage Experimental Unit, MAFF, London.
Beresford, T. (1975). *We Plough the Fields*, Penguin, London.
Best, R.H. (1981). *Land Use and Living Space*, Methuen, London.
Blacksell, M. and Gilg, A. (1981). *The Countryside: Planning and Change*, George Allen and Unwin, London.
Blackwood, J.W. and Tubbs, C.R. (1970). A quantitative survey of chalk grassland in England. *Biological Conservation*, 3, 1–5.
Blaxter, K. (1975). Can Britain feed herself? *New Scientist*, 65, 697–702.
Bowler, I.R. (1979). *Government and Agriculture: A Spatial Perspective*, Longman, London.
Brett, L. (1965). *Landscape in Distress*, Architectural Press, London.
Britton, D.K. (1977). Some explorations in the analysis of long-term changes in the structure of agriculture. *Journal of Agricultural Economics*, 28, 197–209.
Britton, D.K. and Hill, B. (1975). *Size and Efficiency in Farming*, Saxon House, Farnborough.
Centre for Agricultural Strategy (CAS) (1976). *Land for Agriculture*, Report No. 1, University of Reading.
Centre for Agricultural Strategy (1980). *The Efficiency of British Agriculture*, Report No. 7, University of Reading.
Coleman, A. (1976). Is planning necessary? *Geographical Journal*, 142, 411–430.
Commission of the European Community (1981). *The Situation in the Agricultural Markets—Report 1981*, The Commission, Brussels.

Countryside Commission (1977). *New Agricultural Landscapes: Issues, Objectives and Action*, Countryside Commission, Cheltenham.

Countryside Commission (1981). *National Parks: A Study of Rural Economies*, CCP 144, Countryside Commission, Cheltenham.

Countryside Review Committee (CRC) (1978). *Food Production and the Countryside*, Topic Paper 3, HMSO, London.

Davidson, J. and Lloyd, R.J. (Eds) (1977). *Conservation and Agriculture*, John Wiley and Sons Ltd, Chichester.

Dexter, K. (1977). The impact of technology on the political economy of agriculture. *Journal of Agricultural Economics*, **28**, 211–220.

Dexter, K. and Barber, D. (1967). *Farming for Profits*, 2nd edition, Iliffe Books, London.

Edwards, A.M. and Wibberley, G.P. (1971). *An Agricultural Land Budget for Britain, 1965–2000*, Studies in Rural Land Use, No. 10, Wye College, Ashford, Kent.

Feist, M.J. (1978). *A Study of Management Agreements*, CCP 114, Countryside Commission, Cheltenham.

Goode, D. (1981). The threat to wildlife habitats. *New Scientist*, **89**, No. 1237, 219–223.

Green, B. (1981). *Countryside Conservation*, George Allen and Unwin, London.

Hearne, A. *et al.* (1977). The physical and economic impact of motorways on agriculture. *International Journal of Environmental Studies*, **11**, 29–33.

Hooper, M.D. (1977). Hedgerows and small woodlands. In J. Davidson and R.J. Lloyd (Eds), *Conservation and Agriculture*, John Wiley and Sons Ltd, Chichester, pp. 3–32.

Hooper, M.D. and Holdgate, M.W. (Eds) (1968). *Hedges and Hedgerow Trees*, Monks Wood Experimental Station Symposium 4, The Nature Conservancy.

Jones, C.A. (1973). *The Conservation of Chalk Downland in Dorset*, Dorset County Council, Dorchester.

Keenleyside, C. (1977). Voluntary action in conservation. In J. Davidson and R.J. Lloyd (Eds), *Conservation and Agriculture*, John Wiley and Sons, Chichester, pp. 147–171.

King, A. and Conroy, C. (1980). *Paradise Lost? The Destruction of Britain's Wildlife Habitats*, Friends of the Earth, London.

Leonard, P.L. and Stokes, C. (1977). Landscape and agricultural change. In J. Davidson and R.J. Lloyd (Eds), *Conservation and Agriculture*, John Wiley and Sons, Chichester, pp. 121–143.

Mabey, R. (1980). *The Common Ground*, Hutchinson/Nature Conservancy Council, London.

MacEwen, A. and MacEwen, M. (1982). *National Parks: Conservation or Cosmetics?* George Allen and Unwin, London.

MAFF (Ministry of Agriculture, Fisheries and Food) (1970). *Modern Farming and the Soil*, Report of the Advisory Council on Soil Structure and Soil Fertility, HMSO, London.

MAFF (1975). *Food From our Own Resources*, White Paper, Cmnd 6020, HMSO, London.

MAFF (1976). *Wildlife Conservation in Semi-natural Habitats on Farms: A Survey of Farmers' Attitudes and Intentions in England and Wales*, MAFF, HMSO, London.

MAFF (1977). *The Changing Structure of Agriculture 1968–1975*, MAFF, HMSO, London.

MAFF (1979). *Farming and the Nation*, White Paper, Cmnd 7458, HMSO, London.

MAFF (1982). *Annual Review of Agriculture 1982*, Cmnd 8491, HMSO, London.

Mellanby, K. (1975). *Can Britain Feed Itself?* Merlin Press, London.

Moore, N.W. (Ed) (1966). Pesticides in the environment and their effects on wildlife. *Journal of Applied Ecology*, **3** (Suppl.).

Moore, N.W. (1977). Arable land. In J. Davidson and R.J. Lloyd (Eds), *Conservation and Agriculture*, John Wiley and Sons Ltd, Chichester, pp. 23–43.

Munton, R.J.C. (1974). Agriculture and conservation in lowland Britain. In B. Goldsmith and A. Warren (Eds), *Conservation in Practice*, John Wiley and Sons Ltd, Chichester, pp. 323–336.

Nature Conservancy Council (NCC) (1977). *Nature Conservation and Agriculture*, Nature Conservancy Council, London.

Nature Conservancy Council (1982). *Seventh Report: 1st April 1980–31st March 1981*, Nature Conservancy Council, HMSO, London.

National Farmers' Union/Country Landowners' Association (NFU/CLA) (1977). *Caring for the Countryside*, National Farmers' Union (NFU)/Country Landowners' Association (CLA), London.

Newby, H., Bell, C., Saunders, P., and Rose, D. (1977). Farmers' attitudes to conservation. *Countryside Recreation Review*, **2**, 23–30.

Northfield Committee Report (1979). *Report of the Committee of Inquiry into the Acquisition and Occupancy of Agricultural Land*, Cmnd 7599, HMSO, London.

O'Riordan, T. (1981). Farming and conservation—Broads style, *Ecos*, **2(2)**, 19–22.

Parker, D.J. and Penning-Rowsell, E.C. (1980). *Water Planning in Britain*, George Allen and Unwin, London.

Parry, M., Bruce, A., and Harkness, C. (1981). The plight of British moorland. *New Scientist*, **May**, 550–551.

Perring, F.H. and Mellanby, K. (Eds) (1977). *Ecological Effects of Pesticides*, Academic Press, London.

Porchester, Lord (1977). *A Study of Exmoor*, DOE/MAFF, HMSO, London.

Rackham, O. (1971). Historical studies and woodland conservation. In E. Duffey and A.S. Watt (Eds), *The Scientific Management of Animal and Plant Communities for Conservation*, Blackwell Scientific Publications, pp. 563–580.

Reid, I.G. (1981). Farm finance and farm indebtedness in the EEC. *Journal of Agricultural Economics*, **32**, 265–274.

Ritson, C. (1980). *Self-sufficiency and Food Security*, Centre for Agricultural Strategy, Working Paper No. 8, University of Reading.

Royal Commission on Environmental Pollution (1979). *Agriculture and Pollution*, 7th Report, HMSO, London.

Shoard, M. (1980). *The Theft of the Countryside*, Temple-Smith, London.

Webb, N.R. and Haskins, L.E. (1980). An ecological survey of heathlands in the Poole basin, Dorset, England in 1978. *Biological Conservation*, **17**, 281–296.

Weller, J. (1967). *Modern Agriculture and Rural Planning*, Architectural Press, London.

Westmacott, R. and Worthington, T. (1974). *New Agricultural Landscapes*, Countryside Commission, Cheltenham.

Wibberley, G.P. (1976). Rural resource development in Britain and environmental concern. *Journal of Agricultural Economics*, **27**, 1–18.

Worthington, T.R. (1979). *The Landscapes of Institutional Landowners*, Working Paper 18, Countryside Commission, Cheltenham.

CHAPTER 21

Economics and Conservation: The Case of Land Drainage

JOHN BOWERS

An important set of wetland habitats, the grazing marshes located on the floodplains of the rivers of England and Wales are at present under serious threat from large-scale drainage for agricultural purposes.

The two major blocks of these marshes are the Norfolk Broads, along the valleys of the rivers Yare, Bure and Waveney, and the Somerset levels and moors constituting the floodplain of the river Parrett and its tributaries, the Yeo, Tone, Isle and Cary, and, to the North, the basin of the river Brue. Other important areas are the Washes of the Nene and Ouse in Cambridgeshire, the Derwent Ings in Yorkshire—a grade 1 site as listed in *A Nature Conservation Review* (Ratcliffe, 1977)—Amberley Wild Brooks along the river Arun in West Sussex, the river Soar in Leicestershire, parts of the Worcestershire Avon, and the river Stort in Hertfordshire.

Agriculturally these marshes are permanent pasture providing summer grazing for livestock, predominantly cattle, and in some places, a hay crop. They support a range of neutral grassland communities, some of very limited distribution and often species-rich. Many also include small areas of peat, which are often designated as SSSI's. This is so on the Stort and on Halvergate Marshes. Amberley Wild Brooks has a valuable oligotrophic mire. These communities are maintained by a high water table and, in many cases, by regular and frequent flooding. This flooding results either from the river overtopping its banks—and some areas are deliberately kept as washlands—or because of limitations in the capacity of the river system to cope with runoff from the surrounding areas. In addition to the botanical value of the pasture, many of these marshes have wet ditches which contain rare plant and insect communities. Such ditches serving as field boundaries are characteristic of the Somerset levels and of the marshes of Broadland. The latter are a habitat for the dragonfly (*Aeschna isosceles*) which is confined to the sea. Similar dykes on Amberley Wild Brooks contain a rich population of dragonflies, notably *Brachytron pratense, Cordulia aenea* and *Coenagrion pulchellum*. The marshes are

375

botanically and entomologically richest where inundations contain no or little salt water. Salt water seepage is a major factor limiting the ecological diversity of some of the Broadland marshes around Breydon Water.

The grazing marshes are also important as breeding grounds for various species of waders including national rarities such as the blacktailed godwit (*Limosa limosa*) and as winter feeding areas for large numbers of wildfowl, particularly where regular flooding occurs as on the Ouse and Nene Washes, the Derwent Ings and Amberley Wild Brooks. Virtually the whole of the internationally important UK wintering population of the Bewick swan (*Cygnus columbianus bewickii*) is to be found in the grazing marshes that we are discussing.

Drainage of these marshes would usually require alterations to the main river system either to increase its capacity to move water or to reduce the frequency with which it overtops its banks. This in turn is achieved by raising the banks themselves or by other means of protection against tidal surges. The main river system is the responsibility of the regional water authorities. Work is often also necessary to improve the functioning of subsidiary water courses and to create the conditions by which the water table may be lowered. These works are the responsibility of internal drainage boards (IDBs) where they exist; where they do not, this work is also the responsibility of the water authorities, who, under the Land Drainage Act 1976 have powers to take over non-main river functions. Finally drainage requires that individual farmers install field under-drainage. Where works by water authorities and IDBs are not required, field level drainage can be expected to proceed piecemeal, and indeed this is happening on a substantial scale.

The extent of the threat to these areas can be gauged by the fact that there exist schemes by water authorities in various states from formulation to execution for all but two of the areas listed above. The two exceptions are Amberley Wild Brooks where a drainage proposal was defeated at a Public Inquiry in 1978, and the Ouse Washes where most of the potentially improvable land is owned by conservation bodies.

But to describe only these cases understates the problem for two reasons. First, we have not mentioned grazing marshes where reclamation has already been carried out or is in process. An example of the former would be the extensive marshes along the Arun to the north and south of Amberley; the Wild Brooks was the last phase of a comprehensive scheme for the middle Arun. An example of work in progress is Idle Washes on the Yorkshire–Nottinghamshire border, an extensive area of grazing marsh that has now been virtually reclaimed. Second, we have only listed the larger sites and a selection of smaller sites where opposition to schemes is under way. Perhaps a better indication of the scale of the problem is given by an analysis of the 5-year capital works programme of one regional water authority—Anglian Water Authority (Table 21.1). After eliminating expenditure on sea defences (row 1) and what we have

TABLE 21.1   Analysis of Anglian Water Authority capital works programme
1981/82–1985/86.

Division	Lincolnshire	Great Ouse	Essex	Norfolk and Suffolk	Welland and Nene	Total £'000
Total	21 641	12 116	10 207	30 636*	12 161	86 761*
of which:						
Sea defence,	5634	4569	6544	13 774	5825	36 346
Technical	571	1847	17	180	42	2657
Net expenditure	15 436	5700	3646	16 682*	6294	47 758*
of which:						
Potential environmental						
problem	960	4331	1623	13 132*	2873	22 919*
%	6.2	76.0	44.5	78.7	45.6	48.0

Source: Author's estimates.
* Includes £10 555 000 for the Yare Barrier.

called 'technical' (for example, telemetry, flood warning systems, equipment repair and storage facilities) (row 2) we are left with expenditure on agricultural and urban land drainage and flood protection. The Conservation Planning Division of the Royal Society for the Protection of Birds examined the schemes embodied in the residue and identified those where there was potentially an environmental problem in that valuable habitats might be destroyed or damaged; 48 per cent of expenditure constituting nearly £23 million at 1979 prices was so classified representing almost 50 projects. Much of the remaining expenditure would be for urban flood protection.

An indication of the scale of some of the drainage projects can be given by considering the two major areas of grazing marsh.

In the Norfolk Broads, a proposal by the Anglian Water Authority to construct a tidal barrier across the mouth of Breydon Water at Great Yarmouth is moving towards a public inquiry. This barrier would afford protection against tidal surges for the main rivers of Broadland (Yare, Bure and Waveney). The putative benefits are overwhelmingly agricultural. They take the form of the wholesale conversion of grazing marsh to arable and improved grassland. Over the 20-year period for which the project is appraised, conversion of between 9000 and 14 000 ha of marshland is envisaged.

The achievement of these benefits requires drainage investment additional to the barrier: first, by internal drainage boards in lowering the water table on the marshes by deepening dykes and installing better pumps, and second, by the individual farmers in field underdraining. Additionally there would need to be substantial investment by farmers in roads, buildings, plant and machinery.

Whether in anticipation of the Yare Barrier or, more probably, because it is unnecessary to achieve the desired benefits, some of the internal drainage board investment that was expected to follow the Yare Barrier is already taking place.

An application for grant-aid to pump-drain Halvergate Marshes, the largest single block of grazing marsh in Broadland has been made. Over 20 years it is forecast that this scheme will result in the conversion, overwhelmingly to the growing of winter wheat, of 1740 ha. This constitutes about 15 per cent of the benefits from the Yare Barrier. A number of other internal drainage board schemes are also in the process of formulation and a substantial bank-raising project along the Waveney has been proposed by the Anglian Water Authority. It is evident also that the rate of conversion of marsh to arable without internal drainage board investment, although not necessarily without bank-raising by the Anglian Water Authority, is increasing, although precise measurement is not possible.

The second example is the other major English area of grazing marsh, the Somerset levels and moors. A proposal exists for the construction of a tidal barrier on the Parrett near Bridgwater. This has been subjected to preliminary cost-benefit appraisal and is now subjected to search for other agricultural benefits to raise the estimated rate of return. But much more than in the Broads, land drainage in the Somerset levels requires that the capacity of the river system to carry flood water be increased and the control of tidal surges through a barrier is only part of the means of achieving this. The Parrett basin scheme embraces much more than a tidal barrier and interlocks with schemes for improving the drainage of all the rivers' tributaries. The capital works programme of the Wessex Water Authority for 1980–85 contains a commitment of £846 000 for the Parrett basin scheme with an additional £374 000 in associated schemes. There is a further £155 000 for a scheme on the Cary, £60 000 on the Tone and £952 000 on the Yeo. More substantial improvements on the tributaries await progress on the main Parrett basin scheme. Advance commitments for the quinquennium from 1985 include £212 000 on the Tone, £348 000 on the Isle and substantial additional works on the Yeo, all at 1978 prices. This programme of expenditure is aimed at increasing the capacity of river system to permit the drainage of the entire area of the levels and moors that the rivers serve. A similar ambitious programme is the Brue Basin improvement scheme, although this does not at present include a tidal barrier.

As with the Broads, the objective of this major programme is almost entirely to improve agriculture. The main difference is that climate and soils dictate that conversion will be largely to improved grassland that will carry higher densities of livestock than at present rather than to arable. The effect on the environment will be equally disastrous.

While the reclamation of grazing marshes has been continuing for a very long time it is clear that the last decade has seen a marked increase in the rate of their agricultural improvement, to the extent that what we are now seeing is a comprehensive attack upon the surviving marshland. To a degree, the attack is coordinated as well as comprehensive. At root the explanation lies with the agricultural policy of the postwar period which has aimed at maximizing the growth of agricultural output by way of structural improvements such as land drainage. There are two specific triggers for the activities of the drainage authorities: United Kingdom entry into the EEC and the 1973 Water Act. One response to our entry into the EEC was a policy to increase UK agricultural production as a means of countering the adverse balance-of-payments impact and reducing the net contribution to the Community budget. This line of argument is contained in *Food From Our Own Resources* (1973) which presents an expansion programme for agriculture, and is repeated with modifications in a subsequent White Paper, *Farming and the Nation* (1977). It appears that at the time of publication of *Food From Our Own Resources*, a directive was issued to water authorities to consider ways in which their land-drainage divisions might assist the policy.

Section 24 (5) of the Water Act 1973 enjoined the water authorities to carry out surveys of land drainage problems within their boundaries 'from time to time and in any case at such times as the Minister may direct'. The Minister of Agriculture, Fisheries and Food requested that the first survey should be reported on in 1978 while recognizing that further time would be required to complete all the detailed investigations necessary and that therefore further reports would be required to complete the surveys.

The primary purposes of the surveys are to identify and evaluate flooding and land drainage problems within the area concerned, either a water authority division corresponding probably to a main river catchment area or a local land drainage district, to suggest solutions and accord priorities. They provide the basis for the preparation of plans for future works in fulfilment of the authorities' land-drainage functions under the Act (under s 24(6).) The problems identified in these surveys are not confined to those where responsibility lies with the water authority, the so-called main river works. Section 1 of the Land Drainage Act 1976 gives the water authorities a general supervisory role over all matters relating to land drainage. The surveys can, therefore, identify problems on non-main river water courses where responsibility lies with internal drainage boards or local authorities. They encompass drainage problems that affect urban property as well as agricultural land and range from the major to the relatively trivial. Problems may be trivial in two respects: the size of the area affected and, where flooding occurs, its frequency. The comprehensiveness of these surveys is such that there is probably little or no agricultural land in England and Wales of more than 20 or 30 ha in extent subject to fluvial or tidal flooding with a return frequency greater than 1 year in 10 that is not listed as a 'problem' in a s 24(5) survey and for which a solution is not proposed. The surveys are thus, as is intended, a source book for all authorities with land drainage responsibilities: water authorities, internal drainage boards and local authorities.

While we do not known the precise relationship between a s 24(5) survey and a capital works programme of a water authority it is clear that these surveys provided the initial database for the expansion of agricultural drainage to which the Ministry of Agriculture, Fisheries and Food's policy of increased self-sufficiency in foodstuffs was the spur.

Almost all the finance for this drainage programme derives ultimately from taxation, that is it is public expenditure. Grant aid is paid to water authorities at an average rate of 55 per cent, to internal drainage boards at a rate of 50 per cent and to individual farmers at a rate of 37½ per cent. The rest of the expenditure of water authorities and internal drainage boards comes from local rate income. Private money therefore appears only with farm-level investment. Expenditure by water authorities and internal drainage boards on agricultural drainage was about £20 million in 1980–81. The bulk of this related to the improvement of grazing marshes.

If investment is wholly or largely financed by public expenditure then it is public investment. Public investment proposals should be subjected to social cost-benefit analysis to determine their social rate of return (see the bibliography for supplementary reading). Efficiency in the allocation of public investment funds between competing uses is conventionally approached by requiring that, if they are to be financed, projects should achieve a test rate of discount. In land-drainage projects the test discount rate is presently taken to be 5 per cent. In order to attract grant aid therefore, agricultural drainage and flood protection investment of water authorities and internal drainage boards must be subjected to cost-benefit analysis and offer an expected rate of return of at least 5 per cent. No such requirement exists for farm-level investment. Previously, grant-aid applications from individual farmers were scrutinized for financial viability but this practice was dropped in 1981. However, where water authority or internal drainage board investment has consequential farm-level investment which is necessary if the benefits are to be achieved—and this is normally the case—the cost-benefit analysis relates to the entire capital expenditure including this induced investment.

The solutions to the problems identified in the s 24(5) surveys are subjected to cost-benefit appraisal and the benefit-cost ratio is presumably used, together with other criteria, to determine priorities for the capital works programmes of the water authorities. This probably constitutes one of the most extensive uses of cost-benefit analysis in the area of public investment decisions, comparable perhaps with the use made of the technique in the roads programme. The s 24(5) surveys contain the results of hundreds of cost-benefit appraisals. The extent to which they are reported varies greatly between authorities: in some, no more than a code is used to indicate the level of the benefit–cost ratio; in others total costs and benefits are also reported, sometimes with a division into a few broad categories; and in a few, details of the costs and benefits are reported with guidance as to how they were estimated. The analyses are necessarily done to some standard formula and are largely carried out by water authority engineers. MAFF provide detailed guidance notes on the conduct of cost-benefit appraisal and the Agricultural Development and Advisory Service is available to give advice. In addition to the cost-benefit analyses under s 24(5), separate more detailed studies are performed for the larger schemes usually by outside consultants.

I have examined a sample of cost-benefit appraisals embodied in s 24(5) surveys together with a number of more detailed studies by consultants of larger projects. The sample is in no sense a scientific one. The water authorities do not regard the s 24(5) surveys as public documents nor do they consider the consultant's studies as such. Hence the sample is confined to what I have been able to get from various conservation bodies. Nonetheless they are probably reasonably representative and include a number of the larger and more controversial projects including the Yare barrier (Rendell, Palmer, and Tritton, 1977); Halvergate (Dosser, John, and Partners, 1980); the Arun (Middlesex Polytechnic Flood Hazard Research Project (MPFHRP), 1976 and 1978); and various Wessex Water Authority documents on the Parrett Basin Improvement Scheme (the scheme together with that for the Brue Basin is included also in Wessex Water Authority, 1979). They also cover a number of smaller projects including one for Seaton Marshes, Devon.

My conclusion is that the studies contain a number of basic technical defects, all of which have the effect of exaggerating the benefits from agricultural land drainage. Most of the analyses that I have seen, and all of the larger projects, yield rates of return greater than the test discount rate. The Yare Barrier Study (Rendell, Palmer, and Tritton, 1977) was made to yield a rate of return of greater than 10 per cent, which was the level of the test discount rate at that time. When the technical defects are corrected, the true social rates of return on the best projects are of the order of 2–3 per cent, about the same as an afforestation project (Bowers, 1982a). Some of the worst projects have no true positive rate of return at all. If these defects are general to all such appraisals, and I believe that the major ones are, then a number of consequences follow:

(1) The ranking of agricultural projects is too high relative to urban projects, i.e. they impart a bias against urban flood relief.
(2) The ranking between alternative agricultural projects is incorrect.
(3) Given that capital rationing is achieved by use of a test discount rate, too much social capital is devoted to projects for agricultural land drainage. This will arise because agricultural drainage projects which achieve a true rate of return which is less than the test rate of discount are accepted, while non-agricultural non-drainage projects which also fail to meet the test discount rates but offer higher returns than the drainage projects, are rejected.
(4) Given that agricultural drainage projects have deleterious effects on the natural environment and the landscape, the volume of environmental damage done is greater than it would be if there were proper evaluation.

That there are serious consequences for the natural environment is probably beyond dispute (Nature Conservancy Council, 1977). On landscape, professional opinion is also united (for a summary of the case see Shoard, 1980). Both the Broads Authority and the Countryside Commission were sufficiently concerned at the impact of the Halvergate scheme on the landscape to enter negotiations for the payment of compensation for its protection. A technical treatment of these defects can be found in Bowers (1983). In summary they are set out in the following eight sections.

The appropriate level of flood protection

A precondition for cost-benefit analysis is that the proposed investment project be cost-effective in the sense that, given the objective of the project, the lowest cost method of obtaining the benefits is considered. If the project is not cost-effective, in other words there is a cheaper way of getting the same benefits, then cost-benefit analysis is inappropriate and indeed unnecessary since it cannot be socially efficient to carry it out. On its stated objective, the Yare Barrier is not cost-effective in this sense and it is likely that other projects as well are not cost-effective. To explain this we need to consider in more detail the objectives of these schemes.

Agricultural benefits from land drainage and flood protection schemes can take three forms:

(1) Damage avoidance—avoidance of loss of agricultural output that would have resulted from flooding. Most of the benefits from urban flood protection take the form of damage avoidance.
(2) Increased output from existing enterprises as a result of lowered water-tables and longer growing and grazing seasons.
(3) Changes to higher yielding or more valuable enterprises as a result of reduced flood risk or lowered water tables.

Damage avoidance is not normally an important source of benefit from the schemes we are considering, since the agricultural practices are already adjusted to the flood risk that exists—for example, limited summer grazing of livestock. In some cases frequent flooding may be necessary to maintain the grass sward. In the larger schemes damage avoidance is often ignored completely.

Increased yields from existing farming practices are sometimes important in smaller water authority and internal drainage boards schemes. An example is a pumping scheme on North Duffield Carrs, part of the Derwent Ings, where the benefits are said to be increased hay yields and a longer grazing season as a result of a more rapid removal of surface water in the spring and autumn.

The third category of benefit is the major one in most schemes and constitutes virtually the whole of the benefit for the larger ones such as the Yare

Barrier. The typical case would be where reduced flood risk allows farmers to convert from permanent pasture to arable, or to reseed the grassland and increase densities of grazing animals. The benefits are then the enhancement of gross margins—that is, the increase in the value of agricultural output, less any increase in the variable costs (seeds, fertilizers) of attaining it. If changing agricultural practices require increased capital investment by the farmer in the form of underdrainage of fields, farm machinery, roads, etc., then this appears on the cost-side of the cost-benefit calcualtion.

If the benefits of a flood protection or drainage scheme take the form of this third category, then it is possible to calculate what reduction in flood risk is necessary to make it worth the farmer's while to improve his land. Figure 21.1 shows an example of such an exercise for the Yare Barrier. The assumed change is from summer grazing to the growing of winter wheat. The benefits are the increased gross margin less any losses resulting from flood damage. Floods are assumed to give no loss with summer grazing; with winter wheat there is a loss of the crop in the year of the flood and lower yields for some subsequent years. Given the capital cost of the farmer's investment we can plot the rate of return against the flood risk. The two lines refer to different rates of recovery from flooding. All data are from the consultant's report (Rendell, Palmer, and Tritton, 1977).

The level of protection necessary if the farmers are to invest may be read off once we know the required rate of return on the investment—10 per cent is considered normally to be a good rate of return and lower rates are often acceptable; 15 per cent may be put as an upper limit (these are real rates: the data are

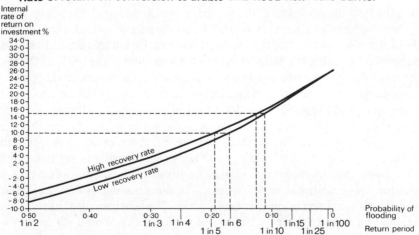

FIGURE 21.1. Based on a model explained in Bowen (1982c).

in constant prices). This suggests that the required level of flood protection is between 1 year in 5 return period and 1 year in 10. But this level of protection is already exceeded over much of the area, and the cost of bringing the whole area up to that level would be much less than that provided by the Yare Barrier which aims at a protection level of 1 year in 100. Thus the scheme is not cost-effective. Other schemes also aim at unjustified levels of flood protection.

This example is worth considering further since it brings out some other important points about agricultural drainage schemes. First, it has the implication that the targeted rate of flood protection falls as the difference between improved and unimproved gross margins increases. The drainage authorities appear to assume the opposite: that higher flood protection is justified because of rising prices for arable crops. This would be true on the avoidance of damage approach if the benefits of crops saved exceeded the costs of enhanced protection levels. It is clear that in general this is not the case.

Second, since the existing level of protection in the Broads, and in some of the other areas that we have mentioned already, exceeds what is required to induce conversion of the marshes one would expect that improvement would be taking place without the drainage improvements. This is indeed happening. Protection of these areas for conservation will thus not be achieved simply by stopping drainage improvements. This will be sufficient in some areas where flood return periods and/or the water table is very high, but the driving force is the high prices of cereal crops, and it is to the structure of farm prices and the support system that generates them that conservation bodies should be looking.

The time-path of conversion

Given the benefits per hectare, the return on the investment depends on the rate at which the benefits are realized—that is, on the rate of reclamation or conversion of the land. In most of the studies I have examined this critical variable is determined by asking the farmers affected about their plans (usually the larger farmers or those indicating greatest interest or enthusiasm) and making *pro rata* assumptions about the rest. No attempt is made to explore the determinants of the decision, and the implicit model is that of inadequate drainage as the principal constraint on conversion. Certainly in the major wetland areas this model appears to be wrong. The striking feature of these areas is the extremely fragmented pattern of land holdings and it is probably truer to say that this factor explains the inadequate drainage, consolidation being a necessary precondition for agricultural improvement. The sheer numbers of holders of small and scattered holdings inhibits consolidation. The schemes seem likely to work by providing a capital gain for holders of small blocks of land which they may realize by selling to those who wish to consolidate holdings. One implication of this is that the social consequences of schemes such as the Yare Barrier and the Parrett Basin improvement scheme are much wider than the proponents seem to realise. Another is that the rate of conversion seems likely to be a slower process than is assumed, the cumulative conversion curve having a positive rather than negative second derivative as is normally assumed (it is typically assumed that 50 per cent of the land is converted in much less than half of the appraisal period). Certainly the case for assuming a rapid conversion of most of the area, the normal assumption, is not established, particularly when many farmers are involved.

The valuation of benefits

A requirement of cost-benefit analysis is that costs and benefits should be counted at their social and not their private valuation. One important source of difference between social and private valuation is the presence of taxes and subsidies. The private capital cost of drainage investment by water authorities and internal drainage boards is less than half of the social cost because these bodies receive grant aid. Similarly the value of agricultural output to the farmer is greater than the social valuation (and the cost of inputs greater than the social valuation) because of the farm price support system. Much of the farmer's receipts is subsidy from consumers. In calculating the social benefits from drainage schemes this subsidy must be deducted just as in calculating social costs grant aid must be added.

Capital grant aid is readily identified and is correctly treated in all the cost-benefit analyses I have seen. Farm price support which comes mainly from the consumer rather than the taxpayer is not easy to measure and in all cases is ignored. That it exists is evident from the large sums levied on foodstuffs entering the EEC and from the rebates paid on exports from the EEC, as well as from the notorious lakes and mountains that the Common Agricultural Policy (CAP) generates (Fennell, 1979). These consumer subsidies are very large; the extreme is the case of dairying where all of the gross margin is effectively subsidy (Black and Bowers, 1981). Failure to take account of them is the largest single source of overstatement of rates of return. Typically the benefits when measured at farm-gate prices (that is, the price that the farmer receives) overstate the social benefits by a factor of two or more.

There exist a number of possible arguments for the use of farm-gate prices for measuring benefits, but they are all unsound (Bowers, 1982c). Perhaps the most seductive is that the subsidy is a social benefit to United Kingdom citizens since it is paid by consumers in other EEC countries. This is in fact not so. A substantial proportion will come from UK consumers (Bowers, 1982b); but in any case the commitment to a common agricultural policy entails a requirement to work in common resource costs. Any other valuations are meaningless.

The problem of production quotas

Another aspect of the confusion between private and social benefits concerns the treatment of crops which are controlled by production quotas. Two enterprises with high gross margins are subject to effective quotas: potatoes and sugar beet. In such cases the crops may only be grown on the land to be drained, at the expense of their being grown elsewhere. The opportunity cost valuation is then the gross margin on the land elsewhere that is released from growing these crops and not the gross margin on the crop itself, plus perhaps an adjustment for differential yields. This problem is not recognized in the studies I have examined.

Agricultural yields

The typical practice in constructing cost-benefit studies of drainage schemes is to take theoretically attainable yields for the improved land. This is plainly inconsistent. The theoretically attainable yields are only reached on the best farms and then not consistently. Actual average yields for the soil type and region are much more appropriate since the theoretically attainable does not constitute a benefit to society.

Flood risk

We have already noted that flood damage resulting from existing agricultural practices on the areas we are considering is probably negligable since the agricultural system is adjusted to the risk that exists. Once agricultural improvements take place however, serious damage can result from flooding. Drainage improvements lower the flood risk but they do not eliminate it. Indeed the return period is usually specified in the scheme design. Deductions from the benefits must be made for losses expected from this lower return period. This is rarely done: the appraisal is often made as though the scheme eliminated flood risk altogether. Not much distortion results if the design return period is 1 year in 100 but where it is 1 in 10 or 1 in 15 considerable overstatement of the benefits results, particularly where the expected flooding has a high salinity.

The project appraisal period

The period over which the benefits and costs of a scheme should be measured is determined by the length of life of the capital works. The decision as to whether to renew the structures is a separate investment decision to be taken when the time comes. Where works by a water authority induce further works by internal drainage boards and investment by farmers, calculating the length of life may be quite complicated. One way of dealing with the uncertainty attaching to benefits occurring many years in the future is to reduce the length of period over which benefits are considered.

In the analyses that I have examined it looks as though length of life is thought of as a variable which may be set to give the desired result of an acceptable rate of return. In one of the internal water authority appraisals (Seaton Marshes) the benefits were summed to infinity. This allowed the project to pass at the test rate of discount, whereas for the period used in that same Authority's s 24(5) surveys (20 years) the project failed. In the case of the Yare Barrier, the length of life was increased between the first and second phases of the study without an explanation being given. This had the effect of course of increasing the rate of return. The Halvergate scheme, which is part of the benefits of the Yare Barrier, is appraised over a longer period than its parent.

Treatment of intangibles

The evaluation of the measurable costs and benefits is only part of a properly constituted cost-benefit analysis. Additionally the unquantifiable effects, the intangibles, should be listed and an attempt made to assess their importance. If a project has substantial deleterious intangible effects, for example on the natural environment or on the landscape then the decision whether to go ahead is essentially a political one involving a subjective weighing of the measurable benefits agains the unmeasurable disbenefits. The more detailed is the information about these unmeasurables, the more informed is the political decision. The subjective evaluation of environmental effects depends, like the valuation of measurable effects, essentially on scarcity. The importance of environmental effects is evidenced by the fact that the two schemes that have gone to public inquiry, Amberley Wild Brooks and Gedney Drove End, were both turned down largely on the grounds of their environmental impact.

While the listing of environmental effects is normal in other uses of cost-benefit analysis such as the roads programme (Nash, 1979), until recently it was unheard of in land-drainage studies. While studies were carried out by conservation bodies objecting to the schemes (Land Use Consultants, 1978; Nature Conservancy Council, South-West Region, 1977; Warburton, 1978), the drainage bodies have taken the line that this was outside their remit. The consequence must be that undue weight is given to agricultural effects. This may be changing with the obligations imposed on drainage authorities by the Wildlife and Countryside Act 1981 and an environmental study has now been produced for the Soar Valley Scheme (Severn-Trent Water Authority, 1982). The attitude beginning to appear is that under pressure from conservation bodies, steps may be taken to protect SSSIs, local nature reserves, etc. Apart from problems with this sort of compromise (for example, the danger of deterioration of sites due to 'island' effects given that SSSIs are not drawn with specific threats in mind), intangibles are still not appearing in the social accounting framework in which the decisions ought to be taken.

CONCLUSION

We have seen that the cost-benefit analyses which underlie the serious and growing threat to an important category of wetland habitats are seriously defective. By way of conclusion I shall endeavour to explain why this state of affairs has come about.

Cost-benefit analysis is a useful aid to decision-making where an authority has a number of competing projects between which, in the absence of the sort of information cost-benefit analysis reveals, it is indifferent. These conditions are not normally satisfied with agricultural land-drainage projects where the proponents have often only one project to which, especially in the case of internal drainage boards, they are firmly committed. The check on the consequent biases should be administered by MAFF which has charge of the disbursement of grants. That this control is ineffective is clear, and the techniques we have criticized receive at least tacit approval. The explanation is that MAFF is not really concerned at all with the socially efficient use of its funds. Its objective is the maximization of the growth of UK agricultural output. For this purpose the existing practices are probably satisfactory. The ultimate arbiter of the national interest in public investment is the Treasury. Grant applications require Treasury approval if they exceed a threshold size. Unfortunately many agricultural drainage projects, including some of those with the most serious ecological impact, do not exceed the threshold. Furthermore, various devices such as phasing of the schemes are used to keep projects below the threshold. Thus what is plainly one scheme at Halvergate is presented as three separate schemes with three separate cost-benefit analyses for this reason (that at least is the view of those I have spoken to; it is otherwise incomprehensible). Even if a proposal does reach the Treasury it is unlikely that a proper examination of its economics and environmental impact will occur in the absence of a public inquiry. The water authorities are for the most part determined to avoid an inquiry and the history of land drainage projects over the last 5 years may be written in terms of a series of battles to bring the schemes under proper public scrutiny.

REFERENCES

Black, C.J. and Bowers, J.K. (1981). The level of protection of UK agriculture, *School of Economic Studies Discussion Paper, 99*, University of Leeds, 42 pp.

Bowers, J.K. (1982a). The economics of afforestation. *Ecos*, 3(1), 4–7.

Bowers, J.K. (1982b). Who pays for the expansion of UK agriculture? *School of Economic Studies Paper 110*, University of Leeds, 11 pp.

Bowers, J.K. (1983). Cost-benefit analysis of wetland drainage, *Environment and Planning*. A, 15, 227–235.

Dosser, John and Partners (JDP)(1980). Three reports for the Lower Bure, Halvergate Fleet and Acle Marshes for land drainage improvement schemes on Seven Mile and Berney levels; Manor House, Halvergate and Tunstall, Acle, Calthorpe, and Tracey levels. (*mimeo*).

Fennell, R. (1979). *The Common Agricultural Policy of the European Community*, Granada Publishing Ltd., London, 243 pp.
Land Use Consultants (1978). *The Landscape Assessment of the Yare Basin Flood Control Scheme*, Countryside Commission, Eastern Region, Cambridge, 28 pp.
Middlesex Polytechnic Flood Hazard Research Project (MPFHRP) (1976). *Proposed Embankment and Drainage Scheme for the Middle Arun: Benefit Assessment*, Enfield, 117 pp.
Middlesex Polytechnic Flood Hazard Research Project (1978). *Proposed Drainage Schemes for Amberley Wild Brooks, Sussex: Benefit Assessment*, Enfield, 96 pp.
Nash, C.A. (1979). The use of economic appraisal in transportation analysis. *Omega*, 7, 441–450.
Nature Conservancy Council (1977). *Nature Conservation and Agriculture*, HMSO, London, 34 pp.
Nature Conservancy Council, South-West Region (1977). *The Somerset Wetlands Project. A Consultation Paper*, Nature Conservancy Council, Taunton, 22 pp.
Ratcliffe, D.A. (1977). *A Nature Conservation Review, 2 volumes*, Cambridge University Press, Cambridge.
Rendell, Palmer and Tritton (RTP) (1977). *Yare Basin Flood Control Study, 3 volumes*, Anglian Water Authority, Norwich, 150, 114, and 360 pp.
Severn-Trent Water Authority (1980). *Land Drainage Survey: Upper Trent Division and Staffordshire*, Birmingham, 204 pp.
Severn-Trent Water Authority (1982). *The Environment (sic) Implications of the Proposed Soar Valley Improvement Scheme*, Birmingham, 22 pp.
Shoard, M. (1980). *The Theft of the Countryside*, Temple-Smith, London, 272 pp.
Warburton, S. (Ed) (1978). *The Yorkshire Derwent. A Case for Conservation*, Yorkshire Naturalists Trust Ltd., York, 73 pp.
Wessex Water Authority (1979). *Land Drainage Survey: Somerset Local Land District*, Bridgwater, 184 pp.

FURTHER READING

Cost-Benefit analysis: elementary texts are Abelson (1979) and Frost (1975). More advanced treatment is given in Mishan (1971) and Pearce and Nash (1981).
Agricultural appraisal: two introductory texts of farm management techniques are Barnard and Nix (1979) and Sturrock (1972). Technical definitions of gross margins etc., and useful statistical material for cost-benefit appraisal is in Nix (annual).

Abelson, P. (1979). *Cost-Benefit Analysis and Environmental Problems*, Saxon House, London, 208 pp.
Barnard, C.S. and Nix, J.S. (1979). *Farm Planning and Control, 2nd edition*, Cambridge University Press, Cambridge, 180 pp.
Frost, M.J. (1975). *How to Use Cost-Benefit Analysis in Project Appraisal*, Gower Press, London, 243 pp.
Mishan, E.J. (1971). *Cost-Benefit Analysis*, George Allen and Unwin, 364 pp.
Nix, J. (annual). *Farm Management Pocketbook*, Wye College, Ashford, approx. 160 pp.
Pearce, D.W. and Nash, C.A. (1981). *The Social Appraisal of Projects*, Macmillan Press, London, 225 pp.
Sturrock, F.G. (1972). *Farm Accounting and Management, 6th edition*, Pitman, London, 228 pp.

Conservation in Perspective
Edited by A. Warren and F.B. Goldsmith
Published by John Wiley & Sons Ltd.

CHAPTER 22

National Parks:
A Cosmetic Conservation System*

ANN MACEWEN and MALCOLM MACEWEN

ORIGINS AND CHARACTERISTICS

National parks originated in the United States, where the opening up of the West revealed the spectacular scenery and natural wonders of the Rocky Mountains. Man was not a newcomer to these regions, where native Americans had apparently lived in harmony with nature for centuries. The trouble was, as a Sioux chief has put it, that 'nature always goes out when the white man comes in'. For the white man combined the technical capacity to destroy nature with ignorance or indifference about the consequences. The overriding motive behind the national park movement was to protect the choicest scenes from the freebooters of private enterprise, who were left free to continue elsewhere their biblical mission to subdue the earth. As the national park concept developed the emphasis shifted from the preservation of natural monuments, wonders or spectacular scenery to the conservation of wildlife and natural systems. It was an anthropocentric concept, invariably justified by the enjoyment of future generations. It was also a democratic one, for the public interest in the enjoyment of the scenery was to override the private interest in its exploitation.

   In the United Kingdom the idea took hold in the 1920s. It was realized in 1949 in a muted form, after two decades of lobbying and four official reports (Addison, 1931; Dower, 1945; Hobhouse, 1947; Scott, 1942) in the National Parks and Access to the Countryside Act. Ten national parks were established in England and Wales between 1951 and 1957 (Figure 22.1; Table 22.1). But they were not, and could not be, national parks in the sense pioneered in the United States and later extended throughout the world. For the United Kingdom has no extensive landscapes more or less untouched by man. The differences between the British and the international meaning of the term 'national

---

* Copyright © 1983 A. MacEwen and M. MacEwen

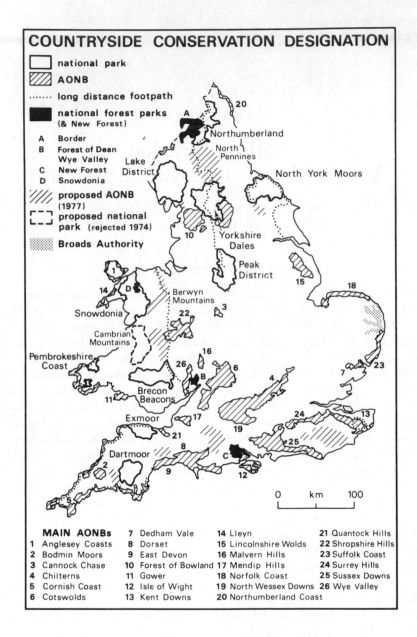

COUNTRYSIDE CONSERVATION DESIGNATION

☐ national park

▨ AONB

······ long distance footpath

■ national forest parks
  (& New Forest)

A Border
B Forest of Dean
   Wye Valley
C New Forest
D Snowdonia

▨ proposed AONB
  (1977)

☐ proposed national
  park (rejected 1974)

▨ Broads Authority

MAIN AONBs	7 Dedham Vale	14 Lleyn	21 Quantock Hills
1 Anglesey Coasts	8 Dorset	15 Lincolnshire Wolds	22 Shropshire Hills
2 Bodmin Moors	9 East Devon	16 Malvern Hills	23 Suffolk Coast
3 Cannock Chase	10 Forest of Bowland	17 Mendip Hills	24 Surrey Hills
4 Chilterns	11 Gower	18 Norfolk Coast	25 Sussex Downs
5 Cornish Coast	12 Isle of Wight	19 North Wessex Downs	26 Wye Valley
6 Cotswolds	13 Kent Downs	20 Northumberland Coast	

FIGURE 22.1. Crown copyright reserved. From MacEwen and MacEwen (1982);
reproduced by permission of George Allen and Unwin.

TABLE 22.1 National park statistics 1981.

	1	2	3	4	5	6	7	8	9	10	11	12
								LAND OWNERSHIP				
National Park	Area (ha)	Population (est. mid-1980)	National Park Authority (ha)	National Trust (ha)	Forestry Commission (ha)	Water authorities of (ha)	Ministry of Defence	Nature Conservancy Council	Other public	Total public and semi-public ownership	% in public and semi-public ownership
Brecon Beacons	134 420	30 000	563	4733	11 122	8624	1006	779	826	27 653	20
Dartmoor	94 535	30 300	1987	1882	1807	4710	3987[1]	302	28 421[2]	43 096	46
Exmoor	68 636	10 000	1979	6605	1340	332	—	—	2560[4]	12 816	19
Lake District	224 300	44 000	8493	50 500	12 734	15 560	461	303	—	88 051	39
Northumberland	103 120	2000	132	756	22 379	1202	22 700	36	—	47 205	46
North Yorkshire Moors	143 200	23 000	865	1200	23 690	271	768	—	5222[5]	32 016	23
Peak District	140 400	38 000	1468	14 200	2600	19 940	1200	139	—	39 547	28
Pembrokeshire Coast	58 350	20 981	1205	2355	720	100	2735	313	308	7736	13
Snowdonia	217 100	26 670	748	19 218	33 883	1955	300	1140	10 130[6]	67 374	31
Yorkshire Dales	176 110	18 500	84	1721	424	221	716	11	66	3243	2
Totals	1 360 000	243 451	17 524	103 170	110 699	52 915	33 873[1]	3023	47 533	368 737[7]	27
%	100	—	1.29	7.63	8.13	3.89	2.5	0.22	3.5	27	

Sources: MacEwen and MacEwen (1982), updated from National Park Statistics, *County Council Gazette*, **November 1981**.

Notes: [1] Excludes 10 000 ha owned by Duchy of Cornwall and used under licence by Ministry of Defence.
[2] Includes 27 592 ha owned by the Duchy of Cornwall.
[3] Excludes Warren Farm (890 ha) bought in 1982, of which 210 ha resold 1983.
[4] Includes land owned by the Crown Estate.
[5] Includes 5032 ha owned by the Duchy of Lancaster excluded from National Park Statistics.
[6] The Vaynol estate, owned by Welsh Office, sales to tenants under negotiation.
[7] There is a small element of double counting in the total, probably less than 1000 ha, caused by one body leasing to another.

park' can readily be seen from the definition adopted by the International Union for the Conservation of Nature and Natural Resources (IUCN) in 1969—20 years *after* the UK legislation:

> A national park is a relatively large area (a) where one or several ecosystems are not materially altered by human exploitation and occupation, where plant and animal species, geomorphological sites and habitats are of specific scientific, educative and recreative interest or which contain a natural landscape of great beauty and (b) where the highest competent authority of the country has taken steps to prevent or eliminate as soon as possible exploitation or occupation in the whole area and to enforce effectively the respect of ecological, geomorphological or aesthetic features which have led to its establishment and (c) where visitors are allowed to enter, under special conditions, for inspirational, cultural and recreative purposes (International Union for the Conservation of Nature and Natural Resources (IUCN), 1975).

Whatever they may be, our 'national parks' do not conform to the IUCN definition and for that reason are not included in the IUCN list of internationally recognized national parks, even though they appear to serve the same broad purposes. They fit rather better, although uneasily, into the IUCN category of 'protected landscapes'. The essential difference between the American or international concepts and our own is that the national parks of England and Wales—there are none in Scotland—are areas in multiple (not state) ownership that *have* been 'materially altered by human exploitation and occupation' for thousands of years and are now changing faster than ever before. There is 77 per cent of their area in private ownership, and little of the land in public ownership (see Table 22.1) has been acquired for the purposes of the national park. Over a quarter of a million people live in them. Their conservation is, therefore, not a question of eliminating the human presence or managing them as reserves, but is part of the bigger problem of integrating the conservation of the living landscape with economic and social life in the wider countryside. Therein lies their significance.

Part 2 of the 1949 Act defined the areas to be designated as

> those extensive tracts of country in England and Wales as to which it appears to the [National Parks] Commission that by reason of (a) their natural beauty, and (b) the opportunities they afford for open-air recreation having regard to their character and to their position in relation to centres of population, it is especially desirable that the necessary measures shall be taken [for] the preservation and enhancement of their natural beauty and the promotion of their enjoyment by the public.

What is common to the national parks is that the designated areas include extensive tracts of 'open country' defined in the 1949 Act as 'mountain, moor, heath, down, cliffs or foreshore'. This definition was extended in 1968 to include woodlands, rivers and canals. So defined, open country, like a park, is accessible to the public. It is also likely to have an above-average scientific interest for nature conservation. It was inevitable that the majority of the national parks would be located in the uplands or on the coast, because they are the most extensive areas left in England and Wales where the scope for changing the natural environment is limited by the physical conditions. They are not so much wild as semi-wild. They offer space and opportunities for recreation. And, although man has left his mark on them, often with a brutal disregard for nature or for scenic beauty, his relationship with nature—in contrast to much of lowland Britain—has been relatively harmonious.

PLANNING AND ADMINISTRATION

The national park system was grafted onto the 1947 Town and Country Planning Act by the 1949 National Parks and Access to the Countryside Act. The system has three distinguishing characteristics. It is cosmetic, being concerned mainly with the appearance of the countryside, and unable to influence to any marked degree the economic or social processes that are changing both the appearance and the ecology of the landscape. It is part of local, not of central government. And it is primarily a planning system, designed to control built development and major changes in land use by ensuring that development conforms to the country development (new structure) plans.

At the local level the national parks are the responsibility of the county councils. Eight of the national park authorities are committees of county councils (in one case a joint committee). They enjoy autonomy in development control (within the policies of the county councils) and in 'countryside' matters (for instance, recreation). They are subject to county council control in finance, staffing, administration and planning policy. They appoint the national park officer, but not the rest of the staff, all of whom are county council employees. Both the Lake District and the Peak District are administered by planning boards, which are not answerable to the county councils, levy a rate, employ their own staff and prepare their own structure (development) plans. One-third of the members on boards or committees are appointed by the Secretary of State for the Environment or for Wales. The appointees are in no sense representatives of government or of the conservation agencies, only one of which (the Countryside Commission) is consulted before their appointments are made. The majority must by statute consist of county councillors or, in the boards, of county council nominees. District councils appoint up to one-seventh of the membership. The national taxpayer provides about 69 per cent of the national park authorities' income, mainly through the National Park Supplementary Grant equivalent to about 75 per cent of 'approved net expenditure' (Table 22.2).

The United Kingdom is, we believe, almost unique in splitting the responsibility for conserving nature at the national level between two government agencies. As we have argued elsewhere (MacEwen and MacEwen, 1982), the

TABLE 22.2   National Park Income and Expenditure 1980–81.

Income	£000	%
From the public		
payments for services, rents, information etc	2040	20
From the taxpayer		
Countryside Commission grants[1]	809	
National Park Supplementary Grant[2]	5249	
Rate Support Grant (Resources Element)[3]	968	7026	69
From the ratepayers[4]		1115	11
		10 181	100

Expenditure	1980–81		(1973–74)	
	£000	%	(£000	%)
Management, administration and planning[5]	3781	37	(842	44)
Recreation[5,6]	2931	29	(490	26)
Information, publicity, national park centres[5]	1939	19	(430	22)
Conservation and estate maintenance[7]	847	8	(57	3)
Land acquisition	501	5	(—	—)
Other	182	2	(97	5)
	10 181	100	(1916	100)

Sources: National Park Statistics, *County Council Gazette,* **November 1981** and **November 1974**. See also Tables 5.5, and 5.7 in MacEwen and MacEwen (1982).

Notes

[1] Countryside Commission pays 75 per cent of information services net costs (after receipts) and 100 per cent of long-distance route costs. The Countryside Commission grant has since been merged in the National Park Supplementary Grant.

[2] National Park Supplementary Grant is 75 per cent (approximately) of approved expenditure, net of receipts, and excluding information services. Grant for moorland conservation at 90 per cent in Exmoor included.

[3] The resources element of the Rate Support Grant varies from county to county; taxpayers' contribution calculated from data supplied by the Treasurer, Somerset County Council.

[4] Rates are payable by all ratepayers in the counties where the national parks are located; residents in national parks pay only a small part of this.

[5] Includes capital expenditure.

[6] Includes car parks, picnic areas, caravan and camp sites, accommodation, catering facilities, access agreements, long-distance routes.

[7] Includes management agreements (£352 000).

historical reasons advanced for the split in 1949 have lost whatever validity they once possessed. The consequence is that the Nature Conservancy Council is responsible for the conservation of 'nature' throughout the United Kingdom. The Countryside (formerly National Parks) Commission is responsible for the preservation and enhancement of 'natural beauty' in England and Wales.

Natural beauty is defined to include flora, fauna, and geographical and physiographical features—precisely the words used in the 1981 Wildlife and Countryside Act to define the interest of the Nature Conservancy Council! But, as the Nature Conservancy took 'nature' for its province, the Commission interpreted 'natural beauty' to mean 'landscape' in the scenic or picturesque rather than the ecological sense of the word.

Both agencies have nationwide responsibilities for advising central and local government on different aspects of conservation. But it is the Commission that designates national parks and areas of outstanding natural beauty (AONBs) and took from the start a protective interest in landscape throughout the countryside, while the Conservancy tended to concentrate in its formative years on research, the management of reserves and the identification of sites of special scientific interest (SSSIs). National parks came to be identified with open-air recreation and the cosmetic aspects of 'landscape'. In Scotland the government failed to implement the postwar reports (Ramsay, 1945, 1947) on national parks. A separate Countryside Commission for Scotland, even more sharply focused on scenic conservation, was set up by the Countryside (Scotland) Act 1967.

In practice the intimate relationship between scenic beauty and nature is recognized by both agencies. The Nature Conservancy Council (1976) has said:

> the natural scenic beauty and amenity of the countryside depends to a large extent upon the maintenance of physical features with their cover of soil, vegetation and animals, these in turn being the expression of patterns of land use evolved by man over the centuries .... National parks, chosen as the outstanding areas of natural beauty, necessarily express this relationship with the basic ecological features to an especially marked degree.

The Nature Conservancy Council went on to say that national parks 'contain some of the least disturbed and developed country in England and Wales and encompass a substantial part of the nation's wildlife resources'. The *Nature Conservation Review* (Nature Conservancy Council, 1977) identified grade 1 and 2 sites that covered 7.8 per cent of the land area in the national parks, compared with 1.9 per cent for the rest of the United Kingdom. But its national nature reserves in national parks extend only to 0.22 per cent of the area. Most of the remaining grade 1 and 2 sites in national parks (although not all) are classified as sites of special scientific interest.

The split between the two agencies, and the cosmetic role of the national park authorities, has frustrated the intimate relationship between landscape and nature conservation in the national parks envisaged by the Dower (1945) report and by the Huxley (1947) report on nature conservation in England and Wales. The Nature Conservancy Council admitted in its evidence to the Sandford Committee (1974) that these hopes had not been realized. A recent study of nature conservation in national parks (Brotherton, 1981) has found that the Nature Conservancy Council treats national parks much as it treats the rest of the country. 'The national park authorities themselves', it adds 'have all but neglected nature conservation purposes .... This lack of interest is consistent with the low priority afforded to nature conservation by local authorities in general and by rural authorities in particular.'

At the local level the system established in 1949 was based on a faulty assessment both of the objective, which was seen mainly as the preservation of scenic beauty, and of the threat, which was seen as the urbanization of the countryside. This assessment led logically to the conclusion that nature or landscape could be conserved easily and cheaply by the local authorities through the new town and country planning system. The legislation refers not to national park authorities but to local *planning* authorities. These have virtually no conservation powers other than development control. The only protection afforded to SSSIs prior to the 1981 Wildlife and Countryside Act (see below) was a provision in the 1949 Act that the local planning authorities had to consult the Nature Conservancy Council before determining any *development* applications that affected an SSSI. Much of the 75 per cent National Park Supplementary Grant provided to national park authorities since 1974, supposedly for 'conservation', is spent on controlling the minutiae of design for insignificant buildings or extensions, or in weighing the pros and cons of new shops or offices in the main urban centres. The main successes of the system have, therefore, been in restraining 'development', although there have also been massive failures (MacEwen and MacEwen, 1982; Sandford, 1977). The government's continuing belief that controls over building are the key to 'conservation' can be seen from the retention of planning controls in national parks and urban conservation areas when they were relaxed elsewhere under the Local Government, Planning and Land Act 1980. At the same time the government resolutely opposed any extension of controls to agricultural or forestry operations, even in national parks.

A fundamental flaw in the system was in fact the exclusion of farming and forestry buildings and operations from development control. The Landscape Areas Special Development Order 1950 gives a measure of cosmetic control over the design and materials of farm buildings in parts of the Lake District, the Peak and Snowdonia National Parks, and the larger buildings are now controlled everywhere, but the smaller buildings and farming or forestry operations are not. The main reason for this sweeping exemption is that in 1949 farming was seen as a wholly benign operation, performing the triple functions of feeding the towns, sustaining the rural population and protecting the countryside. Agricultural operations were, therefore, excluded from the definition of 'development', and a similar exemption was accorded to forestry. The two major land uses, which together affect 96 per cent of the land area of the national parks, were left outside the system set up in 1949 for conserving their scenic beauty, wildlife and natural systems.

A series of official reports in the 1970s and early 1980s (Countryside Review

Committee, 1978; Nature Conservancy Council, 1977a, 1980, 1981; Porchester, 1977; Tourism and Recreation Research Unit, 1981; Westmacott and Worthington, 1976) has, however, clearly identified certain trends in agriculture (and to a lesser extent forestry) as the major threats to nature in the countryside. The effects are both direct (as where agricultural technology affects the soil or the vegetation) and indirect (as when the economic framework and changes in the labour force and social structure inhibit labour-extensive farming or forestry). The direct effects have been most accurately measured in the case of moorland reclamation for agriculture and forestry in the national parks. The Birmingham University Moorland Change Project (Parry, Bruce, and Harkness, 1981) has established that the rate of loss established by Porchester (1977) in Exmoor (approximately 20 per cent between 1947 and 1976) was typical of the losses experienced over comparable periods in the moorland plateaus of the North Yorkshire Moors, Dartmoor, Brecon Beacons and the northen part of Snowdonia. The decline in the area of broadleaved woodlands has not been studied so meticulously, but is general throughout the national parks in the absence of any national policy (as one professor of forestry has put it) other than that of liquidating rather than managing our residual indigenous broadleaved resource (Roche, 1980). SSSIs in national parks are probably at greater risk from tourism or moorland reclamation than SSSIs elsewhere but may be less exposed than the more fertile lowland SSSIs to the agricultural operations that severely damage 4 per cent of SSSIs every year, and possibly as many as 10 per cent (Nature Conservancy Council, 1980, 1981). A study (Tourism and Recreation Research Unit, 1981) of the economies of national parks has firmly established that the rapid fall in the number of farmers and farm workers has been a major factor in the development of more capital-intensive, labour-extensive farming techniques which promote the deterioration of the landscape of the uplands.

The casual visitor to the national parks will be impressed, despite the trends of the past 30 years, by the way in which the beauty and wildlife interest of their richly varied landscapes have, for the most part, survived the adverse pressures exerted on them. This is due in part to the moral (rather than the legal) force exerted by national park designation; to the concern that many farmers and landowners have for considerations other than profits; but mainly to the fact that the changes taking place, although slow, are insidious, and barely noticeable in many cases from one year to the next.

COUNTRYSIDE MANAGEMENT

The yawning gaps in development control, and the difficulty of controlling an insidious process of incremental change, led in the late 1960s and early 1970s to the promotion of the concept of 'countryside management' by the Countryside Commission. The idea was that the national park (and other) authorities would persuade public and private owners or tenants of land to collaborate in positive works to realize the objectives of broad-brush land management plans. The first prototypes for these plans were the national park management plans, which every national park authority was required to prepare by 1977 under the 1972 Local Government Act. The first plans were published in 1977–78 and are due to be reviewed in 1982–83. It is the existence of the national park plans that gives the planning system of the national parks a special significance.

We should not be misled, however, by the seductive language of countryside planning into exaggerating the scope of national park-management plans. They have been described by the Countryside Review Committee as plans 'for the co-ordinated management of large areas of land' (Countryside Review Committee, 1976), or as 'in simple words, getting things done to reduce the conflicts' (Hookway, 1977). They are concerned in fact with recreation and landscape, not with the management of the total resources of an area or resolving major conflicts over land use, and only incidentally with farming, forestry or other social or economic issues. The national park plans are an immense step forward when compared with the void that preceded them. The need to produce a plan obliged every authority to present (for the first time outside the Peak) a com-prehensive picture of the problems, conflicts and opportunities in the park as a whole, an analysis of its physical, financial and human resources, and a phased programme of implementation over 5 years with its associated costs. But the powers of the national park authorities are in inverse proportion to the seriousness of the problems that face them. They have powers and money to provide car parks, picnic sites, toilets, information services and wardens (rangers), and to dictate the siting, materials or design of most buildings. They are powerless when confronted by agricultural or forestry operations inimical to the countryside, or by social and economic trends that deprive rural areas of the labour to maintain a man-managed landscape. The plans tended, therefore, in their first (1977–78) editions, to fudge the more controversial issues.

CONSERVATION ZONING

The plans are not greatly concerned with the conservation of resources, or with sustainable development, apart from some emphasis on protecting good agricultural land and some worries about excessive demands for water and minerals. Nevertheless, a significant change is taking place at least at the level of plan-making and broad-brush land descriptions. Without exception the plans recognize that conservation goes beyond the protection of scenic beauty and of specially designated nature reserves or SSSIs, extending, in the words of the Lake District's Plan, to the conservation and enhancement of 'the ecological resources of the park as a whole' (Lake District Special Planning Board (LDSPB), 1978). Several of the plans identify conservation zones, going beyond SSSIs, the Devon Structure Plan (1981) going so far as to identify the whole of Dartmoor and the Devon portion of the Exmoor National Park as 'nature conservation zones', although the significance of this label in terms of policy is obscure.

The Peak Structure Plan (Peak Park Joint Planning Board (PPJPB), 1980) is the only one prepared exclusively for a national park. When taken together with the Peak National Park Plan it goes further than any of the others (although Dartmoor comes close) in attempting to imbue their policies with a spirit of conservation and to rest them on an ecological base. The natural qualities of the park are seen as the essential qualities to be conserved. To provide a context for land use, land management, conservation and recreation policies, the park (outside the main settlements) is divided into a natural and a rural zone. Broadly speaking, the natural zone is open country where vegetation is almost entirely self-sown. The rural zone consists predominantly of enclosed farmland and plantation woodlands and includes farmsteads, isolated dwellings and hamlets. The interface between the two, where enclosed land can revert to rough grazing, or rough grazing can be enclosed for agriculture or afforested, is regarded as a shifting frontier of potential change for which special policies will be needed. The Secretary of State has deleted these zones from policy statements in the plan.

The principle of zoning for conservation purposes is applied in several parks to woodland management and afforestation. The Lake District Plan (1978) identifies eleven broadleaved woodland conservation zones, where the planting of conifers even as a nurse crop is to be resisted. All the plans focus on the need to remedy the wholesale neglect of small woodlands and the loss of hardwood trees. Eight park authorities are prepared to buy broadleaved woodlands, and all are prepared to enter into management agreements to protect them. But the emphasis is mainly on preventing felling or coniferous planting, or on planting hardwood trees, rather than on positive conservation management. The authorities lack the resources to manage even the woods they already own, let alone to arrest the general decline or to restore natural regeneration in privately owned or National Trust woodlands.

Most of the parks have forestry maps, which represent zoning in its crudest form. These maps were initiated by the National Parks Commission in 1962, when it despaired of any control being imposed on afforestation. They are only concerned with planting on 'bare land', which is assigned to one of three categories: (*a*) where there is a strong presumption that afforestation will be acceptable; (*b*) where there is a presumption against it, but afforestation might be acceptable; and (*c*)where there is a strong presumption against afforestation. These maps are an uneasy compromise that satisfies neither side, and binds neither the landowners nor the authorities. Several of the park authorities regard them as too generous to forestry, and wish to renegotiate them. The Snowdonia map, for example, would allow the afforested area of the national park to be increased from 14 per cent to about 50 per cent of its total area. All the authorities in areas liable to massive afforestation have been pressing for years but without success for powers to control it, and in this they are supported by the Association of County Councils.

The moorland conservation maps (Exmoor National Park Committee (ENPC), 1982) prepared by the Exmoor National Park Committee on the recommendation of Lord Porchester's (1977) report on land use in Exmoor are perhaps the best-known form of zoning for conservation purposes (Figure 22.2). Map 1 defines the total area of moor and heath (19 500 ha), including 1900 ha of intermediate or fragmented areas where seminatural vegetation is not completely dominant. Within map 1 a voluntary system operates by which farmers notify intended changes to the National Park Committee. This is, presumably, the area to which the Secretary of State would apply the compulsory notification procedure envisaged by s 42 of the Wildlife and Countryside Act, should he ever decide to activate it. Map 2 (16 036 ha) defines the area within which the Committee wishes to conserve 'the traditional appearance' of Exmoor. The criteria for map 2 are wholly aesthetic, but the Committee's policies for map 2 comprise 'the strongest possible presumption . . . in favour of the conservation of natural flora, fauna and landscape', and in favour of 'traditional rough grazing compatible with conservation', except where (in a few areas) priority is given to afforestation by an earlier forestry map.

The Committee refused to include an additional 1400 ha that the Nature Con-

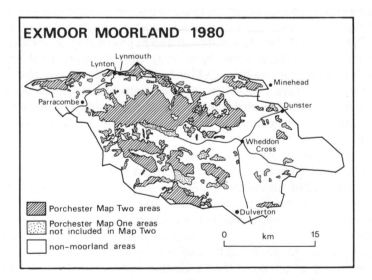

FIGURE 22.2. Porchester's 'Map 1' showed the total area of moorland, and his 'Map 2', the area within which there is the strongest possible presumption against reclamation. From MacEwen and MacEwen (1982); reproduced by permission of George Allen and Unwin.

servancy Council wished to add on ecological grounds. The Conservancy intends, therefore, to prepare a *separate* nature conservation map of Exmoor. This confusion over the purpose of the moorland maps is carried further by s 35 of the Wildlife and Countryside Act, which requires all national park authorities to publish maps defining those areas of moor or heath that the authority thinks it 'particularly important to conserve'—in other words, maps defined on vague subjective criteria.

Yet another form of zoning is introduced by the Act, which imposes a duty on the Nature Conservancy Council *or* the Countryside Commission to notify local authorities of any limestone pavements they believe to be of 'special interest' for their flora, fauna, or geological or physiographical features—but not, apparently, for their beauty. The local authority or the Secretary of State may then make a Limestone Pavement Order, designating the land and prohibiting the extraction of limestone, under penalty of a fine.

IMPLEMENTATION

The implementation of the conservation policies set out in the national park plans depends, essentially, on three factors: the goodwill of farmers and landowners, the negotiation of voluntary management agreements with them, and local management plans or projects (often in collaboration with other authorities). There is a substantial fund of goodwill towards the national parks, but it is continually eroded by the economics of farming and the agricultural and forestry grants and incentives, which push farmers and landowners into operations that are damaging to nature and to landscape. Until the passage of the 1981 Wildlife and Countryside Act (to which we shall return later) the limitations of management agreements were generally recognized and severely inhibited their negotiation.

It is relatively easy to negotiate informal, short-term agreements for purposes such as tree-planting, which impose no serious or long-term obligations or costs on landowners but do offer long-term benefits. The Brecon Beacons National Park Committee, for example, negotiated more than 20 such agreements between 1974 and 1981, although latterly the modest public funds required to finance them (£26 000 in 1979–80) have been drying up. It is easy to negotiate a management agreement where the landowner is obliged to conserve the landscape and to grant public access as a condition of securing relief from capital transfer tax on his land. A case in point is the management agreement for the Bransdale moors negotiated with the Nawton Towers estate on the North Yorkshire Moors. In Exmoor the Government pays 90% grant to support an agreement concluded in 1981 by which farmers and landowners who forego moorland reclamation are compensated by annual payments, which are equal to the profit foregone and indexed against farming profitability. Five 20-year management agreements protecting 813 acres were concluded between 1977 and

1982; others are in prospect. But the park authority protected over 3 300 acres in these years by buying the land. In general, experience has confirmed Feist's conclusion (1978) that 'management agreements cannot remedy deficiencies in the system (such as conflicting government policies, lack of notification machinery or lack of money), or tackle the underlying structural problems of agriculture in marginal areas. . . . Without adequate back-up powers (they) do not offer much in the way of long-term security for the public interest or public investment.'

Local management plans can most easily be developed where the conservation authority has acquired the land (for example, at Nab's Hill farm on the North Yorkshire Moors, or Larkbarrow and Warren Farm in Exmoor). Otherwise, local management plans or projects have been most successful where they have concentrated primarily on recreational objectives, particularly in sorting out acute conflicts at visitor 'honeypots', or in removing friction between visitors and farmers. An outstanding example is the Upland Management Experiment (now the Upland Management Service) in the Lake District (and to a lesser degree in Snowdonia). This service has improved the facilities for visitors by carrying out a host of minor works (mainly to footpaths); it has made life much better for farmers who were suffering badly from damage or annoyance caused by visitors; and it has provided some local jobs. But, as a study of the Hartsop Valley (Feist, Leat, and Wibberley, 1976) demonstrated, upland management could 'only scratch the surface of the real social and economic problems faced by the farming community' that lie at the root of many of the conservation problems in the Lake District. There is a strong conservation element in such local projects as the local plan for Dovedale in the Peak, the Snowdon Management Scheme and the Pembrokeshire Coast long-distance path rehabilitation project. But the cost and the relatively large numbers employed (40 in Snowdon at full strength) preclude their extension widely throughout the national parks.

Two ways of measuring the extent to which the national park authorities are engaged in conservation is to look at their expenditure and their staffing. Table 22.2 shows the distribution of expenditure in the national parks in 1973–74 (the last year before reorganization) and 1980–81. 'Conservation', as used in the official national park statistics, is an all-embracing term that includes the conservation of historic buildings and urban conservation areas. We have, therefore, included under this head in Table 22.2 the money spent on the maintenance of the national park authorities' estates, much of which was acquired for the purpose of conservation.

Expenditure on conservation and estate maintenance has risen from 3 per cent of the total national park expenditure in 1973–74 to 8 per cent in 1980–81 (from £57 000 to £847 000, an increase of about six times in real terms). But the sums spent before reorganization were negligible, and the totals conceal wide

differences. Some 85 per cent of the expenditure of £352 ([...]
ment was spent on the Snowdon management project and [...]
ment service in the Lake District. The eight other parks a [...]
apiece on management schemes and agreements. In contr [...]
planning and administration was £3.781 million (37 per c [...]
£2.931 million (29 per cent) and on information £1.930 mil [...]
Overall expenditure in national parks has been falling steeply since the later
1970s. The Association of County Councils calculated that to achieve the very
modest levels recommended by Sandford in 1974, expenditure would have had
to be increased in 1981−82 by 42.7 per cent from an estimated £8.225 million to
£11.736 million (Association of County Councils, 1981). The National Park
Supplementary Grant for 1982−83 (£6.690 million) implies a cut in real terms
of 11.5 per cent in 2 years.

There is a very long way to go before the priority given on paper to conservation since the Sandford report (1977) is translated into money spent and people
working on the ground. Out of a total of 626 permanent staff employed by
national park authorities in 1980, 256 were engaged in management, administration and planning, 79 were wardens (rangers) whose primary function was
recreational management, and 69 were professionals engaged in project and
estate management (MacEwen and MacEwen, 1982). Only 97 were manual
workers on the ground. The ratio (2.2:1) of professional and managerial staff
to outdoor and ground staff is exceptionally high, and probably increasing as
staff is cut.

One reason why so little is spent on estate maintenance is that only 1.29 per
cent of the land area of the national parks is owned by the national park
authorities (see Table 22.1). The National Trust own 7.63 per cent and the
Forestry Commission slightly more (8.13 per cent). Although public or
semipublic bodies own 27 per cent of the national parks (46 per cent in
Northumberland, 46 per cent in Dartmoor and 39 per cent in the Lake District),
most of this land was acquired for such purposes as water supply, timber production, electricity generation, artillery ranges or other purposes that are often
antagonistic to those of the national park, so that the potential of public landownership for conservation land management is largely frustrated. Consultation between national park authorities and other public landholding bodies has
improved since 1974, but the progress made towards the integration of their
management plans has been slow. Lacking, as the national park authorities do,
both the power of landownership and the ability to protect habitats or landscapes by means of development control or some order-making power, they
have no alternative but to rely on the goodwill of those who do own and manage
the land. It is obviously desirable that, whatever formal powers they command,
the national park authorities should win the goodwill of landholders by example,
advice, persuasion and help with money and labour. But farming, as the

...istry of Agriculture stresses, is first and foremost a money-making business. The difficulty, which no government has yet faced, is that unless the authorities have the ultimate power to control major land-use decisions made for business or other reasons, they are not in a position to 'manage' the countryside. Total reliance on goodwill gives the right to determine policy to those who have property rights in land.

THE WILDLIFE AND COUNTRYSIDE ACT 1981

It is difficult to predict with certainty the consequences of the adjustments made to the national park system by the Wildlife and Countryside Act 1981. The essence of the Act's approach to conservation in national parks is that it excludes all forms of control over changes to habitats and landscape (with a minor exception for limestone pavements); introduces a notification system for SSSIs, and relies almost entirely on the goodwill of farmers and landowners (sweetened in certain circumstances by compensation) to conserve the SSSIs and national parks. It undermines the voluntary nature of the system by imposing one-sided obligations and open-ended financial commitments on the conservation authorities (the Nature Conservancy Council and the national park authorities).

The notification arrangements are inconsistent and confusing. Under section 28, the Nature Conservancy Council must notify landholders of any operations that are likely to damage the flora, fauna or geological or physiographical features of an SSSI. In return, landholders must give 3 months' notice of all such operations to the Nature Conservancy Council. Under the 1980 Agricultural and Horticultural Development Scheme, farmers must also notify the Nature Conservancy Council (in SSSIs) and the national park authorities (in national parks) of all operations for which they propose to claim a capital grant from the Ministry of Agriculture. Outside SSSIs, therefore, there is no legal obligation to notify operations in a national park for which no agricultural grant is claimed. There are derisory fines (up to £500) for breach of the notification procedure in SSSIs, but the authorities have no powers to stop or to deter even the most damaging operations to which they have objected, apart from the Nature Conservancy Council's virtually never-used power of compulsory purchase.

The Act relies exclusively on management agreements between the landholders and the conservation authorities to resolve conflicts. As we have seen, the experience before 1981 had shown that for a number of reasons management agreements had not been successful in resolving acute conflicts of interest or attitudes on a voluntary basis. The Act attempts to get round these difficulties by imposing agreements (and what amounts to a statutory compensation code) on the conservation authorities while leaving the landholders free to accept or to reject them.

Sections 32 and 41 of the Act impose on the Ministry of Agriculture a duty to 'further' conservation when considering grant-aided schemes, but leave the statutory purposes of these schemes (to promote the profitability of the 'agricultural business') unchanged. The duty to further conservation is subject to an overriding obligation to promote the 'agricultural business'. And after the Bill had been enacted, senior Ministry officials asked conservationists not to press them to say what 'to further' conservation means, on the ground that any definition would have to be very restrictive (MAFF, 1981). Sections 32 and 41 also impose a statutory duty on the conservation authorities to offer management agreements (but only in SSSIs, national parks and such other areas as ministers may specify; as we write the Broads have also been specified) should the Minster of Agriculture withhold compensation in response to the authorities' objections. The proffered agreement must restrain the proposed operations and (under s 50) it must offer compensation in accordance with guidance to be given by ministers. But the landholder is free to accept or to reject the terms offered, and to go ahead with his operations.

The precise terms of the guidance had not been announced as this was written (February 1982). But the Act for the first time establishes the farmer's legal right to an agricultural grant in an SSSI or national park, or to compensation for the profit foregone if the grant is withheld on amenity or conservation grounds. The 'improving' farmer can no longer be deterred at no cost to the conservation authority by the withholding of the grant, as had happened occasionally in the past. Ministers have expressed their support for the principles of the guidelines of management agreements negotiated between the Exmoor National Park Committee, the National Farmers' Union and the Country Landowners' Association. This provides, among other things, for annual payments equal to the profit foregone, indexed against the profitability of sheep farming in Exmoor. The Exmoor National Park Committee, which gets a 90 per cent government grant for moorland conservation, has accepted this open-ended commitment. The Broads Authority (which only gets a 50 per cent grant) agreed in 1981 to pay £65 000 annually to conserve some 650 ha on the Halvergate marshes, but refused to index the payments. The constituent local authorities were appalled at the prospect that in 20 years' time the initial £65 000 might have risen to £500 000 a year. The deadlock was broken by the Government's decision not to pay the drainage grants that caused the problem.

Management agreements on these terms can be likened to hot-air balloons which can only be kept aloft by burning money. When the money runs out, the balloon comes down. They ignore the experience of the 1932 Town and Country Planning Act which failed because developers were entitled to compensation if planning permission was refused. The protection afforded to the resource by the 1981 Act is temporary, and if the money runs out all the money already spent will have been wasted. A notional sum of around £150 000 was included in the National Park Supplementary Grant for 1982–83 to implement the

Wildlife and Countryside Act, an invisible drop in the ocean of government spending but a significant measure of its priorities. Governments can be relied on to limit very strictly the funds they will provide, and local authorities (as in the Broads) will not accept open-ended commitments indexed against inflation. The most probable consequence will be that the authorities will exercise extreme caution in objecting to any agricultural proposals, except in Exmoor where the 90 per cent grant enables them to ignore the consequences. The government has refused to extend this grant to other national parks or to other landscapes. Another consequence is that in extreme cases the authorities will have to find the money by cutting down expenditure in other directions.

The government rejected an amendment to the Wildlife and Countryside Bill that would have enabled agricultural grants to be used to encourage conservation and social and economic diversification in national parks. Agricultural policy, incentives and advice therefore remain on their collision course with conservation and other rural interests. National parks in England and Wales were among the first experiments in the conservation of extensive inhabited, man-managed landscapes. But their potential for exploring ways in which conservation can be integrated with social and economic life will not be realized until another government introduces fresh legislation.

REFERENCES

Association of County Councils (ACC) (1981). *Memorandum in relation to House of Commons debate on National Parks, 11 December*, ACC, London.

Addison, C. (chairman) (1931). *Report of the National Park Committee*, Cmnd 3851, HMSO, London.

Brotherton, I. (1981). *National parks and the Achievement of Nature Conservation Purposes*. Landscape Architecture Department, University of Sheffield.

Countryside Review Committee (CRC) (1976). *The Countryside—Problems and Policies*. HMSO, London.

Countryside Review Committee (1978). *Food Production and the Countryside, Topic Paper 3*. HMSO, London.

Dower, J. (1945). *National Parks in England and Wales*. Cmnd 6378, HMSO, London.

Exmoor National Park Committee (ENPC) 1982. Porchester Maps 1 and 2, ENPC, Dulverton.

Feist, M.J. (1978). *A Study of Management Agreements*, CCP 114. Countryside Commission, Cheltenham.

Feist, M.J., Leat, A.M.K. and Wibberley, G.P. (1976). *A Study of the Hartsop Valley*, CCP 92, Countryside Commission, Cheltenham.

Hobhouse, Sir A. (chairman) (1947). *Report of the National Park Committee (England and Wales)*, Cmnd 6628, HMSO, London.

Hookway, R.J.S. (1977). *Countryside Management, the Development of Techniques*. *Proceedings of the Town and Country Planning School*, RTPI, London.

Huxley, Sir J. (chairman) (1947). *Report of the Committee on Nature Conservation in England and Wales*, Cmnd 7122, HMSO, London.

International Union for the Conservation of Nature and Natural Resources (IUCN) (1975). *World Directory of National Parks and Other Protected Areas*, IUCN, Morges, Switzerland.

Lake District Special Planning Board (LDSPB) (1978). *Lake District National Park Plan*, Lake District Special Planning Board, Kendal.

MacEwen, A. and MacEwen, M. (1982). *National Parks: Conservation or Cosmetics?* George Allen and Unwin, London.

Ministry of Agriculture, Fisheries and Food (MAFF) (1981). *Report of Agriculture and Conservation Seminar, 1 December 1981* (circulated to participants), MAFF, London.

Nature Conservancy Council (NCC) (1976). *Memorandum in National Parks and the Countryside*, sixth report session 1975–76, House of Commons Expenditure Committee. HMSO, London.

Nature Conservancy Council (NCC) (1977a). *Nature Conservation and Agriculture*, Nature Conservancy Council, London.

Nature Conservancy Council (NCC) (1977b). A Nature Conservation Review, Nature Conservancy Council, London.

Nature Conservancy Council (1980). NCC calls for further measures to protect wildlife habitats. *Statement*, **12 December**. Nature Conservancy Council, London.

Nature Conservancy Council (1981). New evidence of increasing destruction of wildlife habitats. *Statement*, **11 February**. Nature Conservancy Council, London.

Parry, M., Bruce, A., and Harkness, C. (1981). The plight of British moorland. *New Scientist*, **28 May**.

Porchester, Lord (1977). *A Study of Exmoor*. HMSO, London.

Peak Park Joint Planning Board (PPJPB) (1980). *Peak District National Park structure plan*, Peak Park Joint Planning Board, Bakewell.

Ramsay, Sir J.D. (chairman) (1945). *National Parks: A Scottish Survey*. Cmnd 6631, HMSO, Edinburgh.

Ramsay, Sir J.D. (chairman) (1947). *National Parks and the Conservation of Nature in Scotland*. Cmnd 7235, HMSO, Edinburgh.

Roche, L. (1980). Quoted in *scientific aspects of forestry*, p. 18, report of the House of Lords Select Committee on Science and Technology. HMSO, London.

Sandford, Lord (chairman) (1974). *Report of the National Park Policy Review Committee*, HMSO, London, (for NCC evidence see under NCC, 1976).

Scott, Mr Justice (chairman) (1942). *Report of the Committee on Land Utilisation in Rural Areas*. Cmnd 6378, HMSO, London.

Tourism and Recreation Research Unit (TRRU) (1981). *The economics of rural communities in the national parks of England and Wales*, research report no. 47. Edinburgh University.

Westmacott, R. and Worthington, T. (1976). *New Agricultural Landscapes*. CCP 76, Countryside Commission, Cheltenham.

Conservation in Perspective
Edited by A. Warren and F.B. Goldsmith
© 1983 John Wiley & Sons Ltd.

CHAPTER 23

Environmental Management in Local Authorities

JOHN GORDON

INTRODUCTION

From its conception in the late 1940s the environmental embryo suffered a long gestation until, undernourished, it crawled somewhat limbless into the motherless pouch of the 1960s, when various local authorities established countryside sections and country parks. It was paradoxical that some of the first countryside officers and their troops came from Her Majesty's forces to protect the environment with the same conservative sincerity that was used to propound the philosophy that the only way of protecting anything was to prepare for its annihilation. The same insidious military explosion permeated the county naturalists trust movement. Was the future of the environment to be left not in the hands of those who cared about it but with those who were trained to destroy it?

Until the early 1970s ecologists and nature conservationists were on the fringe of what might be called 'professional activities', that is, architecture, planning, engineering etc. Prior to that time, local authority involvement in 'ecology' had been mainly concerned with development control in relation to sites of special scientific interest (SSSIs) and in the establishing of local nature reserves.

At this time ecologists began to appear on the local authority scene. The first was appointed under the guise of a different job specification—a practice that still continues today; 1972 saw the appointment of the first ecologist *per se* in a county council, and by early 1973 there were about ten to twenty ecologists at work in local authorities in England and Wales.

THE WORKING PANEL OF LOCAL AUTHORITY ECOLOGISTS

This amorphous group met in Milton Keynes in the spring of 1973 and formed itself into the Working Panel of Local Authority Ecologists. Although the interests and responsibilities were wide, the group had many things in common.

411

We were all relatively new, certainly green, faced immense problems, wished to speak to each other to share our hopes, fears and aspirations, but above all had a commitment to the creation or maintenance of a high quality environment. The Panel has five main aims:

(1) To establish an efficient and effective communications network between its members and with other organizations concerned with the natural environment.
(2) To provide a forum for the discussion of matters and problems of mutual interest.
(3) To exchange and disseminate ideas and information.
(4) To standardize (where possible) methods and techniques for ecological surveys, monitoring programmes, etc.
(5) To promote cooperation on matters of common interest.

In order to achieve those objectives the Panel meets twice a year and produces a newsletter. Over the years a number of documents have been produced including a report on the establishment of environmental data banks; a code of practice for the cutting of road verges; and a code of practice for statutory undertakers and site staff.

Membership is restricted to people who are employed in local authorities and spend a considerable amount of their time working on ecological/wildlife conservation matters. 'Local authorities' are loosely defined to include national park authorities and new town development corporations.

Ecology and conservation seem to attract people who like to join organizations, for over the years the Panel has come under pressure to admit landscape architects, planners, officers of county naturalists trusts and other voluntary societies and a miscellany of other local authority personnel. It is unfortunate and a missed opportunity that the membership has been opposed to the admission of ecologists and similar staff from industry, other public bodies and consultancies.

REORGANIZATION OF LOCAL AUTHORITIES, AND THE ECOLOGIST

The reorganization of local government in 1974 and the need to prepare structure plans resulted in a population explosion of local authority ecologists. By the end of 1975 there were about 60, a figure that has remained remarkably constant since (Table 23.1).

In those early days, there was a pioneering spirit; there were no guidelines, job descriptions were written by the incumbent, the challenge and the sense of adventure were exhilarating. It looked like another major step forward for the management of our surroundings but it soom became clear that if local authority ecologists were to survive in this uncharted sea they would have to row their own boat and bale themselves out.

TABLE 23.1 Structure of local government in the United Kingdom (at 1981).

	Number	Number employing an ecologist	Number of ecologists
1. England and Wales			
Greater London Council	1	—	—
City of London	1	—	—
London boroughs	32	—	—
City councils			
Metropolitan	6	4	8
Non-metropolitan	47	19	25
District councils			
Metropolitan	36	1	2
Non-metropolitan	333	3	3
National park authorities	10	7	7
2. Scotland			
Regional councils	9	3	3
District councils	53	5	6
Island authorities	3	—	—
3. New town development corporations			
England	21	2	2
Wales	2	—	—
Scotland	6	—	—

The job titles continued to be varied; planning assistant, planning officer (suitably qualified by 'countryside', 'ecology', etc.) was the most popular, but there remained a plethora of other titles including ecologist, natural resources officer, environmental scientist, biologist, nature conservation advisory officer, landscape ecologist, country parks officer, etc. Whilst in general the job title/description did not seem to inhibit local authority ecologists from roaming across the whole of their authorities' operations there is one title that serves as a constraint on such activities. This contains the words 'nature conservation' or something very similar. In such cases the job holder is restricted to the rather narrow activity of nature conservation rather than being able to range over the whole of the resource-management field.

Whilst the training and qualifications of local authority ecologists varies considerably, there is nevertheless a common thread. Most are trained in one or a combination of the natural sciences, followed by either an MSc in Conservation or Ecology and/or a PhD in Natural Science. In addition, many have gained professional qualifications, for example membership of the Royal Town Planning Institute or the Landscape Institute. Local authority ecologists also need to know a lot about the various aspects of the law, the administration and supervision of contracts and contractors, financial control and the drafting of specifications and bills of quantities.

Ecologists in general (and local authority ecologists are no exception), are equated by many politicians, senior managers and some of the professions with one or both of two extremes: wildlife conservation or ecofreaks. 'Ecological' is, therefore, an unfortunate and singularly inappropriate word to be included in a job title. 'Ecologists' have an increasingly important role in the management of natural (that is, biological) resources but to exploit these opportunities demands a radical reappraisal of the attitudes of politicians (in both local and central government), academic and government ecologists and the professions. It would be more appropriate and relevant for 'local authority ecologists' to be called 'natural (that is, biological) resource officers' (which Merseyside County Council, for example, has been enlightened enough to do). In this context wildlife conservation becomes just another component of a much wider role that 'ecologists' have in the management of air, water and land for industry, housing, recreation and amenity.

Ecologists should be the natural resource equivalent of accountants (*not* economists—for there lies the trouble). As such a natural resource officer may give sound ecological advice that is incompatible with the conservation of nature. For example, should the quality of the air in some cities be cleaned up with the possibility of adversely affecting the conservation of melanistic moths? Should the water quality of an estuary be improved by the installation of a new sewage works when there is the possibility that it may have a detrimental affect on the duck population? Are all schemes for reclaiming derelict land desirable or should some be retained for their visual, historic, wildlife or amenity value (Bradshaw, 1979; Kelcey, 1978)?

The role of the natural resources officer must be to use his or her skills to assess the extent and condition of the natural resources, evaluate them in relation to the opportunities and constraints and to prescribe and advocate how best to exploit and manage them. This requires a concern for people and their condition although I fear few ecologists outside local government are motivated in this way. In spite of what the Nature Conservancy Council says, its kind of wildlife conservation, in practice, is mainly concerned with the preservation of rare or unusual species and habitats in rural areas. As such it satisfies only a relatively small and highly specialized interest group. Unless this narrow attitude can be broadened to include the incorporation of nature into both the urban and the rural environments as a whole and unless it is aimed at preventing common species from becoming rare then it will remain an exercise largely irrelevant to the bulk of the community.

'Ecologists' in local authorities are appointed either as a genuine attempt by the authority to have a more effective resource management policy, or as a luxurious adjunct for public relations and to discharge their conscience, or a mixture of both. Local authority ecologists have few friends; they are too pragmatic for their academic colleagues but too theoretical for their planning, engineering and other professional colleagues, who are still not sure what ecologists can and cannot do. Most people expect ecologists to know everything about everything that lives with the answer normally being required within the hour.

What an ecologist in a local authority can and cannot do depends upon the aspirations and policies of his/her organization, its priorities and the resources that it makes available, all of which can be summarized as the 'political will'.

Although local authority ecologists have to operate under a wide range of diverse and often complex legislation related to planning issues in general (including tree preservation orders, general development orders, The Plant Health Act and a variety of wildlife and conservation legislation), the statutory requirements related to wildlife conservation itself and the efficient use of natural resources are few. The impotent s 11 of the Countryside Act 1968 places a 'responsibility' on public bodies to take conservation into account, but it does not require them to manage the environment effectively. Nevertheless local authorities can and do wield considerable power through land ownership, development control, policy planning and the provision of services. In addition they are answerable to demands from outside the authority, namely from the public and the secretaries of state. As a result of these powers the authorities have at their disposal much greater powers than quangos such as the Nature Conservancy Council who generally only provide advice.

The Nature Conservancy Council has the duty to notify the local planning authority of any site it believes to be of special scientific interest (SSSI). Planning authorities must consult the Nature Conservancy Council before giving permission for the development of an SSSI. Rather late in the day the Department of the Environment (1977) has issued a circular which charges local authorities with taking full account of nature conservation in their structure, local and subject plans, development control, and in the management of their own estates. Whether this becomes as ineffective as s 11 remains to be seen; the indications are that at least at the district council level it is. Local authorities can do various other things such as disseminating information in respect of the Wildlife and Countryside Act 1981. It would be interesting to know how many have done so. Local authorities can also declare local nature reserves.

THE ROLE OF ECOLOGISTS WITH LOCAL GOVERNMENT

What local authority ecologists do varies between the types of authority. As Table 23.2 shows they are employed in a variety of departments, although by far the most are employed in planning departments. Figure 23.1 shows their distribution in Great Britain.

The ecological problems differ enormously from county to county and district to district (Figure 23.2) but the work can be divided into two major roles. The first is education. This seeks to change the values and philosophy of

TABLE 23.2   Summary of the departments employing ecologists (1981).

Engineering	2
Environment	3
Estates	1
Museums	2
National parks	7
Parks	1
Planning	37
Recreation and leisure	3
	56

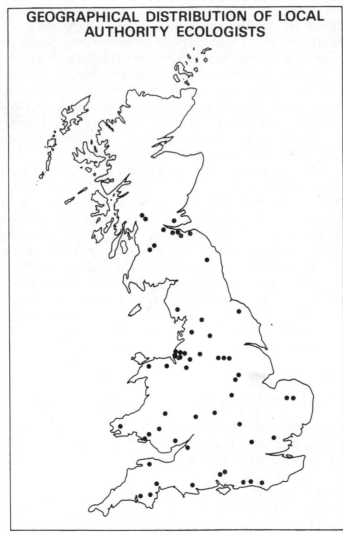

GEOGRAPHICAL DISTRIBUTION OF LOCAL
AUTHORITY ECOLOGISTS

FIGURE 23.1.

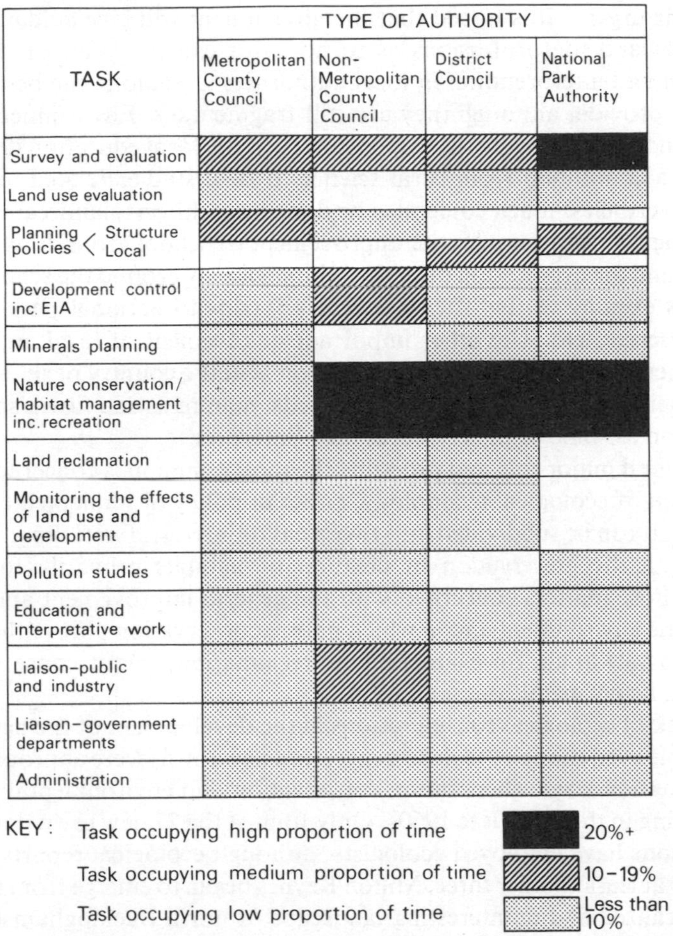

TASK	TYPE OF AUTHORITY			
	Metropolitan County Council	Non-Metropolitan County Council	District Council	National Park Authority
Survey and evaluation	▨	▨	▨	█
Land use evaluation	░	░	░	░
Planning policies ⟨ Structure / Local	▨	░	░ / ▨	▨
Development control inc. EIA	░	▨	░	░
Minerals planning	░	░	░	░
Nature conservation/ habitat management inc. recreation	░	█	█	█
Land reclamation	░	░	░	░
Monitoring the effects of land use and development	░	░	░	░
Pollution studies	░	░	░	░
Education and interpretative work	░	░	░	░
Liaison—public and industry	░	▨	░	░
Liaison—government departments	░	░	░	░
Administration	░	░	░	░

KEY :   Task occupying high proportion of time   █ 20%+

Task occupying medium proportion of time ▨ 10–19%

Task occupying low proportion of time   ░ Less than 10%

FIGURE 23.2. Analysis of the tasks undertaken by local authority ecologists. From Schulte and Conway (1980).

society or certain elements of it. There is still considerable cynicism about, even opposition to, ecology among engineers on the one hand, and the senior politician/manager on the other. The absence of 'ecologists' from many government departments (for example, Countryside Commission and the Agricultural Development Advisory Service), some national park authorities and county councils and most of the district councils indicates that there is still a perception problem at a senior level within public bodies. Much has to be done to educate politicians in central and local government that the natural resources of the nation need to be managed for the long-term benefit of the community and that we cannot continue to live off the capital of the third world. The need to implement the philosophy of the Brandt Commission and the World Conservation

Strategy is urgent. It is hoped that the government will give guidance to local authorities and the professions as to how this can be done. Environmental education and interpretation in local authorities are among the better services that they provide, although they are still fragmentary. Environmental education in schools is generally the responsibility of the local education department. It forms a continuum from rural science to field studies related to examinations. Nevertheless much could also be done by architects, landscape architects and ecologists; for example, the improvement of school grounds as a teaching resource and the exploration of the area immediately around the school, for this can be as exciting as the often unnecessary trips to national parks, national nature reserves, etc. The other important involvement of local authorities in environmental education and interpretation is in the country parks and similar areas, where nature trails, guides to the birds, mammals and flowers, are nearly as common as some of the creatures they 'interpret'.

The second major role and probably the most common is concerned with the application of ecological technology, either at policy or implementation levels (or both). It can be subdivided into two areas of interest. First, those issues concerned with the preservation or creation of habitats primarily for wildlife. Second, those matters concerned with the general improvement and management of the environment and in which nature conservation plays either a minor part or no part at all; for example, land reclamation, energy conservation, air pollution, heavy metal contamination.

The lack of ecological considerations in the development of the new towns is remarkable even though the third generation new towns were approaching or at the designation stage when nature conservation and environmental awareness was peaking in the mid–late 1960s. Only four of the 27 new town development corporations have employed ecologists, although ecological reports were produced for at least another three. Milton Keynes began to emerge from the heavily agriculturalized and uninteresting landscape of north Buckinghamshire at the very end of the 1960s. An ecologist was appointed by the Development Corporation in the autumn of 1972 with the responsibility of monitoring the effects of urban development on the wildlife and its habitats and to advise the various departments of the Corporation on ecological matters.

During the last 9 years a large number of surveys have been carried out at Milton Keynes either of particular habitats (such as road verges, railway embankments, woodland, etc.), particular species (such as small mammals, moths and butterflies) or particular species in particular habitats (such as the invertebrates of selected woodland, fish in the canal). A great deal of effort has been put into the planning, design and management of the wet balancing lakes—impoundment reservoirs built to control flooding and used for a variety of recreational activities: for example, water skiing, sailing and angling. The nature conservation element has involved the protection, where possible, of sites of wildlife value, but probably more importantly the creation of new habitats including woodland, grassland and the restoration of gravel workings to a nature reserve. The work of an ecologist in a new town is described by Kelcey (1975, 1976).

In 1975 Milton Keynes Development Corporation approved the following ecological policy:

1. The Corporation endorses the Council for Europe's Declaration on Nature Conservation for Local Authorities, and encourages all tiers of local government to adopt the principles it contains.
2. The Corporation will ensure that ecological principles and techniques are considered at all stages of the developmental processes, from planning to management.
3. The Corporation will undertake, in cooperation with local wildlife interests, the monitoring of the effects of development, public pressure and management on wildlife and wildlife habitats, including the monitoring of the colonization of newly created environments.
4. The Corporation will endeavour to protect those sites which are listed as being of special wildlife or research value and ensure that they receive sympathetic treatment in the planning, design and management processes.
5. The Corporation will endeavour to establish and maintain a representative collection of plants, animals and ecological specimens found in Milton Keynes in anticipation of the establishment of a City Museum Service.
6. The Corporation will encourage the community (especially children) to understand, participate in, and care for their environment by interpretative schemes, projects and any other appropriate means.
7. The Corporation will exploit to the full, opportunities to create areas or habitats of wildlife value.
8. The Corporation will consult with national, regional and local natural history organisations, and invite their views and recommendations on relevant issues.

Warrington mushroomed up in the mossy prairie of south Lancashire/north Cheshire in the last century to become a sublimely beautiful part of the English industrial landscape. The problems facing the Development Corporation's ecologist were infinitely different from those of Milton Keynes. There was an existing fabric of jumbled buildings, some splendid Victorian ironwork, a vast expanse of derelict peat set in a smoking, boiling, bubbling cauldron and cooled by a river which remains a monument to long-dead municipal benefactors. In this arena of crumbling bricks and shining aluminium, the ecologist could do nothing but good. The input into the planning and landscape proposals has been strong. A report on the ecological approach to landscape design and management of part of Warrington was published in 1980 (Tregay and Moffatt, 1980). Risley Moss has been developed as a Country Park, and a draft management plan for the Moss was produced in 1977 (Moffatt, 1977). A wildfowl refuge has been established in Washington New Town by the coordinated effort of the Development Corporation and other organizations.

It is probable that no other new towns will be designated in Britain and that urban development will be rightly redirected into the cities. The creation by the government of urban development corporations in Liverpool and London is a good start; hopefully they will appear in other cities and major towns. It is disappointing that neither of these development corporations have ecologists or natural resource officers. One would have thought that after the lessons from the new towns, and the immense and valuable contribution that the metropolitan county councils and the Lothian Regional Council have made to environmental management, that we would have seen a further phase in the development of ecological thinking in urban design. But this is the Achilles' heel of ecology in Britain (see Cole, Chapter 16 of this volume), for if little is known and done about the management of the rural environment nothing is known and even less done in the older urban areas. The urban environment is equivalent to Siberia to most ecologists and conservationists; it is sad and salutary that we probably care for and know more about the moon than about the structure, function and processes of urban areas. I fail to understand why politicians are so surprised when the urban population reacts in the way it has been doing recently; it is amazing that it has not done so before.

However, it is satisfying that four of the six metropolitan county councils have ecologists. In most of the local authorities the ecologist's work is concerned with studies and surveys, whilst others spend most of their time on land reclamation. Because most of the ecologists in metropolitan counties are at a senior level, a lot of their work tends to be concerned with policy and the evaluation of studies and surveys. Regional and structure plan work, liaison with the public and industry and the involvement with government working parties also consumes a considerable amount of time. Merseyside County Council has done a considerable amount of work using a variety of remote sensing techniques. A survey of the heavy metals in the soils within the County Council's area has also been undertaken. South Yorkshire County Council has undertaken a county environment study using potential surface analysis (Thomas, 1978).

Only nineteen of the 47 non-metropolitan county councils employ ecologists. There is generally only one person, whereas some of the metropolitan councils have two or more ecologists. The bulk of the work undertaken in non-metropolitan county councils is concerned with nature conservation, recreational management and development control. Liaison with the public and industry occupies a considerable amount of time. Much time was also spent on studies and surveys in relation to the preparation of structure plans (Selman, 1976), but as the enquiries-in-public are completed and the plans receive the approval of the Secretary of State, this work is diminishing. However, the structure plan does not finish with its writing. The policies are continuously monitored and reviewed and enforced where deviation from the approved plan occurs. Subject and local plans arising from the structure plan have to be developed and in the case of difficult issues the 'call in' procedure invoked.

The situation in Scotland

Scotland is covered by nine Regional and three island authorities. Although they operate under a system which is somewhat different from that in England and Wales, their responsibilities are similar to those of the English metropolitan counties, although they cover a larger area. Regional councils are responsible for the production of structure plans in which policies and general proposals for strategical development and land use are stated. They also provide the framework for local plans and development control at a local level. Regional councils are also responsible for highways and water supply and statutory undertakers are obliged to consult their council about proposed routes for major powerlines.

Local authorities own land and many, especially county councils, run colleges of agriculture based on farms, for which wildlife and landscape management plans have been produced, such as for Cheshire County Council's farm at Reaseheath (Crooks, 1976). The work of an ecologist in a non-metropolitan county council is described in more detail by Smith (1976). Only 2 per cent of the district councils have ecologists and most of these are in the midland valley of Scotland, although some receive help from their respective county council or regional ecologist. As with non-metropolitan county councils, most of the work of district councils is concerned with nature conservation and recreation management, with surveys and work on local plans also being important. Development control, land reclamation, liaison with government departments, the public and industry occupy a certain amount of time.

The ecologists in national park authorities are also primarily concerned with nature conservation and recreation management, but probably spend more of their time than the average for local authorities on studies and surveys. Work on structure and local plans also occupies a considerable amount of their time.

Because nature conservation and environmental management issues are considered in so many planning, design and mangement procedures the need for a regional/county overview is essential, especially when the employment of an ecologist by a district council must currently be seen as expensive expenditure unless he or she has other professional qualities. Ecologists therefore need to be at a high enough level to influence the drafting of strategic policies.

LOCAL ECOLOGISTS AND OTHER BODIES

The ecological community and the professions were unprepared for the rise of local authority ecologists. Until then, virtually all the ecologists in Britain were in universities or government agencies, whilst the British Ecological Society (BES) was the only organized body of opinion. Despite its radical and progressive origins, the British Ecological Society has become a closed shop without much of a social conscience. The need for the professional recognition of ecologists was outlined by Polunin (1972) in a leader in *Environmental Conservation*. From this developed a considerable correspondence in future editions of that journal, the *Bulletin of the British Ecologists* and the *Biologist*. The British Ecological Society consistently refused to provide the much-needed leadership towards the professional recognition of ecologists, referring instead to the Institute of Biology's professional register (1978). Apologetically, the Institute of Biology established an environmental division with as much dexterity as a blind man without a stick and with virtually no credibility in the professional world. Even now some Authorities refuse to recognize membership of the Institute of Biology as having the same standing as the Royal Town Planning Institute or the Landscape Institute. The latter (formerly the Institute of Landscape Architects) enlarged its remit, but not without nearly crucifying itself in the process. Nevertheless, the Institute remains the main hope for the professional recognition of ecologists that Nicholas Polunin and others raised a decade ago.

The government's environmental agencies also found it difficult to know how to respond. There was at least another organized ecological voice and another body of opinion with arguably more expertise on how to implement ecological theory than any other group of people in the country, apart from those in industry and consultancies, but these two are not 'organized'. The response was that adopted by most bureaucratic systems when faced with a challenge—to close ranks. But the gaps between theory and practice were obvious, they would not go away, nor could they be avoided. In such circumstances there are two solutions—first, to organize conferences to discuss the issues, and second, to commission studies. So in the early to mid-1970s various conferences were held and studies considered. The British Ecological Society and the Royal Town Planning Institute (RTPI) held a joint conference at University College London in 1973, and another in Liverpool in 1975. The Institute of Terrestrial Ecology (ITE) held a workshop at Monks Wood in 1975 (Holdgate and Woodman, 1975); it is a pity that their plans for further action did not materialize. In 1976 there was a conference in Iceland to discuss the breakdown and restoration of ecosystems (Holdgate, 1978). In 1974 the Institute of Operational Research was given an extension by the Natural Environment Research Council (NERC) to an earlier commission to look at the environmental sciences in relation to regional and structure planning. This resulted in the publication of four reports (Institute of Operational Research, 1976a, b, c and d). The work of local authority ecologists was reviewed by Elkington and Roberts (1977), and again by Schulte and Conway (1980).

It is doubtful whether any of these conferences, reports and the rhetoric of the mid-1970s had any positive effect on the attitude of local authorities. It is still extremely difficult to convince either the executive or elected members that ecologists are good value for money in terms of the contribution they will make to the work of the authority. But this lack of progress both in perception and in technical information is surprising when one considers the two main protagonists are now, and have been for some time, in the best possible position to do something about it.

In a paper by Jeffers (1976) it was suggested that four urgent responses are needed to understand the ecology of Britain:

(1) An urgent need for an extensive survey of the present state of many of the ecological systems.
(2) Basic studies of ecological processes—cycling of nutrients, the regeneration of ecosystems and pathways of pollutants.
(3) Practical experiments to test hypotheses on the management of ecosystems.
(4) Modelling to test the effects of various types of management on various simulated ecosystems.

One wonders how much of this has been achieved and, more pertinently, applied?

Report production was inversely proportional to action, and the consideration of people, for whom we were supposed to be doing it, were omitted as tending to confuse the problems. On rereading the papers and conference proceedings that were produced in the early to mid-1970s, I was struck by how much is still relevant and how little progress has been made in the last decade.

Despite circular 108/77 (Department of the Environment, 1977), the relationship of the Nature Conservancy Council with local authority ecologists is very erratic. In some places it is excellent, in others non-existent, whilst the financial contribution and general support is minimal, especially when compared for example with that given by the inspectorate of ancient monuments to the work of archaeology in local authorities. The major problem facing the Nature Conservancy Council is that it is an interest or pressure group and, therefore, cannot give objective advice or undertake objective assessments of environmental impacts. It is unlikely that the Nature Conservancy Council would produce a report recommending that, having taken all the environmental matters and other factors into account, an SSSI should be developed for some purpose other than nature conservation. It would, therefore, seem much more appropriate for such studies to be undertaken by the Institute of Terrestrial Ecology or its parent the Natural Environment Research Council, neither of which have taken much active interest in the resource management problems of Local Authorites. The National Environment Research Council sees itself as funding its component institutes and universities rather than the type of research needed to provide answers for local authority ecologists. The one major exception is the Ecological Survey of Cumbria (Cumbria County Council, 1978), which resulted from a pilot study of the Lake District National Park and subsequently extended to the whole of Cumbria in association with Cumbria

County Council and the Lake District Special Planning Board (LDSPB) before being applied to the whole country. However, whether this work will be of any value or will be used by the local authorities remains to be seen. There is no doubt that the Institute of Terrestrial Ecology will gladly undertake contracted research for local authorites, although whether they will conduct surveys of wildlife appears doubtful. Local authorities do not have the money nor do they see themselves as being responsible for funding the nation's research.

Some of the strongest ties between local authorities and a government agency is that with the Countryside Commission (for England and Wales). This is because most non-metropolitan county and district councils are concerned about rural landscapes, country parks, outdoor recreation and environmental education and interpretation, whilst metropolitan county councils have considerable urban-fringe problems. Local authorities also administer the Commission's treeplanting schemes and in addition the Commission provides substantial contributions to local authority schemes. The style and scale of the Countryside Commission (England and Wales) is different from the Countryside Commission (Scotland) and the relationship between the Countryside Commission (Scotland) and the Scottish local authorities differs from that in England and Wales. For example, in Scotland the local authorities cannot administer Commission grant-aided treeplanting schemes.

Local authorities are public bodies and ultimately accountable to the public through their selected representatives; they need to take into account the view of the voluntary societies and individuals. A considerable amount of a local authority ecologist's time is taken up in liaison with voluntary conservation interests, but in addition many of them rely on voluntary societies for data and information. It is interesting and extremely frustrating that these pressure groups, including the Nature Conservancy Council, are very good at telling local authorities what to do but rarely how to do it. It is surprising that most local authorities see the need to collect information about housing conditions, employment, traffic flows and the rest, but many fail to understand why they should spend an equivalent amount or even more money on collecting data and information to enable them to manage their natural resources more efficiently. Consider the response if a local authority were to use voluntary labour to undertake a traffic survey or design a bridge! However, the voluntary conservation movement, like the Nature Conservancy Council, cannot provide objective advice and they usually wish to pursue a particular cause vigorously—for example, the fencing off and establishment of a woodland or lake as a nature reserve instead of permitting its development for recreation, in an area where such facilities might be scarce.

CONCLUSION: THE FUTURE OF LOCAL AUTHORITY ECOLOGISTS

Local authority ecologists have been around for virtually 10 years, and this would seem to be a good opportunity to review their achievements, hopes and fears for the future. Because so much of their work is corporate, it is generally difficult for an ecologist to lay *total* claim to any particular achievement. Nevertheless, a great deal has been achieved in influencing policies, directing action and in educating officers, elected members and the public. Landscape schools now include considerably more ecology in their syllabuses so that landscape architects are much more aware. There is much more public awareness and sympathy, mainly because of television, especially through presenters like David Attenborough and David Bellamy. There is also little doubt that international politics, especially in relation to oil, has also played its part. The chief executives' wifes' coffee morning, or an enforced bus or train journey by a chief officer can also have considerable influence.

Decision-making is often ephemeral; sometimes it is hard work, but there is nothing magical about it. The cause of conservation and environmental management could have been much greater over the last 10–15 years had there been greater commitment and leadership from those in a position to give it. Unless there are some radical changes in values, local authorities are, in future, only likely to respond to votes and the fulfilment of their statutory obligations. It is inevitable, regardless of politics, that more power will be devolved and that means the district councils will have greater responsibility for the management of natural resources in spite of the fact that they are the least equipped to deal with it (Beynon and Wetton, 1979). This is particularly alarming because currently many district councils lack the perception, the resources and the expertise to implement the environmental policies devised by the county and regional authorities.

I believe that the ecological issues facing local authorities in the next decade will fall into six categories. There is a desperate need to improve the environment and urban design of our major towns and cities. The cost of energy is likely to cause changes in the structure of society in terms of the distribution and location of housing, industry, transport, leisure and tourism. Agriculture is likely to become more and more intensive and is unlikely to be compromised for wildlife conservation. It is likely that sooner or later some kind of environmental impact analysis will be adopted in the United Kingdom (Goodier, 1978), although the present government would like to see the 'developer' carry most of the cost. Local authorities are going to need greater expertise than most of them have, to undertake the initial negotiations and evaluate the environmental impact analysis when it is submitted. Recreation is likely to have a continuing influence on the environment and will involve both protective and creative solutions. Finally, the maintenance of the landscape of rural and urban areas has received little attention—a lot is based on tradition, a lot on individual prejudices and some of it works but it is likely that in future years landscape will have to be created by management. In addition, maintenance is very costly in terms of energy and labour, although it may be possible to reduce both by the application of ecological theory.

But if local authorities are to take the management of natural resources seriously various things need to happen. There is a need for more *applied* research, the government and its agencies need to provide more leadership and support, and there needs to be a set of national natural resource guidelines, including a wildlife conservation strategy. For example, at a Working Panel of Local Authority Ecologists meeting on estuaries, it became abundantly clear that these systems are very important but that they are subject to *ad hoc* management; responsibility for them is unclearly divided between land owners, the water authority and many district and sometimes several county councils. In view of the lack of interest by research agencies, it may be desirable to establish an Institute of Applied Ecology, funded partly from the rate-support grant and partly from the science vote that will provide information, advisory and analytical services to local authorities. Local authorities are not and never will be research agencies. They will use information if it is available, but will only collect it if it is essential to *their* operations. A local authority will quite rightly not fund a research project or survey which, although it may be of immense value to the nation and to other local authorities, is only of partial interest to itself. Many of the ecological research needs of local authorities are concerned with practical problems, which as far as universities and research institutes are concerned are either mundane or of 'low scientific content'.

If the Nature Conservancy Council and others wish to influence the policies of local authorities they will have to present a reasoned argument based on sound information. One-off surveys are of limited value if the information is not updated. Only by repeated surveys will the reduction in the extent of a habitat or species be established and remedial action taken. Local authorities will never have the funds to undertake this type of work and will rely on the Nature Conservancy Council, the Nature Environmental Research Council and similar agencies to provide it in a useful form.

Most of the authoritative ecological literature is incomprehensible to most engineers, planners, landscape managers, architects and managers (and often to other ecologists!). The local authority ecologist has, therefore, to interpret this information with the inevitable criticism that he is prostituting the science. There are also a number of 'ethical' issues. Should a local authority ecologist be an adviser, advocate or adversary, or stay as close to undertaking surveys and 'research' as he or she can? Does he or she have the duty to tell the truth or the whole truth?

If ecologists are to make any progress in local government and other public bodies, then they cannot afford the luxury of non-involvement in the decision-making process. It is they who must decide the best course of action and then advocate its adoption. Ecologists in local government must be prepared to give 'instant' advice; there is not always a right answer—the solution is often a question of judgement. Mistakes will occur but this is inevitable. However, a good ecologist should be right more often than wrong.

The maintenance of a high-quality environment is dependent upon the simultaneous solution of a great many interacting factors that require a holistic approach and the application of general systems theory. Ecologists are, or should be, trained to do this. Ecologists in local government should pay their way, but unfortunately this will not show up in the profit and loss account. Like education, preventive medicine and the prevention of crime, the efficient management of natural resources is a long-term investment with long-term returns.

ACKNOWLEDGEMENTS

I would like to thank Imperial College and the Department of the Environment for the use of Figure 23.2; John Sheldon, Martin Brocklehurst and Richard Burden for their helpful comments; and Carole Rogers for typing the manuscript—a feat in itself.

REFERENCES

Beynon, J. and Wetton, B. (1979). Planning for nature conservation: are the districts lagging behind? *Planner*, **65**, 6−7.

Bradshaw, A.D. (1979). Derelict land—is the tidying up going too far? *Planner*, **65**, 85−88.

Crooks, S. (1976) (Ed). *Cheshire College of Agriculture. Wildlife and Landscape Management Plan*, Cheshire County Council, Chester.

Cumbria County Council and Lake District Special Planning Board (1978). *An Ecological Survey of Cumbria. Working Paper 4*, Cumbria County Council, Kendal.

Department of the Environment (DOE) (1977). *Nature Conservation and Planning*, Circular 108/77, Department of the Environment, London.

Elkington, J. and Roberts, J. (1977a). Who needs ecologists? *New Scientist*, **65**, 210−212.

Elkington, J. and Roberts, J. (1977b). Is there an ecologist in the house? *New Scientist*, **65**, 276−278.

Elkington, J. and Roberts, J. (1977c). The ecology of tomorrow's world. *New Scientist*, **65**, 411−413.

Goodier, R. (1978). Environmental impact analyses (EIA) and environmental impact statements (EIS): some implications for ecologists. *British Ecological Society Bulletin*, **VII(2)**, 8−9.

Holdgate, M.W. and Woodman, M.J. (1975). Ecology and planning: report of a workshop. *British Ecological Society Bulletin*, **VI (4)**, 5−14.

Holdgate, M.W. (1978). The application of ecological knowledge to land use planning. In (M.W. Holdgate and M.J. Woodman (Eds) *The Breakdown and Restoration of Ecosystems*, Plenum Press, New York, pp. 451−464.

Institute of Biology (1978). Register of environmental biologists. *Biologist*, **25 (4)**, 151−152.

Institute for Operational Research (IOR) (1976a). *The Environmental Sciences in Regional and Structure Planning. (Vol. 1: Main Report)*. Institute for Operational Research, 4, Coventry.

Institute for Operational Research (1976b). *The Environmental Sciences in Regional and Structure planning. Volume 2: A Guide for Planners and Scientists*, Institute for Operational Research, Coventry.

Institute for Operational Research (1976c). *The Environmental Sciences in Regional and Structure planning. Volume 3: Background Papers*, Institute for Operational Research, Coventry.

Institute for Operational Research (1976d). *The Environmental Sciences in Regional and Structure planning. Volume 4: Supplementary Report*, Institute for Operational Research, Coventry.

Jeffers, J.N.R. (1976). The ecological contribution to planning. *Town and Country Planning*, **44 (12)**, 542–545.

Kelcey, J.G. (1975). Ecology in a new British city. *Naturopa*, **22**, 23–25.

Kelcey, J.G. (1976). Planning and ecology in a new town. *Planner*, **62**, 45–47.

Kelcey, J.G. (1978). Why reclaim: a reappraisal of current attitudes. *Reclamation Review*, **1**, 157–161.

Moffatt, D. (1977). *Risley Moss: Draft Management Plan*, Warrington New Town Development Corporation, Warrington.

Polunin, N. (1972). Ecologists and their standards. *Biological Conservation*, **4 (5)**, 321.

Schulte, D. and Conway, G. (1980). *Ecologists in Local Authorities in the United Kingdom*, ICCET Report Series C No. 1, Imperial College, London.

Selman, P.H. (1976). Wildlife conservation in structure plans. *Journal of Environmental Management*, **4**, 149–159.

Smith, R.O. (1976). Planning and ecology in a county council. *Planner*, **62**, 42–44.

Thomas, G.A. (1978). *County Environment Study*, South Yorkshire County Council, Barnsley.

Tregay, R. and Moffatt, D. (1980). *An Ecological Approach to Landscape Design and Management in Oakwood, Warrington*, Warrington New Town Development Corporation, Warrington.

Conservation in Perspective
Edited by A. Warren and F.B. Goldsmith
© 1983 John Wiley & Sons Ltd.

CHAPTER 24

The Voluntary Movement

FRANK PERRING

INTRODUCTION

The voluntary nature conservation movement in the United Kindgon has had a role to play in three distinct areas: the setting up and subsequent influencing of government agencies; the acquisition and management of a series of nature reserves complementary to those run by the central or local authorities; and the changing of public opinion through a programme of education for the young as well as for adults.

INFLUENCING GOVERNMENT AGENCIES

Voluntary organizations have been first the midwives and then the lifelong handmaidens of the statutory conservation organizations in Britain. The first national organization specifically concerned with protecting wildlife was a voluntary one: the Selborne Society for the Protection of Birds, Plants and Pleasant Places which was founded in 1885 (see Lowe, Chapter 19 of this volume).

Having seen the statutory organizations established there is still a need to influence their policies and through them the government to which they are responsible. These activities were coordinated during the 1960s and 1970s by the Council for Nature even though this body included a great many organizations for which conservation was not a primary concern and excluded many others which needed to be represented. In 1979 the Council for Nature was disbanded and its coordinating function taken over by the Wildlife Link Committee of the Council for Environmental Conservation which embraces not only the main voluntary nature conservation organizations such as the Royal Society for the Protection of Birds (RSPB) and the Royal Society for Nature Conservation (RSNC) but also activist bodies such as Greenpeace and Friends of the Earth.

Whilst the voluntary conservation organizations are barred by their charitable status from taking direct political action and canvassing their policies, they are able to inform Parliament, the media and the public through Wildlife Link of the pertinent facts about problems—and these can be expressed more effectively when backed up by the combined membership of the bodies concerned, for these are well in excess of half a million.

There is no doubt that Wildlife Link, chaired by Lord Melchett, had a significant influence on the final shape of the Wildlife and Countryside Act 1981. During its passage through Parliament when over a thousand amendments were tabled, members were made fully aware of the size and influence of the conservation organizations for the first time.

Conservation organizations also have more direct access to members of Parliament through the All-Party Parliamentary Conservation Committee, currently serviced by the Royal Society for Nature Conservation. This meets regularly in the House of Lords where it is addressed by prominent conservationists and its papers go to over 100 MPs. This enables topics of particular importance to be circulated amongst a large number of members of different persuasions and could, for example, lead to a Private Members' Bill brought in under the 10-minute Rule.

Besides having these two forums for political issues, the voluntary bodies and the statutory bodies also meet once or twice a year to consider coordination of technical aspects of wildlife conservation under the aegis of the Conservation Liaison Committee of the Royal Society for Nature Conservation. This has, for example, coordinated the preparation of a series of British Red Data Books published by the Royal Society for Nature Conservation, the production of a series of codes of conduct and a policy on introductions.

Another coordinating body set up through the initiative of the voluntary movement is the Farming and Wildlife Advisory Group (FWAG). This body, born from a conference at Silsoe in Bedfordshire in 1969, has been extremely influential in bringing together members of the key organizations concerned with farming and conservation including leading personalities from the Ministry of Agriculture's Advisory Services, the statutory conservation agencies, the Country Landowners Association, National Farmers Union, Royal Society for the Protection of Birds, Royal Society for Nature Conservation and the British Trust for Ornithology.

As early as 1970 it was apparent that a full-time advisor would be required by the Farming and Wildlife Advisory Group, and this was made possible when both the Royal Society for Nature Conservation and the Royal Society for the Protection of Birds agreed jointly to fund such an appointment. Since then the advisor has been largely responsible for setting up c. 50 county Farming and Wildlife Advisory Groups covering virtually the whole of Great Britain. These county groups have a similar composition at local level to national groups and have generally relied on their own members to advise farmers. However, in two counties, Gloucestershire and Somerset, the Nature Conservation Trusts have taken the initiative and have obtained funds to employ farming and wildlife advisers. In a third county, Suffolk, the local group has set up a consultancy service to farmers on a fee-paying basis with the help of two recently retired agricultural adivsers. These three experiments were the subject of an independent appraisal by the Darlington Amenity Research Trust on behalf of the Countryside Commission which reported very favourably on the reaction of farmers and the high take-up of the advice that was offered.

The voluntary movement sees the need to increase the number of advisers as rapidly as possible, and is seeking the backing of government agencies and charitable trusts to pump prime appointments which, once established, should be able to continue on a fee-paying basis.

In addition to these formal forums the voluntary organizations independently, or in groups relating to the problem, continue to use their influence via the media to protest at inadequacies in the activities of government agencies whenever and wherever these are detected and, by directing their attacks at central government, can try to ensure that organizations like the Nature Conservancy Council (NCC) and the Countryside Commission are adequately funded to meet their own needs as well as having resources to disperse to the voluntary bodies so that they can carry out complementary functions.

ACQUISITION AND MANAGEMENT OF NATURE RESERVES

A measure of success for long used by conservation organizations, statutory as well as voluntary, has been the number of reserves acquired and the total area protected and there is no doubt that the statistics look impressive.

Since its inception in 1949 the Nature Conservancy Council has declared 171 National Nature Reserves (NNRs) which cover over 133 320 ha. The Royal Society for the Protection of Birds, though founded in 1889 and not acquiring its first reserve for 40 years, now owns or manages 80 reserves covering 35 150 ha, the majority of NNR (key site) status.

The Royal Society for Nature Conservation through its 44 associated Nature Conservation Trusts now administer over 1300 reserves covering nearly 44 440 ha which contribute to the protection of 140 of the 735 key sites listed in *A Nature Conservation Review* (Ratcliffe 1977) (Figure 24.1). In addition the National Trust, although not primarily a nature conservation organization, has over 340 properties which are SSSIs, a significant proportion of which are key sites, and the Woodland Trust has, since 1972, begun to purchase woodland, and two of its early acquisitions in Devon were also key sites, whilst another of major national importance was recently purchased in Hertfordshire.

However encouraging these figures may seem there is no cause for optimism, for the following reasons.

(1) The total area of key sites listed in *A Nature Conservation Review* (Ratcliffe, 1977) is about 2,350,000 ha and the combined efforts of the

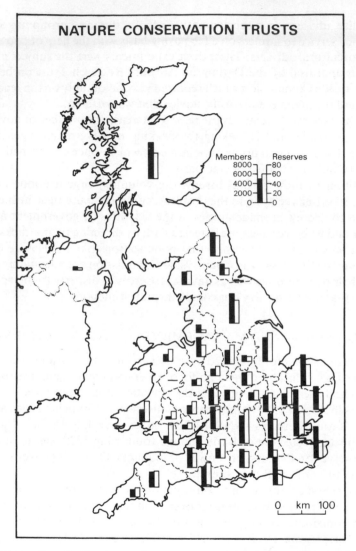

FIGURE 24.1. The distribution of Trusts, their membership and the number of their reserves (as of the end of 1981) in Great Britain and Northern Ireland. Reproduced by permission of Natural World.

Nature Conservancy Council and the voluntary organizations have only resulted in about 20 per cent (181 800 ha) being acquired as reserves.

(2) Likewise only 20 per cent of the sites of regional importance (SSSIs) have any form of protection, and recent evidence suggests that of the remaining 80 per cent, 10 per cent per annum are being destroyed or severely damaged, mainly by agriculture and forestry.

(3) Even areas currently protected by nature reserve status may not be safeguarded. Only a little over a quarter of the National Nature Reserves are owned by Nature Conservancy Council whilst 60 per cent are managed under nature reserve agreements with owners and occupiers, many of which will have to be renegotiated during the 1980s at a cost which could mean that the Nature Conservancy Council will need all its resources for available land purchase to protect what it already holds in some form.

In 1981 the Forestry Commission was instructed by the government to raise £15 million by the sale of forest land over the following 3 years. Amongst the first parcels placed on the market have been several key sites and SSSIs previously held on peppercorn rents as reserves by nature conservation trusts, which they will now have to purchase if these are not to be acquired by private forestry for commercial exploitation.

There is thus an urgent need to marshall all possible resources for the purchase of land. Yet in 1982 the government allocated only an additional £600 000 to meet the extra demands which are likely to fall on the Nature Conservancy Council as a result of the Wildlife and Countryside Act 1981—enough to buy perhaps 2000 ha when nearly 80 000 ha of key sites alone are still at risk.

Whilst the voluntary movement will use its influence on the government to try to increase this meagre sum they continue to demonstrate their will to make a significant contribution themselves. In March 1982 the Royal Society for the Protection of Birds launched an appeal for £1 000 000 with which to purchase woodlands for birds. Between 1971 and 1977 the Royal Society for Nature Conservancy and their associated trusts spent at least £700 000 on acquiring reserves. About 20 per cent of this was donated by the World Wildlife Fund (UK) and the remainder came largely from RSNC/trust resources. In 1980–81 three of the 44 trusts successfully appealed for and raised £300 000 between them to buy reserves. The voluntary movement devotes considerable resources to management: the Royal Society for Nature Conservation has calculated in a recent study (Royal Society for Nature Conservation, 1980) that the Society and the trusts spend £100 000 on their reserves each year. This financial commitment is overshadowed by the value of the voluntary labour that is donated by the trusts' members, which can be conservatively estimated at around 20 000 days per annum.

There is no doubt that, given some government support for a small professional staff, the nature conservation trusts, with their presence in every county, could do even more to mobilize the potential for voluntary work amongst a community that is experiencing a shorter working week and a shorter life of full employment. Indeed in 1978 the Nature Conservancy Council began an imaginative scheme to make 'capacity grants' of £5000 per annum to each of the 38 trusts that were then eligible, one of the purposes of which was to increase their capacity to coordinate the voluntary effort. But the scheme came to an untimely end within a few months through lack of funds from the Department of the Environment. Only 17 trusts received grants, which were not protected against inflation and were terminated after only 3 years. They have been replaced by 3-year grants which contribute only 50 per cent to the cost of a post, and several Trusts have had to turn these down because they cannot meet this extra charge on their resources, thus jeopardizing the great potential for increasing voluntary involvement.

The total amount that was distributed by the Nature Conservancy Council to the trust movement in 1980–81 to buy land, and for management and administration was less than £250 000, yet the trusts contributed around £500 000 to buy reserves and on management alone. In Holland, which has only a quarter of our population, the voluntary movement receives £3 million a year from the government. The Dutch have clearly seen that, at least for the majority of smaller, widely dispersed sites which do not justify the employment of full-time paid wardens or estate workers, the wisest use of limited resources is to assist the voluntary movement so that it has adequate staff to undertake the biological survey of all these local sites, the preparation of management plans and the supervision of their implementation by local voluntary wardens and teams to support them. In this context it might well be argued that it would make economic sense if the government were to make 100 per cent grants to purchase reserves and to provide the necessary tools for management, in exchange for subsequent administration and management by voluntary bodies. This would reduce the escalation of costs that occurs after a reserve has been bought, which makes it so difficult for the Nature Conservancy Council to balance its budget.

EDUCATION

The voluntary bodies have gained the status of charitable trusts because of the educational aspects of their work embodied in their charters; all therefore have education officers or education sections and education is a major activity of organizations such as the Royal Society for Nature Conservation , the Wildfowl Trust and the World Wildlife Fund.

For convenience these educational functions can be fitted into two main areas—general education and on-site education—but, as will be shown, the borderline between the two has become increasingly blurred as nature reserves move into school grounds and audiovisual systems become *de rigueur* at interpretative centres.

General education

Education in nature conservation and wildlife is rapidly growing both in and out of school and the voluntary bodies have had to react to increasing demands while husbanding limited resources in the areas where they produce the greatest effort.

While it might have been ideal for them to employ a team of lecturers to visit schools to promote conservation directly among children, this has not proved cost-effective and the World Wildlife Fund, which operated such a service for many years, has now withdrawn it. It is now seen to be more effective to 'teach the teachers' and voluntary bodies concentrate on the production of packs for teachers' centres, and regional one-day conferences to make teachers more aware of their programmes and the facilities they can provide. These facilities now include a plethora of materials—leaflets, booklets, posters, wildlife 'games' and construction kits, often overlapping in content and produced by 'rival' organizations. One of the problems that has not yet been solved by the voluntary organizations (or the government agencies) is what materials are needed by teachers and schools and how they can collaborate in producing and marketing the best possible product with the least possible overlap. This is an area where a strong lead from a coordinating organization such as the Council for Environmental Education is urgently needed.

The Council can also contribute by seeking to influence the degree to which nature conservation is reflected in the syllabuses of the examination boards. The government is considering the amalgamation of the examinations for General Certificate of Education (GCE) and the Certificate of Secondary Education (CSE) and the reductions in the number of boards; this provides a unique opportunity for the voluntary organizations to press for the inclusion of nature conservation as a major course unit within both the Environmental Studies/Environment Science Panel and the conventional Biology/Integrated Science Panel.

Whilst classroom instruction is important in establishing the principles of the conservation of renewable resources and of wildlife management, education must continue throughout life and will mainly occur outside the classroom and by direct involvement. The brochure for WATCH, the junior arm of the Royal Society for Nature Conservation, a club open to all children and teenagers, says: 'It is a Club with a difference—it's practical, informal and fun, and involves its members in caring for wildlife and the environment as a whole'. Every member receives the club's magazine WATCHWORD which comes out three times a year just before the school holidays. It includes projects which aim to collect information of scientific interest such as the Butterfly Countdown of the summer of 1981, or a census of dragonflies in relation to water pollution in the spring of 1982. But WATCHWORD also includes a newsletter of local activities which gives opportunities for involvement in practical conservation.

The Young Ornithologists' Club, the junior section of the Royal Society for the Protection of Birds, is open to young birdwatchers up to 15 years of age. Though with similar aims to WATCH and an impressive membership of about 100 000, the Young Ornithologists' Club places less emphasis on local involvement. Both WATCH, the Young Ornithologists' Club and other junior wildlife organizations are intended to capture the imagination of the young so that this can later find expression in practical conservation. This is the forté of the British Trust for Conservation Volunteers (BTCV), which attracts and trains its own membership and provides a service in the form of day or residential work parties on reserves run by the Nature Conservancy Council, the Royal Society for the Protection of Birds, the Nature Conservation Trusts, Wildfowl Trust etc., charging a fee to cover its costs. In addition the British Trust for Conservation Volunteers organizes training courses on the theory and practice of management including such arts as coppicing, hedge-laying, wall-building etc., for their own members and those of the other voluntary organizations. This form of education not only helps the voluntary movement, but may provide the experience that a young person needs to lead him to employment as a warden or ranger.

On-site Education

Out-of-school education leads imperceptibly into on-site education. This is the provision of conservation or environmental information in relation to a particular site or area to people who live nearby, or to visitors. The intention is to enhance their awareness and enjoyment of the site in the hope that it may lead to involvement.

On-site nature trails were almost unknown in Britain before 1963, but they are now commonplace and are permanently or temporarily features of most suitable reserves that are managed by the voluntary organizations. Some, however, were created by members who had little experience so that they did not always hold the visitors' attention, particularly that of the young. New types of 'quiz' forms calling for search and observation and are now proving more effective.

Even better, when they can be arranged, are guided walks which allow interaction between the expert and his audience. Voluntary organizations, particularly the Nature Conservation Trusts, often in collaboration with local authorities, are organizing these not only to interpret reserves but to cover the landscape in general and to explain the conflicts between farming and wildlife and how they might be resolved.

But guides cannot always be available and, as the pressures on visiting the countryside continue to grow, information must be provided *en masse*. In this respect the voluntary movement has responded very rapidly in the last decade. In 1970 the Nature Conservation Trusts had no interpretative or information centres open to the public, but by 1982 they had 29, with about six more in the pipeline. Many of them have not been designed to interpret a single site but to guide the visitor to other reserves of interest in the area or county: the Glamorgan

Nature Centre, opened by H.R.H. Prince of Wales in April 1982 is not only in the centre of the county attached to a small reserve and close to one of the Countryside Commission's Demonstration Farms, but provides an introduction to the wealth of wildlife reserves within easy reach along the Glamorgan coast.

CONCLUSION

Though the acquisition, management and interpretation of reserves will continue for some time to be the priorities of the voluntary movement, in time the emphasis will turn to seizing opportunities for actually enhancing wildlife wherever they occur; and this will be particularly so the for the nature conservation trusts with their structure of many hundreds of area groups.

The depth of the penetration of nature conservation within the population will be reflected in how we treat our own backyards, what trees we plant in our streets, whether we find room for a pond and tree nursery in the school grounds, the use we make of city parks for creating habitats, and the work of the community in producing a future for wildlife on derelict land and along abandoned railway lines and canals.

The myriad of opportunities can best be taken by local groups motivated by a personal interest in improving the environment in which they live. They need coordination, encouragement and advice and there are voluntary conservation organizations like the county trusts and the British Trust for Conservation Volunteers who are very ready to give it. What is needed is a government that has the will and the vision to appreciate this potential for conservation as a *social* good, and to make available quite modest sums to allow the voluntary sector to take the lead.

APPENDIX

THE MAIN VOLUNTARY BODIES ACTIVELY ENGAGED IN PROMOTING NATURE CONSERVATION IN THE UNITED KINGDOM

Botanical Society of the British Isles

British Museum (Natural History)
Cromwell Road
London SW7 5BD
Telephone: 01 589 6323

The Botanical Society of the British Isles (BSBI) is the most influential botanical society in Britain and is concerned with all aspects of the study of the vascular plants of the British Isles. It has played a leading part in the conservation of flora. Through its members it has collected data on the distribution of British vascular plants which were the basis of the first British *Red Data Book* published in 1977 by the Royal Society for Nature Conservation (RSNC). Following one of its conferences in 1963 a Wild Plant Protection Working Party was set up which produced evidence which contributed to the passing of the Conservation of Wild Creatures and Wild Plants Act 1975.

The Society publicizes the law by producing posters and leaflets with financial aid from the Nature Conservancy Council and the World Wildlife Fund. Its Conservation Committee, which includes representatives of the British Bryological Society, the British Ecological Society, the British Lichen Society, the Phycological Society, the British Pteridological Society, the Royal Horticultural Society and the Royal Society for Nature Conservation, acts as a forum for matters concerning the protection of threatened British plants.

The Society has two representatives on the RSNC Conservation Liaison Committee (one for higher and one for lower plants) and one on the Council for Environmental Conservation (CoEnCo) Wildlife Link Committee.

The Soceity has over 2500 members from which it appoints a recorder for every County who is expected to collaborate with the local nature conservation trust.

British Trust for Conservation Volunteers

36 St Mary's Street
Wallingford
Oxfordshire
Telephone: 0491 39766

The British Trust for Conservation Volunteers (BTCV) was set up in 1970 to take over the organization of the Conservation Corps previously run by the *Council for Nature*. It is now the largest voluntary organization in Britain that equips and trains volunteers to carry out practical conservation, mainly but not exclusively for wildlife. Through the national Corps it provides work parties with trained leaders to carry out day or residential tasks for many other organizations including the Nature Conservancy Council, the Royal Society for the Protection of Birds, the Nature Conservation Trusts, the Wildfowl Trust, national parks, water authorities, local authorities and private owners; a charge is made to cover costs.

The Trust also helps in the training and equipping of independent conservation corps, notably those of the Nature Conservation Trusts and those attached to universities. In addition the Trust publishes *The Conserver*, a quarterly newsletter and a series of practical conservation handbooks on such subjects as hedge-laying, coppicing, and the use of power-saws.

British Trust for Ornithology

Beech Grove
Station Road
Tring
Hertfordshire
HP23 5NR
Telephone: 044282 3461

The British Trust for Ornithology (BTO) was formed in 1932 with the objectives of seeking greater knowledge of the pesent conditions under which birds live, to marshall and interpret the facts about them and to influence those who can take action to protect them. It maintains a staff of trained biologists who coordinate the work of a network of amateur ornithologists who themselves make observations and collect data throughout the country.

Among its many functions those most pertinent to conservation are: the Common Birds Census which began in 1962 and is designed to detect changes in the population levels of familiar species and to provide an index of changes brought about by new farming and forestry techniques; the publication of *The Atlas of Breeding Birds* in 1976 and the study of the distribution of birds in winter currently in progress. The Trust also regularly compiles censuses of the birds of estuaries and makes enquiries on particular species including sand martin, peregrine and heron, the last of which has been continuously censused since 1928.

A measure of the importance of the Trust's research is that much of it is done at the request of, and with financial support from, the Nature Conservancy Council.

Council for Environmental Conservation

Zoological Gardens
Regent's Park
London NW1 4RY
Telephone: 01 722 7111

The Council for Environmental Conservation (CoEnCo) was formed in 1969 to coordinate 30 national environmental organizations including the Council for the Protection of Rural England, the Royal Society for Nature Conservation, Royal Society for the Protection of Birds, National Trust, and the World Wildlife Fund. Its principal object is to consider matters of national environmental importance and to provide a forum at which concerns that transcend the terms of reference of any one body represented may be expressed and a unified policy developed. It acts through committees on energy, pollution, transport, water and wildlife.

The Wildlife Link Committee of CoEnCo was particularly successful during the passage through Parliament of the Wildlife and Countryside Bill 1981. It coordinated the views of the voluntary conservation lobby and brought pressure to bear on MPs and the Ministries to improve the Bill.

CoEnCo publishes *Habitat*, a monthly newsletter formerly issued by the now defunct *Council for Nature*.

The Youth Unit set up in 1978 promotes and coordinates environmental activities for young people. It has an advisory committee with representatives of all the main youth and environmental organizations.

Farming and Wildlife Advisory Group

c/o Royal Society for the Protection of Birds
The Lodge
Sandy
Bedfordshire SG19 2DL
Telephone: 0767 80551

The Farming and Wildlife Advisory Group (FWAG) was founded in 1969 and is an independent body comprising members of the leading organizations concerned with the countryside both statutory and non-statutory. It aims to encourage understanding between farming and conservation interests by providing a forum for informal liaison and the exchange of ideas, information and experience. To this end it organizes conferences and practical demonstrations on farms. One of its main objectives has been to establish a similar system of liaison and contact between farming and conservation at the local level and 50 county Farming and Wildlife Advisory Groups are now operating covering almost the whole of Great Britain. Whilst there are paid advisors in only three counties, Gloucestershire, Somerset and Suffolk, it is hoped that the number will increase and that there will be close collaboration with the Countryside Commission for England and Wales and their Demonstration Farms project.

Friends of the Earth

9 Poland Street
London W1V 3DG
Telephone: 01 434 1684

An environmental pressure group, Friends of the Earth (FOE) has over 250 local groups in the United Kingdom and is part of a world-wide federation of similar organizations. It pursues campaigns on such topics as safe energy, overpackaging, recycling and endangered species.

Friends of the Earth in Birmingham runs a Greensite Project which aims to put ideas for urban wildlife conservation into practice by developing techniques needed to maintain or enhance the wildlife importance of sites such as church-yards, old railway lines, canal banks and wasteland generally.

National Trust

Head Office:
42 Queen Anne's Gate
London SW1H 9AS
Telephone: 01 222 9251

Conservation and Woodlands Section:
Phoenix House
Cirencester
Gloucestershire GL7 1QG
Telephone: 0285 61818

Founded in 1895 the National Trust is the most important conservation society in Britain with a membership of over 1 million, and is this country's largest private landowner. It owns or protects over 400 miles of unspoilt coastline and over 189 880 ha of estates which include 340 areas of SSSI status, a significant proportion of which are key sites such as the Farne Islands, Box Hill, Wicken Fen and Blakeney Point.

The Trust has established a Conservation and Woodlands advisory team which is carrying out an ecological survey of all its properties with the objectives of preparing management plans for all those of conservation importance.

The advisers represent the Trust on the committees of allied conservation organizations such as the Conservation Liaison Committee of the Royal Society for Nature Conservation.

Royal Society for Nature Conservation

The Green
Nettleham
Lincoln LN2 2NR
Telephone: 0522 752326

The Society, founded in 1912 as the Society for the Promotion of Nature Reserves pioneered nature conservation in Britain. In 1916 it prepared the first list of areas of nature conservation interest, and by 1941 it had acquired eight important nature reserves. It laid the foundations of government action in con-servation, including the creation of the Nature Conservancy Council, and was instrumental in the formation of the International Union for the Conservation of Nature (IUCN). The increasing importance of the Society's role as the national association of the 44 Nature Conservation Trusts which began in 1958, was recognized in the new Royal Charter granted to it in 1976, giving it broader nature conservation objectives and specific powers to promote and assist the Trusts. The Trusts, as corporate members, share in the running of the Society with representatives elected to a majority of the places on its Council; all members of the Trusts are associates. In 1981 it became the Royal Society for Nature Conservation. Total membership exceeds 145 000.

The Society advises and gives assistance to the Trusts and represents their interests with government bodies, particularly the Nature Conservancy Council. Together the Society and the Trusts give information and advice to local authorities, regional water authorities and other official bodies, and to land-owners and farmers about the management of wildlife habitats in the country-side generally.

The Society and the Trusts own or manage over 1300 nature reserves covering nearly 44 440 ha.

The Society produces *Natural World* (a members' magazine) three times a year. It has an active education programme aimed especially at young people, and co-sponsors the WATCH club (see page 443).

Royal Society for the Protection of Birds

The Lodge
Sandy
Bedfordshire SG19 2DL
Telephone: 0767 80551

Founded in 1889 the Royal Society for the Protection of Birds (RSPB) now has over 360 000 members making it the largest voluntary nature conservation organization in Europe. It exists to protect Britain's wild birds and their habitats, and carries out research into the effect of the environment on birds as a basis for advising both government and industry.

The Society owns or manages over 80 reserves covering about 36 360 ha, the majority of key site status with a full complement of full-time wardens and reserve management staff, and a vigorous policy of acquisition.

The Society also undertakes special protection schemes for threatened species such as ospreys, golden eagles and peregrines and employs a small team of in-vestigators working closely with the police to bring prosecutions under the Pro-tection of Birds Acts.

The Society also organizes Beached Bird Surveys whereby hundreds of volunteers patrol our beaches looking for dead seabirds as an estimate of the effect of oil pollution on their populations.

Much of the Society's educational programme is aimed at teachers: a teacher's newsletter is distributed to schools each term through local education authorities. The Film Unit has established an international reputation and its films are regularly seen on television. All members receive a quarterly colour magazine, *Birds*.

The Society also runs a junior section, the Young Ornithologists' Club (page 444).

WATCH

The Green
Nettleham
Lincoln LN2 2NR
Telephone: 0522 752326

WATCH, for children and young teenagers, is an independent educational charity, the Watch Trust for Environmental Education, which is sponsored jointly by the *Sunday Times* and the Royal Society for Nature Conservation (page 442). It is both a national club and the junior wing of the Nature Conservation Trusts. Every member receives the club's magazine *WATCHWORD*, which comes out three times a year just before the main school holidays and includes national projects and competitions as well as information about wildlife and conservation. In addition members receive a newsletter from the local Trust giving information on local events and activities.

The 15 000 members are rapidly being organized into WATCH groups of which more than 300 now exist throughout Britain.

Wildfowl Trust

Slimbridge
Gloucester GL2 7BT
Telephone: 045 389 333

Founded in 1946 the Wildfowl Trust now has over 18 000 members. The Trust maintains a number of collections of wildfowl in refuges in various parts of the country which are open to the public. These contain many rare and endangered species; there are over 120 of the world's 146 species of Anatidae at Slimbridge. In addition some of these refuges are notable for wintering flocks of wild swans, geese and ducks.

The Research Department at Slimbridge, which organizes wildfowl counts, has world-wide links and is the headquarters of the International Waterfowl Research Bureau. The Education Service at Slimbridge, Martin Mere (Lancashire), Washington (Tyne and Wear), Arundel (Sussex), and Peakirk (Cambridgeshire) arranges courses, lectures, tours, exhibitions and films in addition to environmental studies for school children.

Woodland Trust

Westgate
Grantham
Lincolnshire NG31 6LL
Telephone: 0476 74297

The Woodland Trust was formed in 1972 to safeguard existing trees, particularly native broadleaves species such as ash, beech, lime and oak, and to replace some of the millions lost from the countryside in recent years. It plans to acquire at least one woodland in every county and, in its first 10 years has taken 74 woodlands covering 850 ha into its care which are open to its 20 000+ members and, generally, to the public. The Trust is also concerned to create new broadleaved woods by planting land especially acquired for the purpose.

World Wildlife Fund—United Kingdom

Panda House
11–13 Ockford Road
Godalming
Surrey GU7 1QU
Telephone: 04868 20551

Founded in 1961 the World Wildlife Fund is a Swiss-based voluntary organization which administers national organizations in 27 countries including Britain. It aims to create awareness of threats to the natural environment (with emphasis on threatened species and their habitats).

The United Kingdom branch of the World Wildlife Fund plays a major role in funding the purchase and management of nature reserves particularly those being acquired by the Royal Society for Nature Conservation, the Royal Society for the Protection of Birds and the Woodland Trust. Limited resources are also made available for interpretation and educational projects.

An Education Section, replacing the Wildlife Youth Service, was established in 1981. It assists teachers by providing classroom materials on world conservation problems. It produces a quarterly colour magazine *World Wildlife News* which includes a section on the Fund's UK activities.

Young Ornithologists' Club

The Lodge
Sandy
Bedfordshire SG19 2DL
Telephone: 0767 80551

Young Ornithologists' Club (YOC) is the junior branch of the Royal Society for the Protection of Birds and was formed in 1965 to encourage an interest in and inform young people, particularly aged 9–16, about birds. Membership is now over 111 000 as individuals, as family members or as groups in schools or youth organizations. Activities include competitions, holiday courses and field projects. A YOC Leaders' scheme, organized by adult RSPB members, arranges local outings and indoor meetings. Members receive the colour magazine *Bird Life* every 2 months.

BIBLIOGRAPHY AND SOURCES OF USEFUL INFORMATION

Arvill, R. (1969). *Man and Environment*, Penguin Books, London, 332 pp.

Countryside Commission (1972). *The Use of Voluntary Labour in the Countryside of England and Wales*, Countryside Commission, London, 37 pp.

Countryside Commission (1978). *Interpretation in Visitor Centres: A Study by the Dartington Amenity Research Trust*, Cheltenham, 126 pp.

Countryside Commission (1978). *Self-Guided Trails: Report of an Appraisal by the Dartington Amenity Research Trust*, Cheltenham, 112 pp.

Countryside Commission (1981). *County Farming and Wildlife Advisers: Unpublished Report of a study by the Dartington Amenity Research Trust*, Cheltenham, 36 pp.

Dangerfield, B.J. (Ed) (1981). *Water Practice Manuals. Recreation: Water and Land*, Institution of Water Engineers and Scientists, London, 336 pp.

Dennis, E. (Ed) (1972). *Everyman's Nature Reserve: Ideas for Action*, David and Charles, Newton Abbot, 256 pp.

Department of Education and Science (1981). *Environmental Education. Sources of Information*, HMSO, London, 84 pp.

Holliday, F. (Ed) (1979). *Wildlife of Scotland*, Macmillan, London, 198 pp.

Lobbenberg, S. (1981). *Using Urban Wasteland*, Bedford Square Press, London, 42 pp.

Mabey, R. (1980). *The Common Ground*, Hutchinson, London, 280 pp.

Mellanby, K. (1981). *Farming and Wildlife*, Collins, London, 178 pp.

Perring, F.H. and Farrell, L. (1981). *British Red Data Books: 1, Vascular Plants (Edn 2)*, Royal Society for Nature Conservation, Lincoln, 99 pp.

Ratcliffe, D.A. (1977). A Nature Conservation Review, Cambridge Univ. Press, 401 + 320 pp.

Royal Society for Nature Conservation (1980). *SPNC Nature Reserves Study*, Lincoln, parts 1 and 2.

Royal Society for Nature Conservation (1981). *Towards 2000: A Place for Wildlife in a Land-Use Strategy*, RSNC, Lincoln, 25 pp.

Ruck, A. (1980). *Nature Conservation. Why and How?—An Introduction*, Ruck, Hythe, 28 pp.

Seaward, M.R.D. (Ed) (1981). *A Handbook for Naturalists*, Constable, London, 202 pp.

Sheail, J. (1976). *Nature in Trust*, Blackie, Glasgow and London, 270 pp.

Smith, A.E. (1982). *A Nature Reserves Handbook*, Royal Society for Nature Conservation, Lincoln, 176 pp.

Stevenson, D. (Chairman) (1972). *50 Million Volunteers. A Report on the Role of Voluntary Organisations and Youth in the Environment*, HMSO, London, 104 pp.

Wilson, R. (1981). *The Back Garden Wildlife Sanctuary Book*, Penguin Books, London, 152 pp.

Conservation in Perspective
Edited by A. Warren and F.B. Goldsmith
© 1983 John Wiley & Sons Ltd.

CHAPTER 25

Nature Conservation in National Parks in Western Europe

ERIC DUFFEY

INTRODUCTION

Despite the initiatives being taken in North America and New Zealand, where national parks were being created in the 1870s, Europe was slow to follow. Perhaps the concept of vast 'wilderness areas' set aside for all time for the enjoyment of the public and to protect the finest examples of the natural heritage did not seem realistic in smaller industrialized countries with high population densities.

Nevertheless the imagination was stirred, and after the creation of the first five in the arctic north of Sweden in 1909, the idea slowly caught on. The famous Swiss National Park dates from 1914 and two were established in Spain in 1918, but the majority of parks were not designated as such until the years after the Second World War. The number has now increased to 133 (Figure 25.1).

Originally the role of national parks was loosely defined, and not surprisingly their function today varies from country to country depending on national cultures, traditions and attitudes to the natural environment and its wildlife (Duffey, 1970, 1982). In an attempt to establish acceptable common standards for national parks, the Tenth General Assembly of the International Union for the Conservation of Nature and Natural Resources in November 1969 agreed on a precise definition which stated that

a National Park is a relatively large area (1) where one or several ecosystems are not materially altered by human exploitation and occupation; where plant and animal species, geomorphological sites and habitats are of special scientific, educative and recreative interest, or which contains a natural landscape of great beauty; and (2) where the highest competent authority of the country has taken steps to prevent or eliminate as soon as possible exploitation or occupation in the whole area, and to enforce effectively the respect of ecological, geomorphological or aesthetic features which have led to its establishment; and (3) where visitors are allowed to enter under special conditions for inspirational, educative, cultural and recreative purposes.

National Parks in Europe

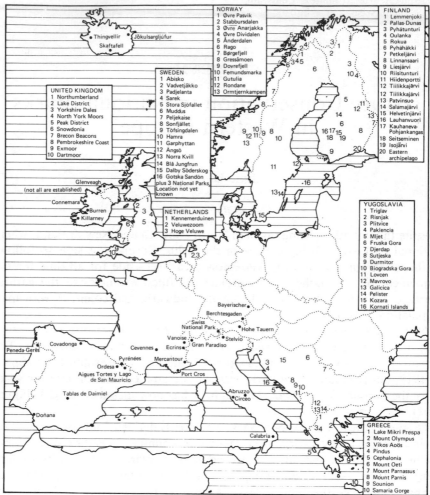

FIGURE 25.1.

In preparing their list of national parks and equivalent reserves for the United Nations, the International Union for the Conservation of Nature (IUCN) (1971) found that in some cases they had to interpret the definition with a degree of flexibility by accepting limited exploitation, but they were firm about the type of area which did *not* meet their criteria, notably

an inhabited and exploited area where landscape planning and measures taken for the development of tourism have led to the setting up of recreaction areas, where industrialisation and urbanisation are controlled and where public outdoor recreation takes priority over the conservation of ecosystems. Areas of this description, which may have been established as National Parks, should be redesignated in due course.

The first two clauses of the IUCN's definition exclude from the UN list all British national parks, and a considerable number of others elsewhere in Europe fail on clause 2, although they are beautiful wilderness areas. These can improve their status by appointing staff and starting an effective management programme. British national parks were all established between 1950 and 1957, well before the 1969 IUCN statement, and their role was defined in different terms. The basis was

an extensive area of beautiful and relatively wild country in which, for the nation's benefit and by appropriate national decision and action, (a) the characteristic landscape beauty is preserved, (b) access and facilities for public open-air enjoyment are amply provided, (c) wildlife and buildings and places of architectural and historic interest are suitably protected, while (d) established farming use is effectively maintained (Dower, 1945).

The value of the IUCN definition is in drawing attention to those national parks or equivalent areas which fall short of what is considered to be the minimum level of care and maintenance and is a constant reminder to those countries who apparently care so little for their natural environment that international opinion is taking note of their shortcomings.

Most definitions of the role of national parks recognize the dual function of providing amenities for the public and a refuge for wildlife. The IUCN definition puts more emphasis on the latter role, but in fact in the majority of countries in western Europe the legislation establishing national parks gives priority to providing enjoyment for the public and to bringing greater prosperity to the region.

There is thus continual pressure on the park director and his executive committee to build more roads, hotels, camp sites, cable cars, ski lifts and holiday homes. A few successfully face up to the struggle but others eventually find it a losing battle and make concessions. Others feel that there is no conflict because parks are for people and the management authority is only limited in what it can do for visitors by lack of funds. The managers of the earliest parks probably assumed that they were large enough to cope with any number of visitors without detrimental effects on wildlife and the landscape. No-one predicted that by 1980, millions would flock to the national parks.

The number of people visiting national parks is very difficult to assess because there are many points of access and tourists cannot be separated from residents. However, accurate counts can be made where there is some control, as at the Plitvice National Park in Yugoslavia, where visitors must pay an entrance fee. The figure for 1980 was 600 000 and has been rising sharply for several years. At the Abruzzo National Park in Italy local circumstances enable the management to make reasonably accurate estimates, and that for 1981 was in excess of 1 million. A similar figure has been estimated for each of the other four national parks in Italy. In France up-to-date figures are not available, but in 1977 the Cévennes National Park recorded 100 000 visitors while in 1975 1 350 000 people visited the Pyrénées Occidentales National Park. The Swiss National Park receives about 350 000 visitors each year.

In Britain considerable efforts have been made to estimate visitor totals in the summer and for the whole year. In the Lake District National Park about 12 million people are thought to have visited the Park during 1980 and of these 2½ million stayed for one or more nights (A. Fishwick, personal communication). In 1977 the Countryside Commission made a visitor census at all ten British Parks during 4 weeks in the summer and in 1980 counts were made of the number of visitors to the National Park Information Centres (Tables 25.1 and 25.2 a, b) (C. Champion, personal communication).

TABLE 25.1   Visitors to British National Parks, based on a 4-week census in 1977 and expressed as number of day-trips (C. Champion, personal communication).

Area	In millions
Pembroke Coast	0.9
Dartmoor	0.8
Exmoor	0.4
Brecon	0.8
Snowdon	1.0
Lake District	1.7
Peak District	2.4
Yorkshire Dales	1.1
Northumberland	0.2
North Yorkshire Moors	0.7

TABLE 25.2   Visitors to information centres at British national parks, April–October 1980 inclusive (C. Champion, personal communication).

	Number of visitors	Number of centres
Pembroke Coast	183 000	7
Dartmoor	154 000	7
Exmoor	45 000	4
Brecon	192 000	4
Snowdon	244 000	9
Lake District	714 000	10
Peak District	231 000	8
Yorkshire Dales	385 000	6
Northumberland	57 000 (1979 data)	5
North Yorkshire Moors	86 000	2

Amost everywhere in Europe it is claimed that the number of visitors are increasing.

Robert Stirling Yard (quoted in Butcher, 1969), referring to American national parks, declared that it was the duty of the people 'to deliver these superb areas unimpaired to the generation following ours'. Unfortunately no-one has yet found out how to assess in scientific terms the human carrying capacity limits of the different ecosystems in national parks before the fauna and flora begin to suffer damage. These studies are urgent as there are signs that the creation of new national parks in Europe has slowed down very considerably and in some countries has ceased altogether.

EUROPEAN NATIONAL PARKS

In the survey that follows the United Kingdom has been omitted because of the more detailed discussion of the subject by MacEwen and MacEwen in Chapter 22.

Scandinavia

The largest concentration of parks in western Europe is in Scandinavia; there are 26 in Finland, nineteen in Sweden, thirteen in Norway, three in Iceland, but none in Denmark. Many are situated north of the Arctic Circle and include four parks, one in Norway and three in Sweden, which share common boundaries and together form what is probably the largest single protected area in Europe. Unlike most other European countries, national parks in Scandinavia are owned by the state and can only be established on state land. The governments therefore have complete control over management and visitor regulation.

Scandinavian countries have together formed the Nordic Council, to examine common problems over many different subjects, including nature conservation and the protection of the environment. Each country takes a sympathetic view of the protection of their natural heritage and this is probably related to the traditional right (*Allemansratten* or 'everyman's right') of people to walk, picnic, pick berries, camp on private land, and swim in, and boat on, private lakes. Thus the creation of national parks in Scandinavia does not necessarily provide access to new areas of countryside, but guarantees that these wild landscapes will not be changed and that the wildlife will be protected. In the north, however, the Lapps are entitled to hunt, graze reindeer and to kill predators on their traditional ranges, even if these lie within national parks.

Finland now has a very comprehensive national park system, the eleven new ones established in 1981 giving it the highest total in Europe. Norway includes within its territory the arctic islands of Jan Mayen, Bear Island and Spitzbergen. The last is by far the largest and has three national parks. There are also two nature reserves, one of which covers 1 ½ million hectares, making it one of the most extensive in the world.

Denmark has no national parks—the country is too small to be able to set aside large areas which would qualify for this status. Moreover, 70 per cent of the country is cultivable land, in contrast to Norway, where only 3 per cent is agriculturally productive, or Sweden with 9 per cent. Iceland is another small country, with virtually no agricultural land apart from coastal pastures, and no forests. It has three national parks which function rather like large nature reserves with a considerable wildlife interest, particularly for birds.

The Republic of Ireland

The Republic of Ireland has four national parks (not all finally established) and others are planned (Anon, 1981). Glen Veigh in County Donegal is as good a wilderness area as one might find anywhere in the British Isles, but so far it has no staff or management. Northern Ireland is perhaps too small to have national parks, but its nature conservation organization is progressive and similar to that in Britain.

The Netherlands and Belgium

Both the Netherlands and Belgium are small countries with very high population densities, 413 km^{-2} and 325 km^{-2} respectively. The Netherlands' achievements for nature conservation are outstanding and rival those of any other country in Europe; Belgium on the other hand has shown much less interest in its natural environment. Neither has national parks in the sense of the IUCN definition, but the Netherlands has three areas (none owned by the state) which are sometimes called national parks—the Kennemerduinen on the coast and the heathlands of Hoge Veluwe and Veluwezoom. In Belgium the vast Hautes Fagnes (3894 ha) state nature reserve of heathland, forest and bog might easily qualify for this status. Very little of the rural area in either country is unsuitable for agriculture or forestry, but whereas Belgium has only 65 km of coastline, nearly all highly developed for tourism, the Netherlands has an extensive coast with numerous islands—including a large part of the Waddensee, which extends into Germany and Denmark and is probably the most important coastal region in temperate Europe for its marine life and as a refuge for migrant and wintering aquatic birds. Parts are protected but the struggle continues for its international conservation status to be recognized by all three governments.

France

France is one of the largest countries in Europe, twice the size of Great Britain, but with approximately the same population total. From Brittany in the north west to the Côte d'Azur and Alpes Maritimes near the Italian frontier, the remarkable diversity of topography, vegetation, soils, climate and wildlife provides many important wildlife areas worthy of protection, but in fact relatively few are safeguarded. There are six national parks (Anon, 1979), the first being established in 1961 (Vanoise) and the most recent in 1980 (Mercantour). Except for the island of Port Cros 10 km off the Côte d'Azur, all are situated in mountainous regions and include superb scenery as well as flora and fauna of great interest. The Vanoise is contiguous with the Gran Paradiso National Park in Italy, and the French Pyrénées Park adjoins the Ordesa in Spain to form two very large transfrontier national parks.

The process of establishing a park in France is complex and time-consuming. Lengthy studies must be made, *départements* consulted and, probably the most difficult of all, the agreement of the communes whose land will fall within the boundary of the park, must be obtained. The refusal of some communes to enter into agreement is shown in the curious irregularity of some park boundaries.

TABLE 25.3   Areas (in hectares) of inner and outer zones of French national parks.

	Inner	Outer
Vanoise	52 839	144 000
Port Cros	694	1800 (marine zone)
Pyrénées Occidentales	47 707	206 000
Cévennes	84 800	237 000
Écrins	91 800	179 000
Mercantour	70 000	(not yet finalized)

French national parks are divided into two parts (Table 25.3), an inner zone which is the park proper, where the principal objective is to protect wildlife and their habitats, and an outer and (larger) zone, where the main purpose is to preserve traditional cultures and ways of life and to provide rural amenities for the public. The inner zone is managed by a director and staff with advisory committee, and the outer by a committee consisting mainly of representatives of local people. The zone managements have separate budgets and are at present funded from Paris.

Zoning within national parks is not uncommon in Europe but no other country has copied the French system perhaps because it appears to have significant disadvantages. It tries to combine, in one geographical unit, two functions which are mutually incompatible—development for tourism and the protection of wildlife, the former being concentrated in a zone which surrounds the wildlife conservation area. National park boundaries cannot be freely selected to include a representative range of habitats because they are influenced so much by political considerations arising out of negotiations with the communes and local authorities. For this reason the inner zone usually includes only the mountain area above the treeline where there is no conflict with economic interests.

Spain and Portugal

Although Spain was one of the earliest countries to establish national parks in Europe (Ordesa and Covadonga National Parks in 1918), there then followed a long period of neglect. It was not until 1955 that another park was formed—the Aigües Tortes–San Mauricio Lake area in the Pyrénées—which, however, did not qualify for inclusion in the 1971 UN list because timber exploitation continued and a hydroelectric works was built within the Park. Spain now has five national parks (Fortuno and de la Peña, 1977), two in the Pyrénées, one in the Cantabrian mountains and two extensive wetlands which are entirely concerned with wildlife conservation. The most famous is the Coto Doñana in the estuary of the River Guadalquivir, and south of Madrid is the remarkable but less well-known marshland of Daimiel. There are three more national parks in the Canary Islands, an administrative province of Spain, and a fourth is proposed for the island of Gomera, also in the Canaries.

Government indifference to wildlife was only gradually influenced by public opinion, and in 1971 the National Institute for the Conservation of Nature (ICONA) was formed and attached to the Ministry of Agriculture. ICONA is a small organization with limited resources and broad responsibilities for forestry, hunting and fishing as well as nature conservation. It has achieved a good deal for the national parks, but if resources were greater and it were backed by effective legislation for the protection of wildlife, more could have been done. The Natural Spaces Act of 1975 was a partial attempt to achieve this. The Act was strengthened in 1977 by regulations which provided for four categories of protected land of conservation value. The first two, national parks and reserves of scientific interest, satisfy the IUCN definition of national parks and equivalent areas, while the other two categories are similar to natural monuments and protected landscapes respectively. The 1975 Act provided an opportunity to reclassify existing protected areas and give them legal status (Viedma, Leon, and Coronado, 1976; Viedma and Ramos, 1978). This process is proceeding slowly; so far ICONA has been able to extend the area of the Coto Doñana National Park from 39 225 ha to 75 765 ha and the Ordesa National Park from 2046 to 15 709 ha.

Spain's neighbour Portugal still has many interesting areas of countryside, particularly along her Atlantic coast, in her hills and in her forests and river valleys. The best known region is the Algarve, the most southerly part of the country, popular with the tourist but also with more nature reserves than the rest of the country. Government support for conservation is provided through the National Service for Parks, Reserves and Heritage Landscapes, in the

Department of the Environment, while the Forest Service also has a section for parks and reserves. Portugese interest in protected areas has grown considerably in recent years but, in general, resources are small and only one national park has been created, the Gerês National Park in the mountains of the northeast bordering the Spanish frontier. In the Portugese Atlantic islands there is a nature reserve on Madeira and two in the Azores.

West Germany, Austria and Switzerland

In Germany, Austria and Switzerland—prosperous countries with a high standard of living—the ecological sciences are as advanced as any in Europe and each has vigorous voluntary nature protection societies. These advantages, together with the common language, might lead one to believe that their policies for the conservation of nature would have developed in a similar, progressive way. In a sense this is true but the development has been curiously one-sided. National parks are few—one in Switzerland, two in West Germany and, very recently, one in Austria. On the other hand nature reserves and nature parks, which have different roles, are very numerous. One reason for this is that the *Länder* in Germany and Austria and the cantons in Switzerland assume the main responsibility for nature conservation, while the federal governments provide guidance and financial assistance but do not take the initiative.

The Swiss federal legislation, for example, has no general provision for establishing protected areas for nature conservation, whether national park, nature reserve or other category. The Swiss National Park was founded on private initiative but was given formal status by a Federal Act in 1914. For many years its maintenance and management was the responsibility of the Swiss League for the Protection of Nature, assisted by the government, and it is only recently that the state has increased its financial aid so that it now contributes half of the total annual cost. Although Switzerland has only one national park, federal legislation defines its function as 'an alpine sanctuary protected from all human interference and influence not serving its purpose and where the entire fauna and flora are allowed to develop freely. The National Park primarily serves scientific research.' (Anon, 1966; Burckhardt, 1980). It is therefore a refuge for wildlife in which the public are guests and expected to behave accordingly. This does not mean that they must suffer irritating restrictions; there is no zonation within the Park, the whole area having the same protection, but no special facilities are provided, and thousands of visitors each year enjoy the wonderful experience of seeing undisturbed nature.

In Zernez, a small town close to the Park, a purpose-built information centre (*Haus der Natur*), with an exhibition hall, lecture room, bookshop and other facilities, describes the Park's objectives, natural history, geology and current research. As with other European national parks outside Scandinavia, the Swiss Park is now owned by the state or the Swiss League but was established by agreement with the communes and landowners. Unfortunately, the creation of other national parks in Switzerland is rather unlikely if the same objectives are followed; since 1914 the tourist trade has grown to such an extent that no commune would now agree to limit its freedom to develop commercial activities.

In Germany the eleven *Länder*, or provinces, are responsible for nature conservation in their own territories, within the framework of the federal constitution, which recognizes five protective categories—National Park, Nature Reserve, Natural Monument, Protected Landscape and Nature Park (Ant and Engelke, 1973). Great emphasis has been given by the *Länder* authorities to the creation of nature parks but there are only two national parks—Berchtesgaden and Bayerischer Wald in the Bavarian Alps. The former is on the Austrian frontier and may later form part of an Austro-German Alpine Park, while the latter borders Czechoslovakia. The high population density and the small proportion of land which cannot be forested or cultivated means that the formation of further national parks is unlikely. However, the growth of nature parks since 1956 has been phenomenal, the total having now reached 59, covering 18 per cent of the land surface of the Federal Republic. Their objectives are defined (Poore and Gryn-Ambroes, 1980) as being similar to protected landscapes but 'intensively developed for outdoor recreation'. Twenty years after the idea was first launched, Selchow (1976) wrote 'the establishment of Nature Parks, which for the most part are situated in areas favoured by nature but economically weak, has exercised a tremendous attraction for those in search of recreation and has improved the infra-structure of the said areas, which in its turn has added to their economic strength'. This role is very different from that of a national park as defined by IUCN, although most nature parks cover large, or even very large, areas and frequently include attractive unspoilt countryside and nature reserves. Their nearest counterparts in other parts of Europe are the French regional parks, and they fulfil the same role as the generally smaller country parks in Britain. The term 'nature park' is occasionally used in Austria and some Scandinavian countries but is applied to different categories of land. In Finland 'nature parks' are strict nature reserves where the public is not allowed, and the term may also be used in other countries by private pleasure garden/safari park enterprises. Transfrontier nature parks have been even more successful than transfrontier national parks and include areas jointly managed with Belgium, Luxemburg and the Netherlands, while proposals with Austria, Switzerland and France are under discussion.

The Austrian federal government has been very slow in protecting its finest landscapes as national parks and did not declare its first until 1981 (part of the Hohe Tauern). Four areas have been proposed for national park status (Wolkinger, 1981), the Hohe Tauern and Niedere Tauern mountain areas, the Neusiedlersee—one of the most important inland wetlands in Europe—and the Danube-Marchauen river valleys. The last two, together with the delta of the Rhine in Lake Constance and a series of reservoirs on the Lower Inn river, were designated as 'wetlands of international importance' by the 1971 Ramsar Convention (Carp, 1980). However, the Austrian federal government has not yet given formal recognition to these proposals.

Italy, Yugoslavia and Greece

The south-eastern part of the region under review is one of contrasts and although the three countries differ greatly in their concern for the natural environment, all have many remarkable areas of great beauty and wildlife interest.

Italy

The cultural heritage of Italy attracts millions of people from all over the world, and perhaps because of this the government has not seen the need to support more than a few national parks. These—two in the Alps, one in the Apennines, one by the coast south of Rome—were all established during the prewar fascist era, possibly for international prestige (Pratesi, 1977). The only postwar national park is that in Calabria but it has no staff. It is said that the parks were forced on local inhabitants, who were against them, and the resentment created has persisted to the present day; gaining their goodwill and cooperation is probably the most difficult of the park authorities' problems. The state own very little of the land within the parks and must therefore be careful not to deprive the local farmers of their traditional rights. Too often the work of the park is regarded as interference, particularly when shooting is stopped and tourist development restricted. These problems can be resolved, however, as is shown by the achievements of the management in the Abruzzo National Park (Tassi, 1979). This park includes a town and several villages but by education, information, and by providing employment and various forms of help to farmers, the Park is now accepted by most of the local people.

The equally beautiful Gran Paradiso National Park in the western Alps is managed by an independent body (as is the Abruzzo)—the Ente Autonomo—which makes its own decisions, although it is financed from Rome. It has recently reorganized its management and is altering the Park boundary in an attempt to meet local objectives. Italian conservationists believe this is a dangerous compromise likely to lead to further concessions and a weakening of the Park status. Poaching is also a serious problem which over 60 armed wardens cannot entirely prevent.

There are several 'proposed national parks' in Italy (Lovari and Cassola, 1975), all areas of outstanding interest, for example Mount Etna in Sicily, Gennargentu in Sardinia, and Pollino in Calabria, but there is at present no government interest in giving formal protection to these sites.

Yugoslavia

Yugoslavia provides a remarkable contrast to Italy because protection of the natural environment, as in many communist countries, is regarded as an important part of their cultural activities. The federal system, which makes wildlife and environmental protection the responsibility of each of the six autonomous republics, has achieved a great deal for nature conservation, and the protection of the natural heritage is taken seriously and promoted conscientiously. The present total of sixteen national parks and about 170 state nature reserves is increasing as further plans are implemented (Godicl, 1981). Each republic has an Institute for Nature Conservation whose function is to advise and promote surveys, including studies for additional protected areas of various categories. Once established, a national park is provided with a director and staff and is responsible to the Environment Committee or Committee for Education and Culture of the Republic. In the early stages the park has financial assistance from the republic government but is expected to become self-supporting as soon as possible.

The parks are zoned with a relatively small area having strict nature reserve status, and about 10 per cent is set aside for tourist development. The remainder—and much the largest part of the park—is managed so that agricultural and forestry activities can be continued under the control of the director and his advisory committee. As elsewhere in Europe, this arrangement appears to combine incompatible policies but seems to work remarkably well due to the very large size of many of the parks and by confining the tourist development to a small part of the whole.

The best example is probably the Plitvice National Park in Croatia, which must be one of the most interesting in Europe. The management formed a company which owns three hotels, camp sites, restaurants, bus and boat services and employs 1300 people—all within the National Park. Each year, during the 4–5 month season, 600 000 visitors pay to enter the Park; but the impact on the environment is not noticeable, because only certain routes can be taken, and the road immediately round the lakes is closed to all but the Park buses. Most visitors want to see the many waterfalls, pools and lakes and, on foot, must follow a route laid out on raised wooden walkways. Yugoslavia is fortunate in still having vast areas of unspoiled landscapes, particularly limestone mountains with magnificent natural forests, and the finest cave system in Europe. Its international importance for nature conservation cannot be overrated.

Greece

The same comment could also apply to Greece, which has a flora exceeding 6000 species, of which 685 are endemic (IUCN Threatened Plants Committee, 1977). The fauna includes the brown bear, jackal, wolf and more species of birds of prey than in other countries in western Europe. It has one of the longest coastlines in Europe, nearly 18 000 km, made up largely by the enormous number of islands in the Aegean Sea.

Unlike Yugoslavia, however, Greece is backward in conserving nature and has probably achieved less than any other country in our region. Public interest in wildlife is slight and alone among countries in western Europe, Greece has no World Wildlife Fund National Appeal, no entomological society, and until very recently, no ornithological society. Nevertheless on the credit side there are ten areas known as national parks, the first of which (Mount Olympus and Mount Parnassus) were established in 1938. In addition there are other categories of protected areas: fourteen aesthetic forests, eleven protected wetlands and 32 natural monuments (Duffey, 1982). But these sites receive little or no protection so their status is 'on paper' only. At least two national parks have been so changed by uncontrolled development that proposals have been made to disestablish them. The main difficulty is that no park (except one) has any staff specifically concerned with wardening and management. The exception is the Samaria Gorge in Crete, where staff of the Forest Service patrol the Park, management work is done and information panels have been installed by the entrance. The gorge is one of the most remarkable natural features in Europe, falling from 2000 m to sea level in 18 km through superb scenery, and the whole length is traversed by a public path.

The Greek Forest Service has a small section responsible for national parks and other categories of protected land, but its resources are very limited. The voluntary organization, the Hellenic Society for the Protection of Nature, has achieved a great deal for conservation by persuasion and eduation, but does not own land as nature reserves. The international interest in the Greek fauna and flora is clearly demonstrated by the large number of study visits by scientists and university students and the concern is shown by the International Union for Conservation and the World Wildlife Fund in the loss of outstanding ecosystems. One hopes that as a member of the EEC, Greece will qualify for help to preserve its unique natural features as well as for economic development.

DISCUSSION AND CONCLUSIONS

The conservation of nature in western Europe takes many forms, each an expression of different cultures, ways of life and traditions. The struggle towards a common understanding of how best to achieve an effective wildlife policy is making progress, although slowly. In several countries national parks are the principal means of protecting the wildlife heritage and are managed accordingly. A few do not attract many visitors, for example, those in arctic Scandinavia where access may be difficult, or flat marshlands such as the Coto Doñana and Daimiel National Parks in Spain, so that visitor control is easy and there is little conflict with wildlife conservation.

In scenic mountainous areas of southern Europe, however, where most national parks are situated, the number of visitors may be so great that management is directed solely to their needs and to the wishes of the local residents who depend on the tourist for a living. Whether consideration is given to the effects of high intensity use on the fauna and flora appears doubtful; research is a luxury few can afford and consequently evidence of conflicts is anecdotal or absent.

In well-managed parks with a record of scientific study such as the Abruzzo in Italy, some evidence of the effects of public disturbance to wildlife is beginning to accumulate. This Park has one of the best western European populations of brown bears outside Yugoslavia and is visited by over 1 million people each year. Disturbance causes the bears to wander more frequently outside the Park and although protected by law they may be shot by hunters or farmers. The Swiss National Park authorities also realize the potential dangers to wildlife, the number of annual visitors having increased tenfold in 10 years (Schloeth, 1974). The IUCN guidelines of March 1980 (Poore and Gryn-Ambrose, 1980) include a recommendation that all national parks and equivalent areas should have management plans, and if scientifically based this would be a great step forward in reconciling public use with wildlife protection. However, there is little evidence that this advice is being followed.

In small countries such as Britain and the Netherlands there is a clear distinction between the strict nature reserve, where the main objective is wildlife conservation, and national parks or equivalent areas, which are primarily for people. In southern Europe the assumption that national parks fulfil a major protective role has resulted in nature reserves being given a low status in the hierarchy of protected areas. In Table 25.4 the highest numbers of nature reserves (*sensu lato*) are scored by the more northern countries and of the others many receive no management or surveillance and are only reserves 'on paper'. There is little evidence that the governments concerned are increasing their support to speed up the establishment of reserves.

Although the number of visitors to national parks continues to increase, there are several other categories of land which welcome the tourist. State forests in most countries are open to the public and in some instances, for example, Denmark, they have signed footpaths, nature trails and information leaflets. In France there are 24 regional parks in areas of great interest, and West Germany has its 59 nature parks. Britain has 167 country parks and also the unique National Trust, unparalleled elsewhere in Europe. The distribution of these land categories is very uneven and outside the three countries mentioned the regional nature park idea is not well developed. The use made of them, and whether they succeed in attracting some of the people who would otherwise go to national parks, has not been studied.

The close economic, political and scientific links between the countries of western Europe through the EEC, Council of Europe and other agencies should provide the means for a detailed study of the conservation and recreation needs

TABLE 25.4 The numbers of national parks and nature reserves (*sensu lato*) in each of 19 countries in western Europe.

	National parks	Nature reserves[a]	Source
Austria	1 (in part)	180	Wolkinger (1981)
Great Britain and Northern Ireland	10	1642[b]	
Ireland	4	12	Personal information; Anon (1981)
Belgium	—	62	Personal information; Poore and Gryn-Ambroes (1981)
Denmark	—	410	Dahl (1979)
Finland	26	814	Personal information; Poore and Gryn-Ambroes (1981)
France	6	589[c]	Anon. (1979)
Germany	2	1321	Ant and Engelke (1973); Erz (1979)
Iceland	3	32	Personal information; Poore and Gryn-Ambroes (1981)
Italy	5	166[d]	Personal information; Lovari and Cassola (1975); Poore and Gryn-Ambroes (1981)
Luxembourg	—	17	Poore and Gryn-Ambroes (1981)
Netherlands	(3)	1215	Personal information; Anon., (1980); Hesse (1981)
Norway (with Spitzbergen)	18	207	Holt-Jensen, 1978; Poore and Gryn-Ambroes (1981)
Spain (with Canaries)	8	87	Personal information; Fortuno and de la Peña (1977)
Sweden	19	1591	Personal information; Poore and Gryn-Ambroes (1981)
Switzerland	1	300	Personal information; Burckhardt (1980) Poore and Gryn-Ambroes (1981)
Portugal (with Madeira and Azores)	1	19	Personal information; Duffey (1982)
Greece	10	25	Personal information; Duffey (1982)
Yugoslavia	16	170	Personal information; Godicl (1981)
Total	133	8859	

[a] The National Park total is reasonably accurate at the time of writing. The total of nature reserves is based partly on published figures and partly on personal information, but the accuracy is very variable. In addition the categories of land selected is arbitrary—including nature reserves (private and state), bird sanctuaries, wildfowl refuges, game reserves, nature parks and regional parks. Nature monuments, protected landscapes, country parks, SSSIs, etc., are excluded.

[b] National nature reserves, Trust reserves and RSPB reserves.

[c] Including 24 regional parks and 503 wetlands where shooting is banned or restricted.

[d] 101 Reserves in state forests, 15 regional forest reserves, six regional parks.

of each country and how well they are being met by the existing protected areas. This calls for an international research effort which, if adequately supported, would provide basic information to enable governments and conservation agencies to direct their limited funds to those areas where there is the greatest need.

REFERENCES

Anon. (1966). *Through the Swiss National Park; A Scientific Guide*, Committee for Scientific Research in the National Park, Switzerland.
Anon. (1978). *Scotland's Scenic Heritage*, Countryside Commission for Scotland, Perth, Scotland.
Anon. (1979). *Guide de la nature en France*, Bordas, Paris.
Anon. (1980). *Hanboek van natuurreservaten en wandelterreinen in Nederland*, Natuurmonumenten, 's-Graveland.
Anon. (1981). *Areas of Scientific Interest in Ireland*, An Foras Forbatha, Dublin.
Ant, H. and Engelke, H. (1973). *Die Naturschutzgebiete der Bundesrepublik Deutschland*, Bundesanstalt für Vegetationskunde, Naturschutz und Landschaftspflege, Bonn-Bad Godesberg.
Burckhardt, D. (1980). *Die schönsten Naturschutzgebiete der Schweiz*, Ringier, Zofingen, Switzerland.
Butcher, D. (1969). *Exploring our National Parks and Monuments*, 6th edn, revised, Houghton Mifflin, Boston.
Carp, E. (1980). *Directory of Wetlands of Interntional Importance in the Western Palearctic*, United Nations Environment Program and International Union for the Conservancy of Nature, Gland, Switzerland.
Dahl, K. (1979). *Kort over Danmark. Tekst og detailkort over fredede områder*, Danmarks Naturfredningsforening, Copenhagen.
Dower, J. (1945). *National Parks in England and Wales*, Cmnd 6628, HMSO, London.
Duffey, E. (1970). Wildlife conservation in Europe. In *The 1969 Handbook of the Society for the Promotion of Nature Reserves*, pp. 1–36.
Duffey, E. (1982). *The National Parks and Nature Reserves of Western Europe*. MacDonald and Co., London.
Erz, W. (1979). Katalog der Naturschutzgebiete in der Bundesrepublik Deutschland. *Naturschutz Aktuell*, **3**, 103. Bonn.
Farnett, G., Pratesi, F. and Tassi, F. (1977). *Guide alla natura d'Italia*, 4th edn. Arnoldo Mondadori, Milan.
Fortuno, F. and De La Peña, G. (1977). *Reservas y cotos nacionales de caza*, 5 vols. Incafo, Madrid.
Godicl, L. (1981). The protection of rare plants in nature reserves and national parks in Yugoslavia. In H. Synge (Ed), *The Biological Aspects of Rare Plant Conservation*, John Wiley, Chichester, pp. 491–502.
Hesse, J. (1981). *Mécanismes institutionnels et économiques de la conservation des espaces naturels aux Pays-Bas*. Ecole Polytechnique, Laboratoire d'Econométrie, Paris.
Holt-Jensen, A. (1978). *The Norwegian Wilderness National Parks and Protected Areas*. Tanum-Norli, Oslo.
International Union for the Conservation of Nature (IUCN) (1971). *United Nations List of National Parks and Equivalent Reserves*, 2nd edn. Hayez, Brussels.

International Union for the Conservation of Nature (IUCN) Threatened Plants Commitee (1977). *List of Rare, Threatened and Endemic Plants in Europe*. Council of Europe, Nature and Environment Series, no 14, Strasbourg.

Lovari, S. and Cassola, F. (1975). Nature conservation in Italy: The existing National Parks and other protected areas. *Biological Conservation*, **8**, 127–142.

Poore, D. and Gryn-Ambroes, P. (1980). *Nature Conservation in Northern and Western Europe*. International Union for the Conservation of Nature, Gland, Switzerland.

Pratesi, F. (1977). *Parchi Nazionali e zone protette d'Italia*. Musumeci, Aosta.

Schloeth, R. (1974). Problems of wildlife and tourist management in the Swiss National Park. *Biological Conservation*, **6**, 313–314.

Selchow, E. (1976). The Nature Park idea has triumphed. *Nature and National Parks*, **14(51)**, 19–20.

Tassi, F. (1979). *Parchi Nazionali*. La Nuova Italia, Florence.

Viedma, M.G. and Ramos, A. (1978). A commentary on Spain's 1975 Protection of Natural Spaces Act. *Biological Conservation*, **14**, 13–23.

Viedma, M.G., Leon, F. and Coronado, R. (1976). Nature conservation in Spain: a brief account. *Biological Conservation*, **9**, 181–190.

Wolkinger, F. (1981). Die Natur- und Landschaftsschutzgebiete Österreichs. *Österreichische Gesellschaft für Natur und Umweltschutz*, **7**, 154.

Index